Industrial Applications of Green Solvents

Volume I

Edited by

Inamuddin[1,2,3], Mohd Imran Ahamed[4] and Abdullah M. Asiri[1,2]

[1]Centre of Excellence for Advanced Materials Research, King Abdulaziz University, Jeddah 21589, Saudi Arabia

[2]Chemistry Department, Faculty of Science, King Abdulaziz University, Jeddah 21589, Saudi Arabia

[3]Department of Applied Chemistry, Faculty of Engineering and Technology, Aligarh Muslim University, Aligarh-202 002, India

[4]Department of Chemistry, Faculty of Science, Aligarh Muslim University, Aligarh-202 002, India

Published by **Materials Research Forum LLC**
Millersville, PA 17551, USA

Published as part of the book series
Materials Research Foundations
Volume 50 (2019)
ISSN 2471-8890 (Print)
ISSN 2471-8904 (Online)

Print ISBN 978-1-64490-022-2
eBook ISBN 978-1-64490-023-9

Distributed worldwide by

Materials Research Forum LLC
105 Springdale Lane
Millersville, PA 17551
USA
http://www.mrforum.com

Manufactured in the United States of America
10 9 8 7 6 5 4 3 2 1

Table of Contents

Preface

The concept of green chemistry emerged as a natural pollution preventive measure in chemical industries. A wide range of products including chemicals used for protecting crops, improving yield and manufacturing of medicines are using many substances and thus intentionally or unintentionally release harmful end products that pollute the environment and at the same time are dangerous for the human health. It is well-known that around the world billions metric tons of hazardous waste is produced per day the by chemical industries. Therefore, it becomes very important to control the pollution and generation of hazardous waste at this pace. In 1970, The United States formed the Environmental Protection Agency (EPA) which set several environmental regulations/guidelines for the safeguarding of human and environment health. Since then, EPA encourages researchers and chemist to synthesize and design new processes in order to reduce the production of toxic and hazardous waste. With the help of green chemistry, the focus is on the synthesis of environment-friendly products, low energy requirement processes, alternatives of hazardous substances and also preventing pollution in its first place. For this purpose, various research grants/funds are allocated. In 1995, President Clinton started the Presidential Green Chemistry Challenge Award by recognizing the improvements in the field of green chemistry. Geoffrey Coates of Cornell University won this award for developing a metal complex that is used in making plastic by using very cheap starting material carbon monoxide and carbon dioxide and in these plastics, no hazardous chemicals are present as it is free from bisphenol-A. Green chemistry is not limited to industries only but also helpful in pharmaceutical companies. Yi Tang, another Presidential Green Chemistry Challenge winner, developed a new method for making cholesterol drugs by using cheaper starting material and less hazardous solvent. In a similar way, ethyl lactate manufactured from corn starch and soybean oil is used to replace hazardous cleaning solvents as it biodegrades to carbon dioxide and water and is also available at low cost as compared to petrochemical solvents. These are the very few examples of applications of green chemistry. Green chemistry is helpful for the protection of human health and the environment by reducing pollution, eliminating, reducing or replacing the hazardous waste and are beneficial for the economy.

Industrial Applications of Green Solvents Volume 1 is intended to explore some of industrial applications of green solvents (water, ionic liquids, supercritical carbon dioxide, terpenes) in the industrially important areas such as chemical synthesis including lipase-catalyzed reactions, organic synthesis, esterification reactions, heterocycles as well as paint industry, leather industry and gas separation membranes.

Inamuddin[1,2,3], Mohd Imran Ahamed[4] and Abdullah M. Asiri[1,2]

[1]Chemistry Department, Faculty of Science, King Abdulaziz University, Jeddah 21589, Saudi Arabia.

[2]Centre of Excellence for Advanced Materials Research, King Abdulaziz University, Jeddah 21589, Saudi Arabia

[3]Department of Applied Chemistry, Faculty of Engineering and Technology, Aligarh Muslim University, Aligarh-202 002, India

[4]Department of Chemistry, Faculty of Science, Aligarh Muslim University, Aligarh-202 002, India

doi: https://doi.org/10.21741/9781644900239-1

Chapter 1

Plant Cell Culture Strategies for the Production of Terpenes as Green Solvents

Mubarak Ali Khan[1*], Tariq Khan[2], Huma Ali[3]

[1]Department of Biotechnology, Faculty of Chemical and Life Sciences, Abdul Wali Khan University Mardan (AWKUM), Mardan 23390, Pakistan

[2]Department of Biotechnology, University of Malakand, Chakdara Dir Lower, Pakistan

[3]Department of Biotechnology, Bacha Khan University, Charsadda, KP, Pakistan

makhan@awkum.edu.pk, write2mubarak@gmail.com

Abstract

Green chemistry implies the synthesis and design of fine chemicals of biological origin which can be used in industrial applications. Petroleum-based solvents such as n-hexane and toluene, etc. used in different extraction procedures have shown adverse effects on human health and the environment. The alternatives are the terpenes derived from essential oils have represented their potential as green solvents. The practical use of d-limonene and α-pinene in the extraction of oil from various substances has recently, been investigated as an alternative to petro based solvents. These volatiles are mainly produced in different parts of medicinal and aromatic plants. However, their limited productivity due to geographic variability and environmental fluctuations in natural plants does not meet the emerging industrial demands. In this chapter, the potential of different plant cell culture approaches for the enhancement of production of important terpene volatiles in different plants have been elucidated. Biotechnological methods which can improve the yield of essential oils through genetic engineering of the metabolic pathways responsible for the biosynthesis of terpene volatiles have been described with recent examples.

Keywords

Essential Oils, Volatiles, Green Solvents, Terpenes, Cell Cultures, Callus Cultures

Contents

1. Introduction

Green chemistry employs all those techniques, methods and principles, which are environmentally friendly in terms of reduction of hazardous substances generally used during the production of chemical products. Greener synthesis of starting materials for different industrial applications, including organic synthesis has been recognized globally for their non-hazardous effects on the environment and human health. Plant oils and fats from animals have diverse industrial applications of anthropogenic interests. During the extraction of oils and fats from plants or animals, n-hexane an important petroleum-derived organic solvent has been used as the most commonly used solvent. It was preferred for many years during traditional Soxhlet extraction procedures. Being a non-polar solvent and with low boiling point, n-hexane has shown high efficiency and simple recovery during extraction of oils [1]. However, during the extraction and recovery procedures, the leakage of n-hexane to the environment can cause negative effects on human health and environment. Therefore, it's crucial to explore alternative options and choice of solvents for extraction which can minimize or prevent the associated adversities.

Plant-derived natural compounds such as essential oils have diverse potentials and can be used as alternatives to petroleum-based solvents in various industrial applications. The bioactive volatiles extracted from the essential oils such as α-pinene, d-limonene, linalool and eugenol etc. have been used as natural solvents in many extraction procedures, dyes, and the aroma and as cleaning and degreasing agents. Essential oils could be obtained from different parts of medicinally important plants such as roots and rhizomes, leaves, bark and branches, flowers, fruits, and seeds. Essential oils derived terpene solvents can have many benefits as solvents of choice in the different extraction procedures. Application of terpene solvents can significantly reduce treatment time and energy. Like n-hexane, toluene is also a recommended solvent for distillation processes, in the cosmetics and pharmaceutical industries. Being a petroleum based solvent, the excessive use of toluene can cause severe environmental and health complications on its leakage to the environment [2]. The lab scale potentials of toluene can be replaced by the eco-friendly monoterpene, d-limonene as an alternative to toluene in the food industries [3].

However, production and yield of the important volatiles are limited in the natural plants by factors including but not limited to climatic conditions, cultivation parameters, harvesting time and soil [4]. Biotechnological methods can provide promising means for the efficient production of useful natural volatiles through eco-friendly methods [5]. Plant cell cultures hold tremendous potential in the production of higher yield of secondary products including terpenes. Some novel volatiles which are not present in wild plants, may also be produced through biotransformation during *in-vitro* cultures [6]. Different types of cell cultures callus and cell cultures, adventitious roots culture, shoot culture, hairy roots culture and embryo culture, etc. can be used for the production better yield of important secondary volatiles [7]. Secondary metabolites in plants are generated as a part of metabolism and act as defensins for the plants [8]. Different manipulation and elicitation strategies can be adopted to enhance the yield of these secondary metabolites through *in vitro* cultures of plants [9]. The different biotechnological approaches can manipulate the terpene biosynthetic metabolic pathways in the plant cell for enhanced biosynthesis of terpene volatiles. Generally, terpene biosynthesis in the plant cell is regulated through the deoxyxylulose phosphate and mevalonate-dependent (MVA) [10].

In this chapter, the potential of cell cultures of plants to produce enhanced quantity of essential oils in medicinal and aromatic plants will be reviewed and discussed in the light of current innovations in cell culture technology and biotechnology. We will deal with compounds such as monoterpenes and sesqui-, di- and triterpenes which are classified as green solvents for their potential industrial applications.

2.1 Essential oil as a source of green solvents

Essential oils are the plant-derived natural products and are called "essential" due to the essence of the distinctive fragrance they contain, specific to the fragrance of the respective plant from which they are derived. A large portion (10%) of the known plant essential oils has important commercial applications [1, 11]. Aromatic plant species grown in different geographic zones may differ in the composition of these essential oils. Generally, essential oils are mixtures of aromatic compounds mainly composed of terpene hydrocarbons, a product of secondary metabolism in many aromatic plants [12].

Terpenes are organic compounds assembled from five-carbon isoprene units. Isoprenoids are simple hydrocarbon molecules which form the basic unit of terpenoids. Isoprenoids constitute to form monoterpenoids (C_{10}) and then sesquiterpenoids [13], which, ultimately, form other metabolites including carotenoids, chlorophyll pigments, phytohormones, rubber and sterols, turpentine [14]. Among the terpenes hydrocarbons, monoterpenes, for instance, α-pinene, d-limonene, and p-cymene etc. represent a large proportion of essential oils [1] as indicated in Fig. 1.

The monoterpene hydrocarbons are of interest in many industrial applications, including food, fragrance, cosmetic and pharmaceuticals. In the food industry, the monoterpenes are attributed to their diverse applications as natural organic solvents with an eco-friendly trademark and prominent diversity. Large varieties of monoterpene organic solvents obtained from essential oils can efficiently replace the use of conventional solvents such as petroleum or halogenated solvents in a variety of extraction procedures. Their physical properties are different due to the presence of monocyclic, bicyclic and acyclic hydrocarbon chains. One of the significant monoterpenes, d-limonene, harvested as a byproduct from citrus peel is a low cost and eco-friendly green solvents having potential applications as degreasing and cleaning properties with optimal industrial performance [5]. The practical use of d-limonene in the extraction of oil from various oleiferous substances has recently, been investigated as an alternative to organic solvents. Similarly, the analysis of α-pinene through qualitative and quantitative measures showed similar results in extraction procedures as obtained using n-hexane as a chemical solvent. Obtained through steam distillation or fractionation of pine oleoresins, α-pinene represents the major component of oils derived from confers, rose, basil rosemary, etc. and is considered as an important potential green solvent in many industrial applications [3]. The monoterpene, p-cymene derived from leaf essential oils of many woody plants has also shown positive results as an ingredient of the perfumes (musk), fragrance or as a solvent for dyes and varnishes in important industrial applications [2].

Industrial Applications of Green Solvents I   Materials Research Forum LLC
Materials Research Foundations **50** (2019) 1-20   doi: https://doi.org/10.21741/9781644900239-1

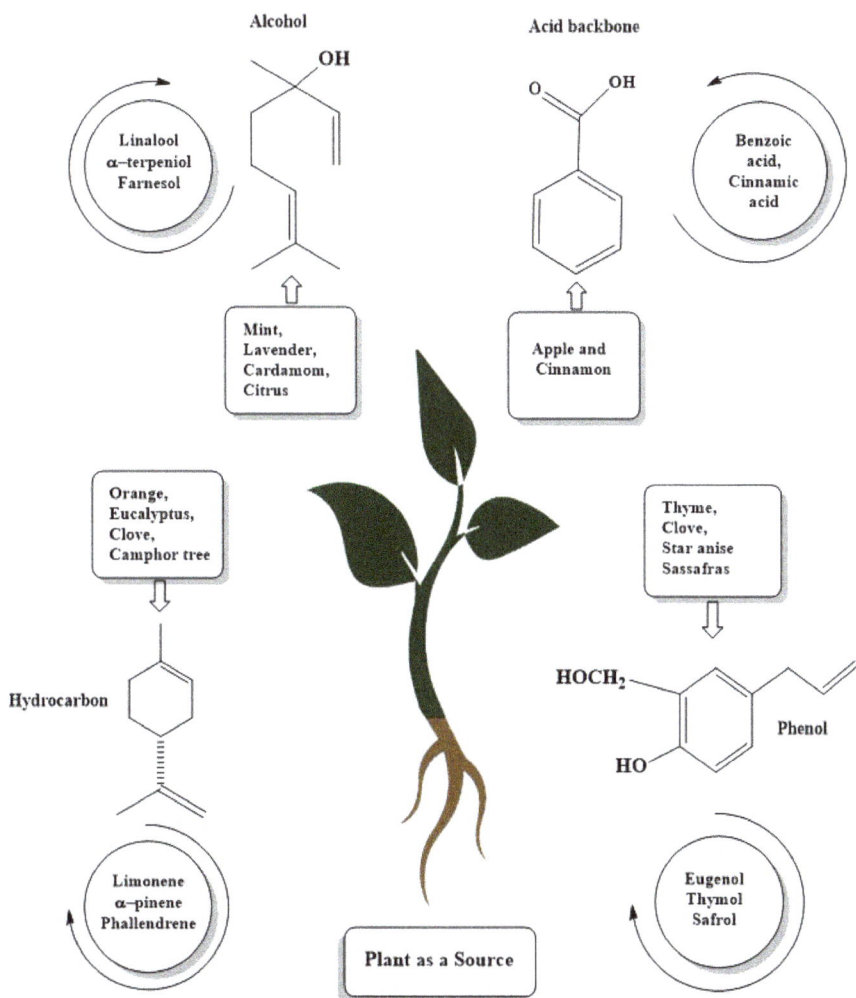

Fig.1 Different type of familiar compounds in essential oils.

2.2 Green solvents as an alternative to chemical solvents

Generally, for extraction of oils and fats from plants or animals, n-hexane has been used as the most commonly used solvent. A non-polar solvent with low boiling point i.e. 69 °C, n-hexane was preferred for many years during traditional Soxhlet extraction procedures as an important petroleum-derived organic solvent. Its two main advantages include high efficiency and simple recovery during the extraction of oils [1]. However, due to leakage of n-hexane during the extraction and recovery procedures, the prospective negative effects on human health and environment urged for the exploration of alternative options in order to prevent the associated adversities. Among the monoterpene hydrocarbons, α-pinene and p-cymene have been used recently in the extraction of vegetable oils. Both have shown promising results in oil extraction and have tested for their optimal performance as good alternatives to n-hexane and alcoholic solvents [5]. Application of terpene solvents in the process of green extraction can significantly reduce treatment time and energy. During several distillation processes, toluene is the recommended solvent for extraction in the cosmetics and pharmaceutical industries. Being a petroleum based solvent, the excessive use of toluene has to be reduced due to its environmental and health complications on its leakage into the environment [2]. Recently, the monoterpene, d-limonene has been used as an alternative to toluene in the food industries for the determination of moisture content in the food products. The organic solvents such as acetone, dichloromethane, and chloroform etc. have routinely employed during extraction of food colors such as carotenoids. These solvents are very efficient because of their versatile volatility and dissolubility [3]. However, these solvents can contaminate the end products due to the mixing of the residues and can eventually affect human health and the environment. Reduction of these volatile solvents can be achieved by the application of d-limonene during the extraction procedures, which can significantly enhance the extraction yield and may also protect the environment [5].

2.3 Essential oil yield and limiting factors for production in wild grown plants

Medicinal and aromatic plants are the major sources for essential oils. Most of these plants synthesize the essential oils in different parts through distinct developmental stages in their life cycle [15]. Since the major proportions of all the essential oils are plant-based they are called as green solvents [16]. Many plant species of genera Abies and *Ocimum* are reported to synthesize essential oils in different parts of the plants. They are employed for treating arthritis, bronchial asthma, bronchitis, cancer, diarrhea, dysentery, insect bites, malaria, eye disorders, and skin disorders [12]. The bioactivity of these plants is suggested to be mainly due to essential oils richness in terpenoids found in their leaves.

Different parts of plants such as seeds, roots, peels, leaves, fruits, and barks produce several essential oils [17]. Essential oils are found in bark and branches in case of camphor and cinnamon, leaves in case of oregano, eucalyptus and mint, roots and rhizomes in case of ginger, flowers as violet, lavender, rose, and jasmine and fruits and seeds of lemon, orange, nutmeg, and pepper. Generally, 5% of the dry matter of vegetables is formed by essential oils. In addition to the part of plant, environmental conditions for example climate, cultivation, soil and harvesting time also influence the composition of the essential oil [4]. Due to these multiple factors, the essential oil volatiles like other natural products are found in limited quantities in different plant tissues. Confinement of specific plant tissues for instance seeds or flowers for production of volatiles limit the yield of essential oils from the cultivated plants. Since the seeds or flowers represent a small weight compared to the whole plant in a harvesting season, eventually result in the low yield of useful volatiles and the operational procedures might be expensive. The higher market demand of the potent volatiles has been fulfilled by the production of chemically synthesized compounds, for instance, the synthetic product of natural vanillin is traded in the market for consumers [1]. However, the chemical synthesis of the natural products, although a cheaper option but is not preferred by consumers due to its negative impacts on human health and the environment. Biotechnological methods can provide promising means for the efficient production of useful natural volatiles through eco-friendly methods [5].

3.1 Biotechnological production of terpenes

Compared with the wild counterpart, plant *in vitro* cultures hold tremendous potential in the production of higher yield of secondary products including terpenes. However, in some instances, the volatiles detected in the intact plants were not found in the *in vitro* culture and vice versa. Besides, some novel volatiles have been reported to be produced as a consequence of biotransformation in *in-vitro* cultures, those which were not present in the wild plants [6]. *In vitro* cultures include all the methods such as callus and cell cultures, adventitious roots culture, shoot culture, hairy roots culture, embryo culture etc. which have been used for the production of the better yield of important secondary metabolites including essential oil [7] as indicated in Table 1 [15, 18-35]. Plants produce chemical compounds as a response to any type of stress exerted upon them. These compounds have been produced as a part of secondary metabolism and thus called secondary metabolites. Secondary metabolites which act as defensins for the plants are useful medicinal compounds for different animal disorders [8]. *In vitro* cultures can yield substantial amounts of secondary metabolites through different manipulation and elicitation strategies to enhance the yield [9]. Altogether, the different biotechnological

Materials Research Forum LLC
doi: https://doi.org/10.21741/9781644900239-1

methods can enhance the metabolic pathways in the plant cell for enhanced biosynthesis of terpene volatiles. Generally, terpene biosynthesis in the plant cell starts with the metabolic regulation of two important precursors known as isopentenyl pyrophosphate (IPP) and dimethylallyl diphosphate (DMAPP). It is worth mentioning that IPP in higher plants is biosynthesized through two different metabolic pathways including mevalonate-dependent (MVA) pathway and mevalonate-independent pathway also known as deoxyxylulose phosphate pathway, schematized in Fig. 2 [10].

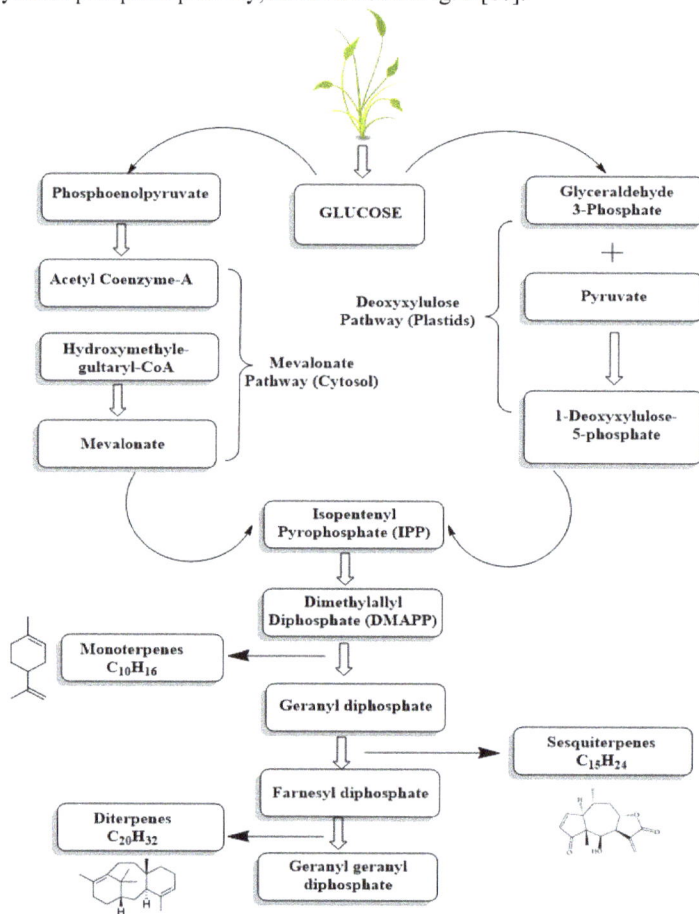

Fig.2 Terpenoids biosynthetic metabolic pathways in plants.

Table 1 Plant cell cultures of different medicinal plants with essential oil profiles

S. No.	Plants Name	Type of cell culture	Total volatile compounds	Abundant terpene volatiles	References
1	*Thymus moroderi*	Micro propagation	42	1,8-Cineole, camphor	[22]
2	*Mentharotundifolia* L huds	Micro propagation	18	Carvone	[23]
3	*Salvia dolomitica*Codd	Micro propagation	83	α–Pinene, β-phellandrene, borneol	[21]
4	*Salvia officinalis*	Micro propagation	73	Camphor, cis-Thujone	[24]
5	*Thymus vulgaris* L	Micro propagation	54	Thymol, γ-terpinene, p-cymene, carvacrol	[25]
6	*Anthemisobilis*	Shoots culture	30	o-farnesene, 2-hexenol	[26]
7	*Saturejakhuzistanica*	Shoots culture	14	Carvacrol, Γ-terpinenes, P-cymene	[27]
8	*Ajuga bracteosa*	Shoots culture	34	α-Pinene, β-pinene, β-ocimene	[18]
9	*Lallemantiaiberica*	Shoots culture	55	limonene, α-Pinene,	[20]
10	*Agastacherugosa*	Shoot culture		α-Pinene, estragole	[19]
11	*Anthemisobilis*	Shoots culture	30	o-Farnesene, 2-hexenol	[26]
12	*Ocimumbasilicum*	Shoots culture	37	Estragole, ethyleugenol 8-Cineole	[28]
13	*Origanumacutidens*	Callus culture	38	α –pinene, carvacrol	[29]
14	*Lallemantiaiberica*	Callus culture	9	Thymol, octane, carvacrol	[30]
15	*Rosa hybrid* L	Callus culture	11	Geraniol, Citronellol	[31]
16	*Daucusgenotypes.*	Callus culture	30	α –Pinene, carotol, α –Bergamoten	[32]
17	*Ajuga bracteosa*	Cell culture	29	β-Pinene,1-Terpinene-4-ol	[15]
18	*AgastacherugosaKuntze*	Cell culture	14	2,3-butanedione, limonene 2,6-nonadienal	[33]
19	*Cupressuslusitanica*	Cell culture	10	b-Thujaplicin. Camphor	[34]
20	*Tripterygium regelii*	Adventitious roots culture	12	Clastrol, diterpenoids	[35]

3.2 Approaches to improve the yield of terpenes produced by plant cell culture technology

Micropropagation is a robust and reliable technique used for multiplication of plants through *in vitro* cultures; it produces a large number of homogeneous plants in a short period of time. Besides, production of bioactive secondary metabolites can be enhanced in medicinal plants with this technique. During micropropagation, tiny parts of the plants commonly called as explants excised from different plants species can be micro propagated under optimized growth condition of culture media, temperature and photoperiod [36]. To engineer secondary product metabolic pathways through genetic manipulation of plants, the establishment of *in vitro* plant regeneration systems facilitates these efforts [37]. Conventional breeding for the enhanced production of high-quality plant products is still faced with many challenges. Nevertheless, plant genomics and biotechnology research have produced more knowledge to a better understanding of the complex genetics and biochemistry involved in the biosynthesis of these plant secondary metabolites including terpenes [38]. Although plant cell cultures don't provide those specialized glandular structures as found in natural plants for the accumulation of essential oils and related products, however, they provide good plant materials for accumulation and harvesting of non-poisonous terpenes. Different approaches can be employed through *in vitro* plant cell cultures for the stimulation of terpene biosynthetic pathways to influence the production of elevated levels of essential oils [6] as indicated in Fig. 3. Similarly, plant cell cultures can be manipulated for the upregulated biosynthesis of essential oils through:

1. Changing the composition of chemical reagents in the culture medium

2. Induction of differentiation in plant cell cultures.

3. Creating unique artificial sites for accumulation of volatiles.

4. Altering the physical conditions for *in vitro* growth, such as light, temperature and gaseous environment.

3.3 Terpenes in callus and cell cultures

Plant cell suspension cultures compared with wild plants and other types of cultures, have the advantage of being (1) less prone to various environmental variations (2) stable production platforms of homogeneous and uniform yield (3) rapid growth (4) reproducible (5) able to synthesize novel products that do not normally exist in the native plants [39]. In addition to medicinal products, cell suspensions have been employed to produce compounds used as fragrances, food flavors, and additives, dyes, coloring agents

[9]. There are, however, many limitations to cell suspension culture technology including slow growth and scale-up hurdles, the instability of cell lines and subsequent lower yield of some important metabolites [36]. A lot of important aromatic plants have been exploited for the production of useful volatiles such as industrially important monoterpenes through callus and cell cultures [15, 18]. Considerable levels of monoterpenes along with carotenoids were detected in the callus cultures of *Tanacetum vulgare* L. The promising green solvents, for instance, α-pinene and d limonene have been produced in the cell cultures of Pelargonium variants. Callus and cell cultures in Mentha species have shown variable trends in the production of monoterpenes within the different cell lines. In some studies the cell cultures were found to accumulate only the precursors of the volatile compounds; however, few Mentha cultures have been recommended for the production of characteristic monoterpenes, those found in the intact menthe plants. Likewise, callus cultures of *M. Piperita* have been reported for the accumulation of monoterpenes in special secretory organs. These volatiles were found in correspondence to the intact plants [6]. Callus and cell cultures of three aromatic plants species *Taiwan Aoshiso*, Akachirimen, and Aochirimen accumulated higher levels of monoterpenes than the wild grown respective plants.

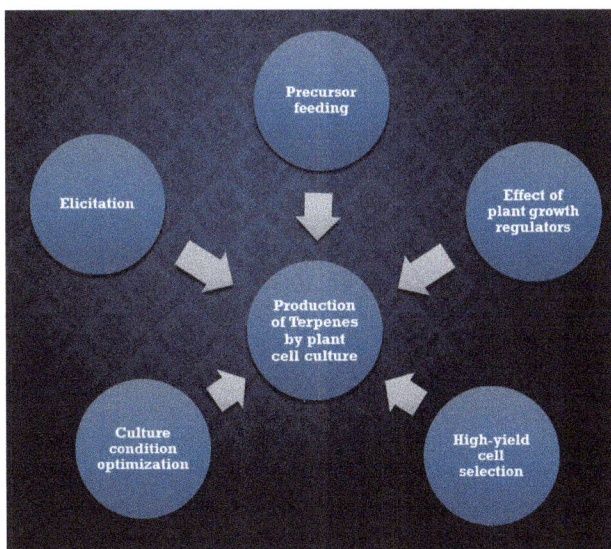

Fig.3 Plant cell culture strategies for enhancement of production of terpenes in plants.

Light intensity or quality during *in vitro* cultures can influence the plant cell's physiological and hormonal status through the initiation of distinct metabolic pathways that can influence and regulate the biosynthesis of important essential oils [15, 40]. In cell cultures of *Ocimumbasilicum* L, constant light illumination produced higher total essential oil yield including the potent volatile linalool than the cell cultures grown under complete darkness. The process of elicitation by application of chemical elicitors e.g. phenylacetic acid and methyl jasmonate and under the effects of physical elicitors such as the absence of light illuminance in the cultures has positively influenced the production of monoterpene *Ajuga bracteosa* cell culture [Fig.4]. Higher levels of monoterpene hydrocarbons such as α-pinene, β-pinene, β-limonene, β-ocimene, 1-terpinene-4-ol, caryophyllene, β-farnesene, myrtenal, citronellyl acetate, and β-element were detected in the cell cultures grown under the influence of methyl-jasmonate and constant dark [15] [Fig. 5]. In another study, the important monoterpenes such as limonene and terpinolene were elicited by methyl jasmonate under dark in higher amount in *Rosa damascene* cell cultures [41]. The process of elicitation is directly linked with the biosynthesis of essential oils in the plant cell. Several factors are responsible for the regulation of volatiles biosynthesis. These factors include genetic makeup of the explant used in cell cultures, type of culture media and the *in vitro* developmental phase of plant cells [42].

3.4 Terpenes in shoot cultures

In vitro regenerated shoots have been found to accumulate higher levels of monoterpenes. The role of exogenous application of plant growth regulators in the culture media has been reported to stimulate the biosynthesis of potent organic volatiles having industrial applications as green solvents [6, 15, 19, 20]. In a recent study by Ali et al. [18], it was observed that *Ajuga bracteosa*, (a high valued medicinal plant) accumulated higher levels of monoterpene hydrocarbons including limonene (3.4%), α-pinene (5.3%), camphene (4.45%), α-thujone (9.4%), 1,8 –cineole (14.3%), borneol (11.4%), camphor (12.2%), and nerol (9.2) in the shoots raised *in vitro* in response to application of thidiazuron (TDZ), an important plant regulator [18] [Fig. 5]. In another similar study, TDZ supplementation into the MS media produced a substantial amount of monoterpenes and sesquiterpenes through shoot cultures in medicinally potent plant *Lallemantiaiberica* [20]. The higher production of the important terpene volatiles in the regenerated shoots can be attributed to the different attributes of shoot cultures, such as the juvenile stage of the differentiated shoots, as the monoterpenes biosynthesis is directly linked to the young and immature shoot with higher metabolic potential [21]. Biosynthesis of terpene metabolites generally takes place in epidermal cells of shoot or leaf and store in special glandular structures called as leaf trichomes [18]. In another study, compared with callus cultures, the *in vitro* raised shoot cultures in medicinally important plants *L. angustifolia* and *R. officinalis* were found to accumulate higher levels of monoterpenes hydrocarbons [43].

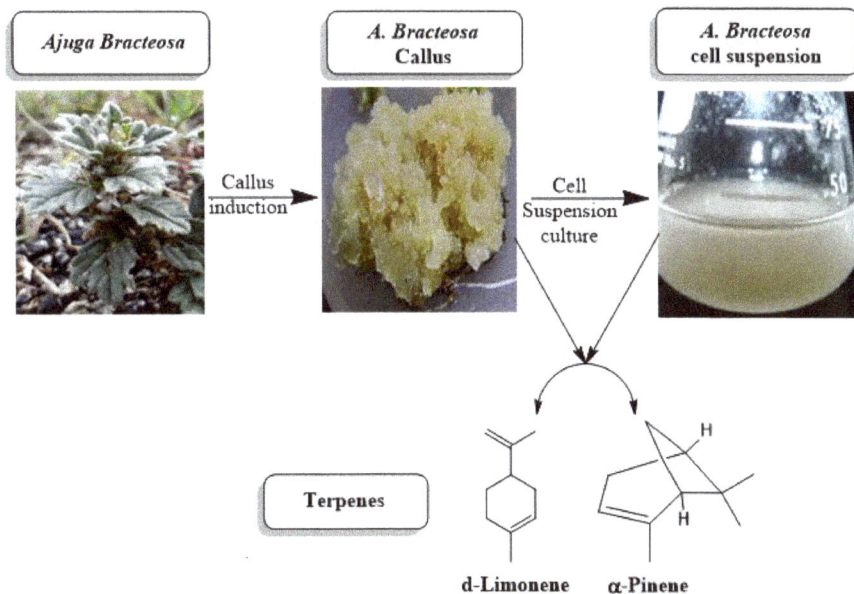

Fig.4 Development of plant callus and cell suspension cultures for production of terpenes in Ajuga bracteosa.

As the growth and development during *in vitro* shoot cultures are highly influenced by the effects of different plant growth regulators, so the biosynthesis of terpenes is positively correlated to the *in vitro* growth and developmental. Besides, the ontogenetic changes in the shoots as a result of plant cell growth in an artificially maintained growing environments and the accelerated but controlled secondary metabolism during *in vitro* cultures are the other important reasons which influence and regulate the biosynthesis of secondary volatiles [19, 21].

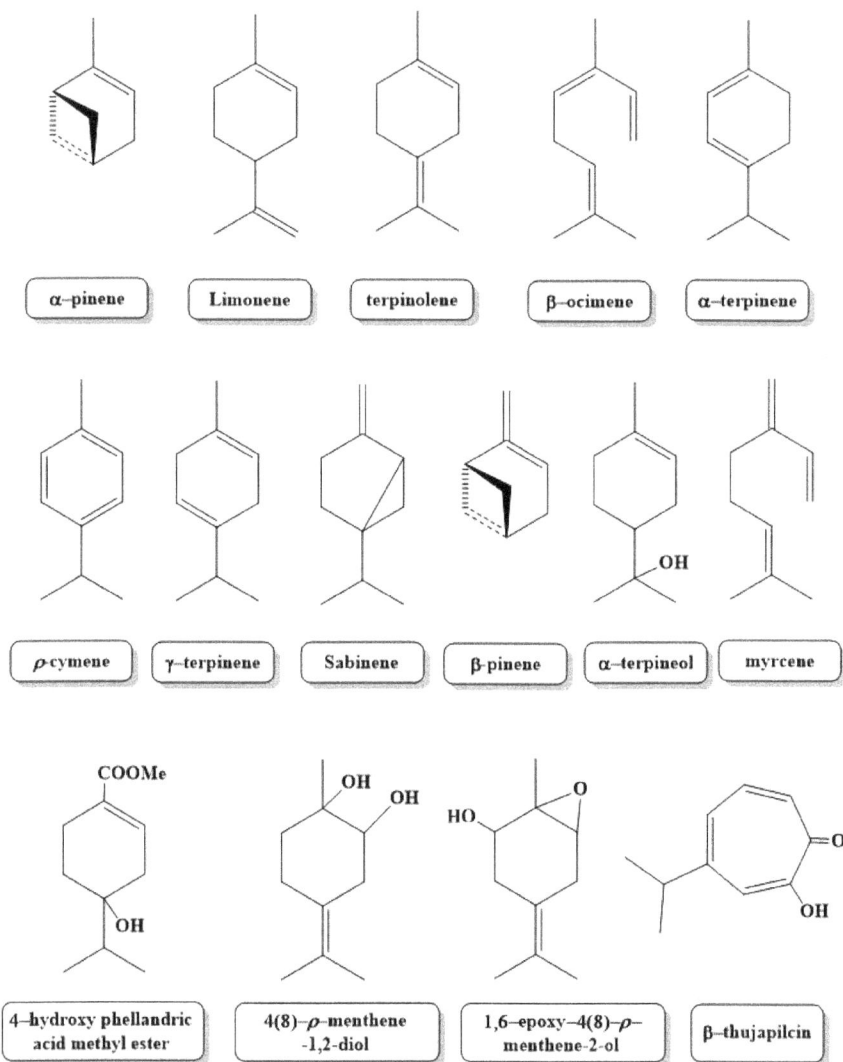

Fig.5 Chemical structures of different monoterpene volatiles detected in shoot cultures of Ajuga bracteosa.

3.5 Terpenes in hairy roots

Generally, the potential of plant cell cultures for the production of bioactive secondary metabolites can be enhanced by the induction of cell differentiation. Within the different cell culture approaches, hairy root cultures hold tremendous potential for the biosynthesis of volatile organic compounds besides other classes of potent secondary metabolites. When plant tissue is genetically transformed by *Agrobacterium rhizogenes* which inserts its T-DNA though Ri plasmid, results in the formation of hair like small and fine roots. The advantage of hairy root culture technology is that it does not require further media supplementation of cell cultures with plant growth regulators, because the inserted T-DNA carries the genes responsible for indigenous biosynthesis of auxins. Lacking the property of geotropism, hairy roots are highly branched and can grow faster than normal roots. They not only produce the metabolites at levels similar to the normal roots but also the metabolites which are produced in the aerial parts of the natural plants. Further, the hairy roots are phytochemically and biochemically stable like any other cell culture technologies. Hairy root cultures have been focused on their potential in the biosynthesis of notable natural products including volatile organic compounds [43].

An excellent review article, [44] has analyzed the different aromatic plant species which have been capable of the production of considerable levels of essential oils including terpenes through hairy root culture technology. Among the different plants, the hairy roots of *P. anisum and A. millefolium* resulted in producing essential oils, like the essential oil profiles of the mother plants, also some new metabolites were detected those were not found in the native plants. In certain cases such as hairy roots of *D. carota* and *L. alpinum*, the essential oil profiles of the volatiles were found in elevated levels, compared with the respective callus and cell cultures. The incremented yield of essential oil can be attributed to the application of elicitation strategies during hairy roots cultures [Fig. 6]. Nonetheless, the metabolic pathways for the biosynthesis of volatiles can be manipulated by using more effective transgenes that can be inserted into the T-DNA region.

4.1 Genetic engineering of plants for enhanced terpenoid biosynthesis

Few reports are available on genetic engineering of different plant species through transformation with the candidate genes responsible for terpenes biosynthesis. Particularly the metabolic pathways responsible for the production of mono and sesquiterpenes were tailored for enhanced production of these important organic volatiles. In these studies, the Cauliflower mosaic virus promoter (CaMV 35S) was used for the overexpression of the reductoisomerase DXR of the mevalonate MEP pathway in peppermint and a significantly higher (50%) increase in total essential oil production was observed. The yields of cyclic monoterpenes were enhanced by overexpression of limonene synthase enzyme in the plastid. The overexpression of the rate-limiting factors enhanced significantly the specific yield of monoterpenes [37]. It is crucial in some instances to enhance the yield of specific compounds of interest such as the monoterpenes α-pinene and d-limonene which are suitable alternatives to the hazardous chemicals. Thus the *in vitro* cultures through a genetic transformation in plants can boost the production of the desired compounds [45]. For instance, the production of monoterpene alcohols can be accelerated by the over expression of linalool synthase, the enzyme responsible for the profound production of glycosylated forms than the free form. Likewise, overexpression of prenyl transferase has been found to increase the yields of the linear as well as some cyclic sesquiterpenes [43].

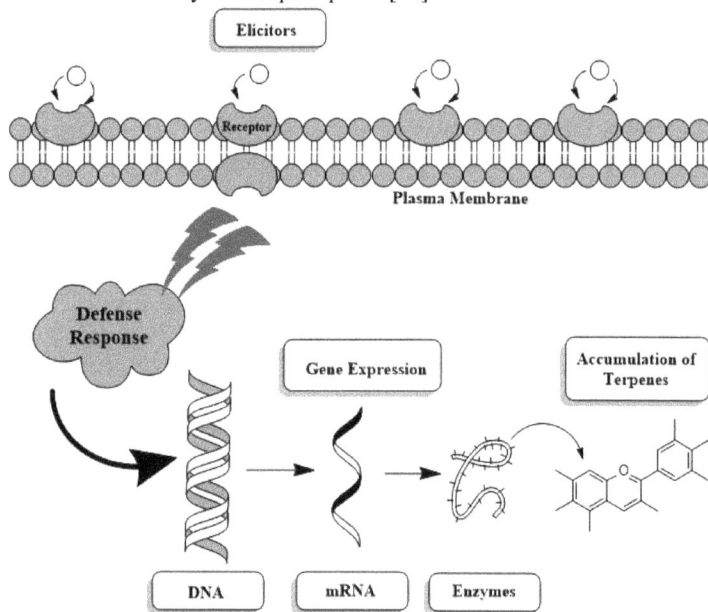

Fig.6. Impacts of elicitors on gene expression, secondary metabolism and production of terpenes in plant cell cultures.

Conclusion

The major monoterpenes derived from essential oils have proven their potential as green solvents of becoming alternatives to petroleum-based solvents in various industrial applications. The limited productivity of these volatiles in the natural plants could not meet the higher market demand of the potent volatiles in industrial applications. Biotechnological methods through manipulation of plant cell cultures in many medicinal and aromatic plants have improved the yield of essential oils. The recent advancements through genetic engineering of the metabolic pathways have resulted in the enhanced biosynthesis of terpene volatiles. However, further research studies should focus on the application of metabolomics and transcriptomic approaches to increment further the yield of essential oils in plants.

References

[1]   Y. Li, A.S. Fabiano-Tixier, F. Chemat, Essential oils as green solvents, in: Y. Li, A.S. Fabiano-Tixier, F. Chemat, Essential oils as reagents in green chemistry, Springer, 2014, pp. 55-61.

[2]   S. Bertouche, V. Tomao, A. Hellal, C. Boutekedjiret, F. Chemat, First approach on edible oil determination in oilseeds products using alpha-pinene, J. Essent. Oil Res. 25 (2013) 439-443.

[3]   M. Virot, V. Tomao, C. Ginies, F. Visinoni, F. Chemat, Green procedure with a green solvent for fats and oils' determination: microwave-integrated Soxhlet using limonene followed by microwave Clevenger distillation, J. Chrom. 1196 (2008) 147-152.

[4]   E.M. Napoli, G. Curcuruto, G. Ruberto, Screening of the essential oil composition of wild Sicilian rosemary, Biochem. Systemat. Ecol. 38 (2010) 659-670.

[5]   S. Chemat, V. Tomao, F. Chemat, Limonene as green solvent for extraction of natural products, Green Solvents I, Springer, 2012, pp. 175-186.

[6]   T. Mulder-Krieger, R. Verpoorte, A.B. Svendsen, J. Scheffer, Production of essential oils and flavours in plant cell and tissue cultures. A review, Plant Cell Tissue Organ Cult. 13 (1988) 85-154.

[7]   T. Khan, B.H. Abbasi, I. Iqrar, M.A. Khan, Z.K. Shinwari, Molecular identification and control of endophytic contamination during in vitro plantlet development of Fagonia indica, Acta Physiologiae Plantarum 40 (2018) 150.

[8]   R. Rani, M.A. Khan, W.K. Kayani, S. Ullah, I. Naeem, B. Mirza, Metabolic signatures altered by in vitro temperature stress in Ajuga bracteosa Wall. ex. Benth, Acta physiologiae plantarum 39 (2017) 10.

[9]   S. Saeed, H. Ali, T. Khan, W. Kayani, M.A. Khan, Impacts of methyl jasmonate and phenyl acetic acid on biomass accumulation and antioxidant potential in

adventitious roots of Ajuga bracteosa Wall ex Benth., a high valued endangered medicinal plant, Physiology and Molecular Biology of Plants 23 (2017) 229-237.

[10]   M. Zuzarte, L. Salgueiro, Essential oils chemistry, Bioactive essential oils and cancer, Springer, 2015, pp. 19-61.

[11]   L.S. Nerio, J. Olivero-Verbel, E. Stashenko, Repellent activity of essential oils: a review, Bioresour. Technol. 101 (2010) 372-378.

[12]   M.A. Khan, T. Khan, A. Nadhman, Applications of plant terpenoids in the synthesis of colloidal silver nanoparticles, Adv. Colloid Interface Sci. 234 (2016) 132-141.

[13]   N. Sangwan, A. Farooqi, F. Shabih, R. Sangwan, Regulation of essential oil production in plants, Plant Growth Regul. 34 (2001) 3-21.

[14]   S. Zwenger, C. Basu, Plant terpenoids: applications and future potentials, Biotechnol. Mol. Biol. Rev. 3 (2008) 1-7.

[15]   H. Ali, M.A. Khan, N. Ullah, R.S. Khan, Impacts of hormonal elicitors and photoperiod regimes on elicitation of bioactive secondary volatiles in cell cultures of Ajuga bracteosa, J. Photochem. Photobiol. B Biol. 183 (2018) 242-250.

[16]   Y. Li, A.S. Fabiano-Tixier, F. Chemat, Essential oils: from conventional to green extraction,in: Y. Li, A.S. Fabiano-Tixier, F. Chemat, Essential oils as reagents in Green Chemistry, Springer, 2014, pp. 9-20.

[17]   P. Masango, Cleaner production of essential oils by steam distillation, Journal of Cleaner Production 13 (2005) 833-839.

[18]   B.H. Abbasi, H. Ali, B. Yücesan, S. Saeed, K. Rehman, M.A. Khan, Evaluation of biochemical markers during somatic embryogenesis in Silybum marianum L, 3 Biotech 6 (2016) 71.

[19]   L. Daviet, M. Schalk, Biotechnology in plant essential oil production: progress and perspective in metabolic engineering of the terpene pathway, Flavour and Fragrance Journal 25 (2010) 123-127.

[20]   J. Díaz, A. Bernal, F. Pomar, F. Merino, Induction of shikimate dehydrogenase and peroxidase in pepper (Capsicum annuum L.) seedlings in response to copper stress and its relation to lignification, Plant Science, 161 (2001) 179-188.

[21]   T. Khan, B.H. Abbasi, M.A. Khan, M. Azeem, Production of biomass and useful compounds through elicitation in adventitious root cultures of Fagonia indica, Ind. crop. Prod. 108 (2017) 451-457.

Industrial Applications of Green Solvents I   Materials Research Forum LLC
Materials Research Foundations **50** (2019) 1-20   doi: https://doi.org/10.21741/9781644900239-1

[22]   H. Ali, M.A. Khan, W.K. Kayani, T. Khan, R.S. Khan, Thidiazuron regulated growth, secondary metabolism and essential oil profiles in shoot cultures of Ajuga bracteosa, Ind. Crop. Prod. 121 (2018) 418-427.

[23]   C.C. Giri, M. Zaheer, Chemical elicitors versus secondary metabolite production in vitro using plant cell, tissue and organ cultures: recent trends and a sky eye view appraisal, Plant Cell Tissue Organ Cult. 126 (2016) 1-18.

[24]   P. Olgunsoy, S. Ulusoy, U.Ç. Akçay, Metabolite production and antibacterial activities of callus cultures from rosa damascena mill. petals, Marmara Pharmaceutical Journal 21 (2017).

[25]   J.K. Holopainen, Can forest trees compensate for stress-generated growth losses by induced production of volatile compounds?, Tree physiology 31 (2011) 1356-1377.

[26]   S. Zielińska, E. Piątczak, D. Kalemba, A. Matkowski, Influence of plant growth regulators on volatiles produced by in vitro grown shoots of Agastache rugosa (Fischer & CA Meyer) O. Kuntze, Plant Cell Tissue Organ Cult. 107 (2011) 161.

[27]   N. Pourebad, R. Motafakkerazad, M. Kosari-Nasab, N.F. Akhtar, A. Movafeghi, The influence of TDZ concentrations on in vitro growth and production of secondary metabolites by the shoot and callus culture of Lallemantia iberica, Plant Cell Tissue Organ Cult. 122 (2015) 331-339.

[28]   L. Bassolino, E. Giacomelli, S. Giovanelli, L. Pistelli, A. Cassetti, G. Damonte, A. Bisio, B. Ruffoni, Tissue culture and aromatic profile in Salvia dolomitica Codd, Plant Cell Tissue Organ Cult. 121 (2015) 83-95.

[29]   Y. Gounaris, Biotechnology for the production of essential oils, flavours and volatile isolates. A review, Flavour and Fragrance Journal 25 (2010) 367-386.

[30]   J.A.T. da Silva, Floriculture, ornamental and plant biotechnology, Global Science Books 2006.

[31]   S.C. Roberts, Production and engineering of terpenoids in plant cell culture, Nat. Chem. Biol. 3 (2007) 387.

[32]   A. Marco-Medina, J.L. Casas, In vitro multiplication and essential oil composition of Thymus moroderi Pau ex Martinez, an endemic Spanish plant, Plant Cell Tissue Organ Cult. 120 (2015) 99-108.

[33]   S. Bhat, P. Maheshwari, S. Kumar, A. Kumar, Mentha species: in vitro regeneration and genetic transformation, Mol. Biol. Today 3 (2002) 11-23.

[34]   P. Avato, I.M. Fortunato, C. Ruta, R. D'Elia, Glandular hairs and essential oils in micropropagated plants of Salvia officinalis L, Plant Science, 169 (2005) 29-36.

[35]   V.R. Affonso, H.R. Bizzo, C.L.S. Lage, A. Sato, Influence of growth regulators in biomass production and volatile profile of in vitro plantlets of Thymus vulgaris L, J. Agr. Food Chem. 57 (2009) 6392-6395.

[36]   M.L. Fauconnier, M. Jaziri, M. Marlier, J. Roggemans, J.-P. Wathelet, G. Lognay, M. Severin, J. Homes, K. Shimomura, Essential oil production by Anthemis nobilis L. tissue culture, J. Plant Physiol. 141 (1993) 759-761.

[37]   F. Sadeghian, J. Hadian, M. Hadavi, A. Mohamadi, M. Ghorbanpour, R. Ghafarzadegan, Effects of exogenous salicylic acid application on growth, metabolic activities and essential oil composition of Satureja khuzistanica Jamzad, Journal of Medicinal Plants, 3 (2013) 70-82.

[38]   L.E.F. Monfort, S.K.V. Bertolucci, A.F. Lima, A.A. de Carvalho, A. Mohammed, A.F. Blank, J.E.B.P. Pinto, Effects of plant growth regulators, different culture media and strength MS on production of volatile fraction composition in shoot cultures of Ocimum basilicum, Ind. Crop. Prod. 116 (2018) 231-239.

[39]   M. Sökmen, J. Serkedjieva, D. Daferera, M. Gulluce, M. Polissiou, B. Tepe, H.A. Akpulat, F. Sahin, A. Sokmen, In vitro antioxidant, antimicrobial, and antiviral activities of the essential oil and various extracts from herbal parts and callus cultures of Origanum acutidens, J. Agr. Food Chem. 52 (2004) 3309-3312.

[40]   S.M. Razavi, H. Arshneshin, A. Ghasemian, *In vitro* callus induction and isolation of volatile compounds in callus culture of Lallemantia iberica (M. Bieb.) Fisch. & C.A. MEY, Journal of Plant Process and Function 5 (2017).

[41]   K.P. Singh, M. Bala, S.P. NAMITA, S. Kumari, Mutation breeding in roses: A review, Journal of Ornamental Horticulture 19 (2016) 55-74.

[42]   D. Jawdat, H. Al-Faoury, A. Odeh, R. Al-Rayan, B. Al-Safadi, Essential oil profiling in callus of some wild and cultivated Daucus genotypes, Ind. Crop Prod. 94 (2016) 848-855.

[43]   T.H. Kim, J.H. Shin, H.H. Baek, H.J. Lee, Volatile flavour compounds in suspension culture of Agastache rugosa Kuntze (Korean mint), J. Sci. Food Agr. 81 (2001) 569-575.

[44]   R. De Alwis, K. Fujita, T. Ashitani, K.i. Kuroda, Volatile and non-volatile monoterpenes produced by elicitor-stimulated Cupressus lusitanica cultured cells, J. Plant. Physiol. 166 (2009) 720-728.

[45]   F. Inabuy, J.T. Fischedick, I. Lange, M. Hartmann, N. Srividya, A.N. Parrish, M. Xu, R.J. Peters, B.M. Lange, Biosynthesis of diterpenoids in Tripterygium adventitious root cultures, Plant physiol. 175 (2017) 92-103.

Industrial Applications of Green Solvents I
Materials Research Foundations **50** (2019) 21-60

Materials Research Forum LLC
doi: https://doi.org/10.21741/9781644900239-2

Chapter 2

Ionic Liquids as a Green Solvent for Lipase-Catalyzed Reactions

A.A. Elgharbawy[1], M. Muniruzzaman[2*], H.M. Salleh[1], M.D.Z. Alam[3]

[1]International Institute for Halal Research and Training (INHART), International Islamic University Malaysia, PO Box 10, Kuala Lumpur 50728, Malaysia

[2]Centre of Research in Ionic Liquids (CORIL), Chemical Engineering Department, Universiti Teknologi PETRONAS, Bandar Seri Iskandar, Malaysia

[3]Department of Biotechnology Engineering, Kulliyyah of Engineering, International Islamic University Malaysia, PO Box 10, Kuala Lumpur 50728, Malaysia

* m.moniruzzaman@utp.edu.my

Abstract

Enzymes are important contributors to the current industrial development as they activate the reactions through enormous pathways. In recent years, ionic liquids (ILs) have become qualified media for clean extraction, photochemistry, green processing, electro-synthesis, and pharmaceutical applications. Although many enzymes have been studied in ILs media, lipases showed exceptional stability, selectivity and production yields. This chapter briefly outlines some merit as well as the downsides of current ILs applications in lipases reactions including the production of biodiesel, esterification and other established applications.

Keywords

Ionic Liquids, Lipase, Biocatalysts, Biodiesel, Transesterification

List of abbreviations

Cations

1-heptyl-3-methylimidazolium	$[C_7MIM]$
1-butyl-3-methylimidazolium	[BMIM]
1-hexyl-3- methylimidazolium	$[C_6MIM]$
1-ethyl-3-methylimidazolium	[EMIM]
1-dodecyl-3-methylimidazolium	$[C_{12}MIM]$
1-penty-3-methylimidazolium	[PMIM]
1-ethyl-2,3-dimethylimidazolium	[EDMIM]
1-hexadecyl-3-methylimidazolium	$[C_{16}MIM]$

1-hexyl-3-methyl-imidazolium	[HMIM]
N-octadecyl-N',N'',N'''-trimethylammonium	[C_{18}tma]
N-methyl-N-propanolpyrrolidinium	[C_1C_3OHPyr]
vinyl-3-ethylimidazolium	[veim]
3-methyl-2-(1-sulfobutyl)-1H-imidazolium	[BSO_3HMIM]
1-hexyl-3-methylimidazolium	[C_6mim]
1-methyl-3-octylimidazolium	[OMIM]
1-butyl-3-methypyridinuim	[Bmpy]
1-alkyl-3-methylimidazolium	[RMI]

Anions

tetrafluoroborate	[BF_4]
trifluoromethanesulfonate	[CF_3SO_3]
chloride	[Cl]
bromide	[Br]
hexafluorophosphate	[PF_6]
acesulfamate	[Ace]
saccharinate	[Sac]
bis(trifluoromethylsulfonyl)imide	[Tf_2N]
methylsulfate	[MS]
trifluoromethanesulfonate	[TfO]
bis(trifluoromethyl sulfonyl)imide	HTf_2N
acetate	[OAc]
methanesulfonate	[$MeSO_3$]
bis(trifluoromethylsulfonyl)imide	[TFSI]
hydrogensulfate	[HSO_4]

Other abbreviations

Heteropoly anion-based Brønsted acidic ILs	HPA-ILs	
Poly (1-vinylimidazole-based)	P(VB-VS)	
N,N-dimethyl-N-(3-sulfopropyl) cyclohexylammonium hydrogen sulfate ([Ps-N-Ch(Me)$_2$][HSO$_4$])		
N,N-dimethyl-N-(3-sulfopropyl) cyclohexylammoniumtosylate ([Ps-N-Ch(Me)$_2$][p-TSA])		
Burkholderia cepacia lipase	BCL	
porcine pancreatic lipase	PPL	
Rhizopus delemar lipase	RhDL	
Penicillium expansum lipase	PEL	
Thermoanaerobacter thermohydrosulfuricus lipase	TTL	
Candida rugosa lipase	CRL	
Candida antarctica lipase B	CALB	
free fatty acids methyl esters	FAMEs	
deep eutectic solvents	DESs	
sugar fatty acids ester	SFAE	

3-amino-propyltriethoxysilane   (APTES-Fe_3O_4)
Ionic liquids   ILs
Ethanol   EtOH
Tetrahydrofuran   THF

Contents

1. Introduction

Enzymes are biocatalysts which are widely contributing to enhancing the yield and speeding up the reaction rates in many industrial processes. Enzymes have the potential to function efficiently in supercritical fluids, organic solvents and any other aqueous environment which extends the technology perspectives of enzymes [1]. Organic solvents are not being a matter of concern due to their reasonable price, however, their contribution to the pollution due to toxicity, environmental concerns, volatility, and physical hazards, should be diminished. TThere is an urgency for researchers and industries to find alternatives to the conventional solvents used in industrial applications [2]. Considering these issues, ionic liquids (ILs) became a competitor with organic solvents as alternatives and attractive eco-friendly substitute due to their featured physicochemical merits [3].

The question of what an ionic liquid (IL) actually is, has been a topic of debate for a long time. The simple definition could be given as 'a liquid salt consisting of ions and ion pairs'. However, when ILs are mentioned today, they are often looked upon as organic salts with a low melting point of $100\ ^\circ C$ or lower and also known as room temperature ionic liquids (RTILs). RTILs possessing highly desired properties have attracted a broad range of developments as well as research [4].

ILs are ammonium, cholinium, imidazolium, pyridinium, pyrrolidinium, phosphonium and morpholinium derivatives combined with mineral or organic anions [5]. The physicochemical properties of ILs such as viscosity, density, hydrophobicity, and solubility can be tuned by proper selection of their components to adjust for certain applications. Therefore, ILs are regarded as "designer solvents" [1].

ILs have been applied in many fields including electro-synthesis, green processes, catalysts, fuel cells membrane batteries, lubricants, separation agents, nanomaterials synthesis, nucleophilic substitution reactions and biochemical transformations [6]. Accordingly, due to the adaptable properties of ILs, enzymes display better activity and prolonged stability. For illustration, *n*-heptane was substituted with IL in *Candida antarctica* lipase B system. The lipase was stabilized by the ionic liquid, 1-heptyl-3-methylimidazolium bis((trifluoromethyl)sulfonyl) imide $[C_7MIM][Tf_2N]$ due to the low nucleophilicity of the anion and the hydrophobic asset of the IL [7]. Enzymes such as lipases and proteases were reported to be thermally stable in ILs. Few suggestions on the stability were demonstrated. For instance, Sivapragasam et al. [5] conveyed that the chemical and structural stabilities of DNA and proteins may be boosted by IL, which can be reflected by the capability in the processing of biomass, transformation and catalytic processes [8]. Additionally, a wide range of enzymatic processes has been achieved in IL media, for example, aminolysis, alcoholysis, hydrolysis, esterification, and polymerization [5].

Enzymatic-aided reactions and whole-cell reactions have been performed in ILs. In many occasions, it was denoted that exceptional enzyme stability, yields, selectivity were achieved in IL media. Novozym® 435 (an immobilized lipase B from *Candida antarctica* (CALB) was stable in IL. Likewise, few oxidoreductases, such as D-amino acid oxidase, chloroperoxidase, peroxidase, and laccase were reported [9]. Lipases are very effective in chemical reactions as they are tolerant, capable to catalyze varied reactions such as epoxidation, esterification, and transesterification, and easier to obtain over other enzymes [10]. The current applications of lipases in IL media have established many advantages, such as better enzyme recyclability, high selectivity and excellent conversion rate [11]. However, most lipases display certain drawbacks when applied in specific reactions due to their sensitivity to the reaction media and the ability to operate in mild conditions [12]. In recent years, reports to explore the features that can be applied in serving the technology and the research interest are expanding (*Figure 1*).

In this context, lipases demonstrated improved stability in ILs in contrast to traditional solvents [5]. This chapter briefly outlines some merits as well as the downsides of recent applications of ILs in lipase-catalyzed reactions. The limitations and the recommendations for the ILs to be regarded as 'green solvents' in chemical and biochemical reactions are also discussed.

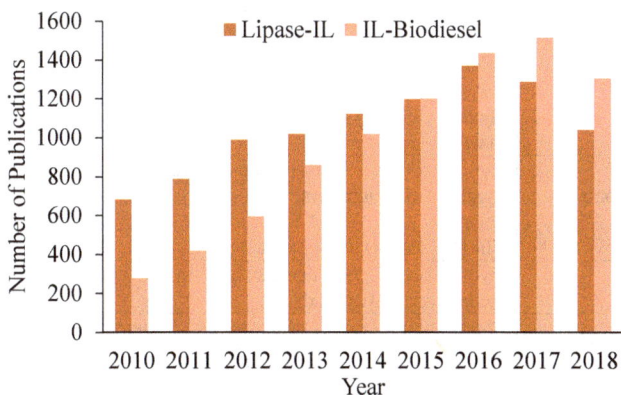

Figure 1. Number of articles published from "Scopus" by using keywords "Ionic liquids and lipase" as well as "Ionic liquids and biodiesel" during the 2010-2018 (September) period.

Industrial Applications of Green Solvents I   Materials Research Forum LLC
Materials Research Foundations **50** (2019) 21-60   doi: https://doi.org/10.21741/9781644900239-2

2.   Lipases: An overview

Enzymes catalyze a wide range of processes best in aqueous media and at particular pH with exquisite selectivity and stereospecificity. Among enzymes families, lipase is a vital component in the modern industry due to the capability of lipases from diverse sources to catalyze synthesis and hydrolysis reactions such as transesterification and esterification, depending on the thermodynamic conditions [13]. Lipases are, therefore, broadly distributed among microorganisms, plants and higher animals, where they are involved in the lipid's metabolism and can be extracted intracellularly or extracellularly. Lipases' catalytic mechanism involves a nucleophilic attack by the hydroxyl group of serine towards the substrate's carbonyl group, which generates an acyl-enzyme intermediate stabilized by an oxyanion hole. [14].

Table 1. Lipase applications in industry.

Category	Application	Ref.
Food industry	Bread enhancers to increase bread texture, color volume and shelf-life. Oil and fat restructuring Milk fat's hydrolysis, modification of butter fat, cheese ripening Synthesis of flavoring agents and emulsifiers Improving meat and fish flavors.	[15]
Chemical and Fuel industry	Synthesis of esters and emulsifiers Biodiesel production Formulation of detergents (oil stains removal for cleaning purpose). Transesterification of oils Reagents for lipid analysis. Hydrolysis of oils and fats to obtain monoacylglycerols, diacylglycerols and fatty acids.	[14] [15]
Cosmetics and perfumes	Esterification for Skin and sun-tan creams, bath oil. Synthesis of fragrances Digestive aids, sugar-based surfactants, specialty lipids Chiral synthesis intermediates	[14]
Pharmaceutical	Hydrolysis of polyester alcohols in the manufacturing of medicine; digestive aids	[15]
Other industries	Improving the quality of papers by removing the pitch from the pulp produced for papermaking	[16]

The excellent stability and capability of lipases to catalyze synthesis coupled with the catalytic polyvalence (catalytic promiscuity), have generated many industrial applications and a lot of research and development work. Certainly, lipases may function on broad range of substrates and can be introduced in few forms (native enzymatic powder, immobilized on supportive materials, liquid or solid). Consequently, they are used in various applications as presented in *Table 1*.

3.   ILs in enzymatic reactions: Advantages and merits

ILs are generally regarded as 'green solvents' although some have been reported to be toxic. Regardless, the green feature enabled the substitutional potential for conventional solvents. As they have negligible vapor pressure, ILs do not evaporate, and therefore, they abolish the environmental complications that are caused by volatile solvents [17]. Moreover, ILs can serve as media for the synthesis of several types of inorganic and organic materials. The advantages of ILs are shown in *Table 2*. With exceptional merits and properties, ILs can serve many applications due to the tunable structure with regards to the anions, cations and the side chain on the cation [18].

Table 2. Advantages of ionic liquids (ILs) in lipase-catalyzed reactions

Property	Merits	Ref.
Almost infinite possibilities of structural alteration	A 'designer solvents' as the cation and anion groups can be tuned as desired based on the application E.g., promote extraction efficiency, alter the reaction rate, increase substrate solubility, enhance lipase stability and reactivity. Adjustable physical and chemical properties.	[21]
Low melting point (<100 °C)	Extraordinary solubilization power of very hydrophobic substances in water. Facilitates product separation. Facilitates a new class of non-aqueous solvents with improved polarity.	[22]
Non-volatile with negligible vapor pressure	ILs do not evaporate at ambient conditions. ILs are non-flammable They are less toxic and readily recyclable.	[1]
High thermal stability	ILs remain liquid at wide range of temperatures ILs do not decompose at high temperatures.	[23]
Water-immiscible ILs	Provide a polar substitute with non-aqueous nature for biphasic system. Decrease the rate of deactivation.	[24]
Dissolve a wide range of natural and synthetic materials	Enhanced products, enzyme and substrates solubility. Allow lipase reactions in highly polar systems. ILs are suitable solvents for several kinds of inorganic, organic, ganometallic polymeric substances, small molecules.	[25]
Positive effect on enzymes	Enzyme activation and stabilization. Constructive influence on the reaction equilibrium or the enzyme's specificity Enhance lipase performance in solvent	[26]
Recyclability and reusable	Reduce energy use and cost of operation. Eco-friendly. Allow the reuse of lipases.	[27]

The uses of ILs have been conveyed from various industrial applications. For instance, ILs are known to serve as a potential catalyst in biodiesel production due to their merits, such as ease of separation, safe handling, fewer effluents to the surrounding, apt for continuous reaction and recyclability [19]. Moreover, in biodiesel formation system, ILs offer protection from deactivation induced by methanol. Generally, short-chains; 1,3-dialkylimidazolium cations such as 1-butyl-3-methylimidazolium bis(trifluoromethylsulfonyl)imide [BMIM][Tf$_2$N] or 1-butyl-3-methylimidazolium hexafluorophosphate [BMIM][PF$_6$], are used in the biocatalysis synthesis of biodiesel in a biphasic system [20].

4.   ILs properties and featured characteristics in lipase stabilization

ILs have been signifying their capabilities as a medium for many enzymatic processes, as a result of their merits. Enzymatic reactions are environmentally friendly processes which require lower temperature, high selectivity, and specificity. In this context, industrial processes that involve enzymes are auspicious resources towards clean manufacturing [28]. Nevertheless, lipolytic synthesis requires the use of non-aqueous solvents to dissolve a sufficient quantity of the substrate for the reaction to occur. Moreover, there is no complete understanding and basis of the enzyme's activity and stability predictions in ILs.

The choice of ILs may not be as simple as it looks, as they may have a negative impact on the enzymes. For instance, most enzymes could face an irreversible deactivation due to the strong hydrogen bonds (H-bonds) formation with the anions of ILs [28]. Hence, an alteration in cation or anion is reported to affect the IL's physiochemical properties which influence the enzymatic process through either stabilizing of the enzyme, activation, or structural modification [29]. Furthermore, the enzyme's stability or activity may be adapted by either tuning the ILs to diminish the deactivation effect or improving the enzyme tolerance in the ILs. It is therefore important to understand the enzyme's activation approach in ILs.

The IL's specificity towards stabilization, unfolding, refolding of various proteins and their activities have demonstrated by several studies. ILs were described as biotransformation media for enzymes such as cellulases and lipases, which occur if the enzymes are thermostable. Based on the Anfinsen hypothesis, the sum of interatomic interactions and the sequences of amino acids determine the three-dimensional (3D) and thermodynamically-stable native protein's structure in its regular physiological state [30].

ILs are superior compared to organic solvents due to the enzyme's stability improvement in the enzyme-catalyzed reactions [30]. This particularly applies in regard to the thermostability of the created system. For the synthesis of biodiesel, recycling the enzyme is also desirable when using IL to substitute organic solvents [31]. This is alongside with the low volatility characteristic that prevents the evaporation of ILs at the room temperature [31]. Quite a few enzymes have been recognized by their compatibility with certain types of ILs. For instance, *Candida rugosa* lipase (CRL) and *Candida Antarctica* lipase B (CALB) exhibited higher thermostability in the presence of ILs and therefore, the bioprocesses can occur at elevated temperatures if required [32].

4.1   Stabilization and activation of lipases in ionic liquids (ILs)

The novelty of using lipases in ILs media to enhance the catalytic activity is a promising and advanced approach, however, it comes with several encounters. On the other hand, a variety of ILs was compatible with lipases regardless of their microbial source [31]. This was also supported by the fact that hydrophilic ILs do not deactivate lipase but rather dissolve the lipase and form a biphasic phase that confines the retrieval of the enzyme and ILs for successive recycle, as ILs possess high polarity [33].

The interactions between *Candida rugosa* (CRL) and imidazolium or cholinium-based ILs have been investigated by Guncheva et al. [34] where the anions were based on amino acids (Leu, Trp, Thr, Val, Met, Ile, Gly). They suggested that these ILs resulted in structural rearrangements of the protein molecule which enables the enzyme's active site to be available for the substrate. Moreover, CRL displayed higher thermal stability in 1-methyl-3-octylimidazolium hexafluorophosphate [OMIM] [PF$_6$] than in hexane [35].

Zhao et al. [36] have prepared ether-functionalized ILs with acetate or formate anions; that are capable to dissolve a variety of substances and are also compatible with lipase. The prepared ILs dissolve oils at 50 °C; where lipases upheld its activity even in high concentration (50%, v/v) of methanol. The soybean oil transesterification in IL-methanol mixture showed the likelihood of ILs for lipase-stabilizing and oil-dissolving in the effective biodiesel production.

Another study reported by Schöfer et al. [37] revealed that the IL-lipase suspensions were stable and may as well be recycled for three cycles while losing only less than 10% of activity. Upon assessment, lipase (C*andida antarctica*lipase B) had 10-fold increment in the activity in [BMIM][Tf$_2$N] and1-butyl-3-methypyridinuim tetrafluoroborate [Bmpy][BF$_4$]. Furthermore, this established the argument that the enantioselectivity of lipases, generally, has a lesser value in hydrophilic IL than in hydrophobic ILs.

It has been recognized that long alkyl chains resulted in the increment of hydrophobicity of ILs. Various ILs with assorted cations and anions have shown the IL inhibition capability when the hydrophobicity of ILs increased. For instance, 1-butyl-3-

methylimidazolium tetrafluoroborate [BMIM][BF$_4$] exhibited the minimal inhibition of lipase activity, where 1-butyl-3-methylimidazolium trifluoromethanesulfonate [BMIM][CF$_3$SO$_3$] had a strong inhibitory effect, while 1-butyl-3-methylimidazolium chloride [BMIM][Cl] acted as medium inhibitor. It was concluded that for the ILs with the same cation but different anions, the hydrophobicity, and hydrogen bond ability influence their inhibiting action, while the hydrophobicity increment is associated with the inhibition capacity if the same anion was used [38]. It can be suggested that more stable and active biocatalysts can be obtained by integrating IL with enzymes, owing to the protective effect of IL towards lipases to encounter the deactivation in many cases.

4.2   Methods of stabilization of lipases in ILs

Several methods have been to stabilize enzymes in IL media including chemical and physical alteration of the enzymes and the modification of the IL itself, as summarized in *Figure 2*. Examples of stabilization are elaborated in *Table 3*.

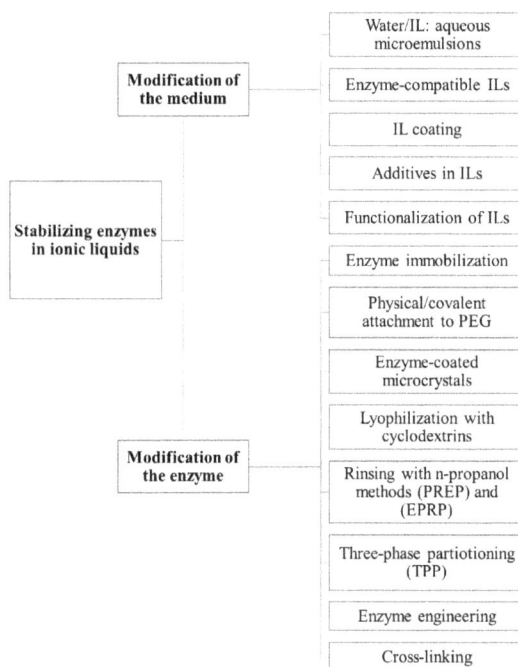

Figure 2. Methods for enzymes stabilization in ILs, summarized from [32].

Table 3. Examples of lipase activation and modification in ILs for stability enhancement.

Lipase	Method of stabilization	Ionic liquid	Enzyme performance	Ref.
Candida rugosa lipase	IL-Coating	Tetraethylammonium *L*-asparaginate and tetraethylammonium *L*-histidinate	Catalysis esterification of fatty acids and oleyl alcohol in hexane. The coated lipase displayed an improved activity than the free lipase.	[39]
Candida rugosa lipase (VII)	Functionalization of IL	*N*-methyl-*N* propanolpyrrolidiniumbis (trifluoro-methanesulphonyl)imide $[C_1C_3OHPyr]Tf_2N$	By prolonging the reaction time from 2 to 6 h, high yield of biodiesel could be obtained for the recovered lipase.	[40]
Burkholderia cepacia(BCL)	Physical adsorption on aerogel modified with protic ionic liquid	N-methylmonoethanolamine pentanoate	BCL immobilized on aerogel-IL showed an improved thermostability than free and BCL immobilized on original support material. Formed a stable and active biocatalyst.	[41]
Candida rugosa lipase	IL-compatibility	[BMIM][MS] and [BMIM][BF$_4$]	5% of [BMIM][MS] enhanced the enantioselectivity (E) of lipase by a factor of 50	[42]

Porcine pancreatic lipase (PPL)	Immobilization on Fe_3O_4-Chitosan nanocomposites functionalized with ILs	Imidazolium-based ILs	Improved activity, stability, and reusability of immobilized PPL	[43]
Candida rugosa lipase	Micro-encapsulation	(1-Vinyl-3-ethylimidazolium bis (trifluoromethylsulfony) amide) ([veim][Tf₂N])	Exellent stability of IL-encapsulated-lipase	[44]

Stabilization and activation of enzymes in ILs through immobilization on solid supports via a covalent linkage, cross-linkage, physical contact or encapsulation, are the most common approaches. For instance, *Burkholderia cepacia* lipase (BCL) immobilized with hydrophobic IL (N-methylmonoethanolamine pentanoate) at 1.0% (w/v) demonstrated an increase of 35 times in comparison with lipase without IL. The IL-lipase system yielded 46.2% ethyl esters production from triglycerides [45]. IL can also act as an agent for the immobilization of enzymes to enhance their activity. A significant improvement in the lipase activity was recorded when the enzyme was coated by RTSPILs (room-temperature solid phase ionic liquids) [46]. The RTSPILs are solid ILs below or near the room temperature and these can be used as a support matrix for immobilizing the enzymes via physical adsorption. The enzymes immobilized on RTSPIL may offer better effectiveness than immobilizing on the regular support materials, due to the specific influence of ILs' charge on the enzyme molecule. Hence, the hydrophobic RTSPILs have been auspicious support in immobilization technology [47].

Miao et al. [48] have synthesized Fe_3O_4 nanoparticles by chemical co-precipitation using the template as 1-butyl-3-methylimidazolium tetrafluoroborate [BMIN][BF_4]. The Fe_3O_4 nanoparticles treated with 3-amino-propyltriethoxysilane (APTES-Fe_3O_4) were used as carriers for the lipase (*Candida antarctica*), in combination with glutaraldehyde as a coupling reagent. The lipase displayed good diffusion in rapeseed oil and methanol. Moreover, an external magnetic field can quickly separate the magnetic nanoparticles from the solution for reuse. Likewise, porcine pancreatic lipase (PPL) immobilized on magnetic chitosan nanocomposites modified with IL preserved 91.5% of the initial activity after ten repetitive cycles [43].

Coating of the lipase with IL can induce significant alteration in the secondary structure. Stabilities and catalytic efficiencies of *Candida rugosa* lipase (CRL) and *Rhizopus delemar* lipase (RhDL) were reformed in IL containing a non-nutritive sweetener as an anion, such or acesulfamate or saccharinate (5% (w/v) 1-butyl-3-methylimidazoliumacesulfamate [BMIM][Ace] and 1-butyl-3-

methylimidazoliumsaccharinate[BMIM][Sac] [49]. Also, tuning the ILs components; the cations and anions seem to improve the stability of the enzyme. A study of hydrophilic IL by Solhtalab et al. [50] highlighted the effect of the length of alkyl chain of [C_nMIM][Br] on the activity of *Thermoanaerobacter thermohydrosulfuricus* lipase (TTL). The maximum activity of TTL was recorded in 1.0 M 1-butyl-3-methylimidazolium bromide [BMIM][Br], 0.3 M 1-hexyl-3- methylimidazolium bromide [C_6MIM][Br] and 0.3 M 1-butyl-3-methylimidazolium bromide [EMIM][Br], whereby no noteworthy response was detected in 1-dodecyl-3-methylimidazolium Bromide[C_{12}MIM][Br], which showed that hydrophilic ILs have positively affected the activity.

5.   Factors influencing IL-lipase reactions

The proper selection of the IL for the application as solvents or co-solvents for enzymatic reactions is crucial. Some key factors related to ILs such as composition, polarity, viscosity, temperature, pH, and water content have been documented in the literature and explored by many researchers. Some of these factors are discussed below.

5.1   IL composition

ILs anions have nucleophilicity properties and can form hydrogen bonds (H-bonds) based on their properties and structures. Therefore, the anions could influence enzymes' stability and activity by the strong interaction with the enzyme and resulting in conformational changes [51]. For instance, the stability of CALB was investigated by He and coworkers [52] in imidazolium-based IL with various cations and anions. Their study showed that lipase could retain the activity after reusing for five times in hydrophobic ILs such as [EMIM][Tf_2N], 1-penty-3-methylimidazolium hexafluorophosphate [PMIM][PF_6] and 1-ethyl-2,3-dimethylimidazolium bis((trifluoromethyl) sulfonyl) imide [EDMIM][Tf_2N] whereas lipase's activities in organic solvents and hydrophilic ILs were severely reduced. They observed that the highest lipase's activity was detected in [EDMIM][Tf_2N], where no activity was detected in imidazolium-based IL with long chain cations.

On the other hand, a mixture of hydrophobic and hydrophilic ILs positively affect both the stability and activity of an enzyme. Lipase from *Candida rugosa* was co-immobilized with a mixture of two ILs whereby the hydrolysis and esterification activities of the immobilized lipase were greater by 14-fold than in silica gel without IL, respectively. The highest stability of immobilized lipase was attained in 1-hexadecyl-3-methylimidazolium bis((trifluoromethyl)sulfonyl) imide [C_{16}MIM][Tf_2N], the hydrophobic IL [53]. The effect of IL's structures based on [BF_4]$^-$, [Tf_2N]$^-$ and [PF_6]$^-$ anions and 1-alkyl-3-methylimidazolium ([C_nMIM]$^+$) cation, on the CALB was investigated by Wang et al. [54]. They demonstrated that the occurrence of fewer than six carbons in the alkyl chain of the cation would correspond to the IL's hydrophobicity and lipase activity. They found that additional increment in the chain length impaired the activity of CALB. The enzyme activity depending on the nature of anion was in the order $[BF_4]^- < [PF_6]^- < [Tf_2N]^-$.

5.2   IL polarity

All probable intermolecular interactions, either specific or nonspecific, between the solvent and the solute that is not resulting in chemical alteration, have been found to be dependent on the polarity of ILs [55]. The basicity of hydrogen bonds is an additional factor that may influence the activity of enzymes. The polarity of an IL is similar to formamide or the alcohols with a fewer number of carbon atoms [56]. In addition, ILs with non-coordinating anions such as $[PF_6^-]$ and $[Tf_2N^-]$ have lower polarity than alcohols with fewer carbon atoms as shown by the solvatochromic study. High polar solvents promote the polar substances' solubility and result in more selective and faster reactions [51].

Several studies have investigated the enzyme stability and conversion process in terms of polarity influence. Zhao et al. [57] have reported that the stability of enzymes by IL may be described from enzyme dissolution and substrate steady-state stabilization by IL. They have tested a group of imidazolium-based ILs such as [OMIM][BF$_4$], [HMIM][Cl], [BMIM][Br], [EMIM][PF$_6$], [BMIM][BF$_4$] [EMIM][Br]. Their study revealed that the polarity variances between the ILs did not directly affect the lipase activity.

In general, the enzyme's activity can be enhanced by increasing the IL's polarity. For example, the *Pseudomonas cepacia* lipase exhibited higher activity in the racemic 1-phenyl ethanol acetylation with vinyl acetate in IL. The initial reaction rate in more polar IL, [EMIM][BF$_4$], was 3-fold faster than in the less polar IL, [BMIM][PF$_6$] [58].

Ha et al. [52] on the other hand, observed that [BMIM][BF$_4$] had a less stabilizing effect on Novozym 435® (lipase acrylic resin from *Candida antarctica*) in comparison with other hydrophobic ILs. Few more studies have suggested that hydrophilic ILs such as [EMIM][BF$_4$], [BMIM][BF$_4$] and [BMIM][BF$_4$] comparatively displayed high activities. The anion polarity rates acetate as the highest and $[PF_6^-]$ as the lowest. Meanwhile, the cation polarity depends on the chains or branches attached to the charged center, with less polarity for octyl (8C) than a methyl (1C) group. The declining polarity for cations can be ordered as ammonium > imidazolium > pyridinium > pyrrolidinium, with identical substituents on the cations [59].

Studies have shown that enzyme activity can be possibly linked with the strength of the anion's hydrogen-bond acceptor whereby enzyme stabilization could be a result of low basicity of hydrogen-bond anions [60]. Nevertheless, the cations role cannot be neglected on the overall performance and properties of ILs. Concurrently, we may deduce that the lipase activity and stability may not depend only on the IL polarity, but also on the structure of the enzyme and the amino acids in the active site which may display different charges in different polarities and H-bond strength.

5.3   IL viscosity

ILs have viscosities that are higher than regular organic solvents due to comprehending van der Waals interactions, the H-bonding and/or ion-ion interactions. In this respect, ILs' viscosities have shown to generally rise with the elongation of the cation's alkyl chain for the same anion due to the stronger force of van der Waals [51]. IL's viscosity influences the enzymatic reaction mechanism. The reaction rate may be influenced by the viscosity in terms of the limitation of mass transfer. It is rational that high IL's viscosity affects the structure and the activity of the enzyme by slowing down the conformational transformations. However, this does not apply to all biocatalytic reactions that take place in IL [26]. In general, ILs with less viscosity, hydrophobic in nature, chaotropic cation and kosmotropic anion have shown better enzymatic activity and stability. However, there is no universal rule due to many contrary reports to generalize the trend. Despite some contrary results, CALB has catalyzed the transesterification of *n*-butanol and ethyl butyrate in more than 20 ILs which endorsed the effect of the IL's viscosity on the rate of reaction [57].

The high viscosity results in a great mass transfer tolerance, which hampers the contact between the enzyme and substrates or among the substrates, which result in a slower reaction rate [61]. For instance, Novozym 435 was used in ILs "1-ethyl-3-methylimidazolium methylsulfate [EMIM][MS], 1-ethyl-3-methylimidazolium trifluoromethanesulfonate [EMIM][TfO], [OMIM][Tf$_2$N],", with a solvent-free medium and *tert*-butanol, whereby free fatty acids methyl esters (FAMEs) conversion in *tert*-butanol was about 65.8% but decreased after that. The conversion in [EMIM][TfO] was 80% after 12 h, which is 15% greater in IL than in *tert*-butanol, although the viscosity of [EMIM][TfO] (42.7 cP) is ten times higher in contrast to *tert*-butanol (4.312 cP) [62].

It is denoted that long long alkyl chains with imidazolium ring have a larger value of viscosity, though an influence may result from the anions as well.

5.4   pH

The pH value plays a vital role in lipase activity in non-aqueous enzymatic catalysis, by ruling the amino acid residues located at the active site of the enzyme. Amino acids are neutral, positively charged or negatively charged molecules, hence, change in pH of the medium affects the charges and thereby the enzyme activity and the rate of the reaction. Therefore, enzymes require a certain pH to function properly.

In this context, Rios et al. [63] have reported a better yield of ethylene glycol esters during synthesis by CALB and IL system at pH 7.0. On the other hand, low pH values resulted in the protonation of amino acids residues in *Candida rugosa* lipase at the positions (His449 and Glu341), which enabled the nucleophilic attack by Ser209 oxygen from the substrate's carbon, consequently leading to lipase activity reduction [63]. Generally, enzyme-organic reactions depend much on pH of the medium where the enzyme reaction occurs. The pH varies upon varying the water content of the reaction system and solvents type, in which both affect the local polarity and the enzyme's active site. An IL-reaction medium would act similarly, which might be a tedious issue due to the IL's polar and ionic nature. Due to the changes in IL composition, IL may have potentials to alter the pH of the medium. Moreover, the availability of water in the reaction system might result in the decomposition of some ILs and hence, lower the enzymes' activity and stability [64].

The bis(trifluoromethyl sulfonyl)imide (HTf_2N-IL) has higher acidity than both [BMIM][Tf_2N] and [BMIM][BF_4]. The pH of the medium not only depends on the nature of the enzyme but rather on the IL composition. The occurrence of [BMIM][Tf_2N] as an additive was assessed with immobilized *Bacillus sp.* ITP-001 lipase, at 37 °C in the pH range of 3.0–9.0 during a hydrolysis reaction. The IL addition significantly stabilized the optimal pH for the enzyme. CRL has an optimal pH of 7.2 which dropped to 7.0 in IL medium. The enzyme activity dropped to 30 – 40% in contrast with the optimum value of pH is acidic (lower than 6.0). On the other hand, when the pH was higher than 8.0, no activity was detected [65].

In actual fact, the IL systems with different cations and same anion changed the enzymatic activity similarly as the pH changed. Rather, IL systems with the same cation and different anions behave contrarily towards the enzyme activity as pH shifted. This implies that the pH influence on lipase activity in hydrolytic reactions highly relies on IL's anion.

5.5   Water content

The amount of water in IL systems is known to impact the activity and efficiency of the enzymatic reactions. Enzymes are suspended in reaction media at low water content as they are not soluble in most of the ILs [66]. It is an established fact that to facilitate the enzyme activity, a certain amount of water is required to increase the enzyme's active structural sites. Nevertheless, an additional amount of water has a destructive impact on the enzyme, whereby it facilitates the aggregation of the enzyme which reduces the diffusion of the substrate and results in enzyme inactivation. It has been observed that addition of a proper amount of water positively affected the yield of biodiesel, where the highest yield was attained when water content was in the range of 0.4–0.8 g; and an additional increment in water decreased the yield. The initial water content is critical in reactions involving enzymes. For instance, in biodiesel production, high content of water leads to the hydrolysis of ester instead of the synthesis. To illustrate the critical role of water content in an enzymatic reaction, the drying of the feedstock has been found critical to avoid oils and fats hydrolysis to free fatty acids (FFAs). The occurrence of FFAs resulted in the formation of soap and hindered in the purification of the products [67].

Enzymes are mainly inactive in dry media. However, the enzymatic reactions are likely to be enhanced if a small water quantity is introduced into the non-aqueous medium. In this context, the major role for water is to ensure optimal activity of the enzyme. For instance, lipase displays higher activity in solvents with less polar nature and more hydrophobic property in most cases. Therefore, regulating the water level in enzymatic reactions is mandatory. The water content is usually recognized as thermodynamic water activity (a_w) [68].

For *C. rugosa* lipase, Ulbert et al. [35] ranked the solvents with optimal water content as hexane < toluene< [ONIM][PF$_6$]< [BMIM][PF$_6$] <[BMIM][BF$_4$] which is in harmony with reducing the solvent's polarity. A solvent has higher propensity to strip off the water from the enzyme's surrounding when it becomes more polar. ILs have a better capacity to hold water as they possess high polarity. Two common approaches have been used to maintain the water activity in the system; (i) pre-equilibrium stage separation of the enzyme and substrate solution using a saturated aqueous mineral solution, or (ii) an aqueous buffer preparation by adding a hydrated salt to the reaction. When a hydrated salt occurs in the non-aqueous system, it may influence the activity of the enzyme through water buffering and specific effects, which are vital for the enzyme to function optimally. The solvent's requirement of water determines the controlling effect of each role. Therefore, using hydrated salt to regulate the water activity in the biocatalytic IL system should be applied with extreme caution [68].

The initial activity of lipase (CALB) was lower in 1-hexyl-3-methylimidazolium [C_6mim][Tf_2N] in almost anhydrous medium (a_w=0.17), while the highest enzyme activity was obtained at a_w=0.28. Further increase in the water activity leads to a reduction in enzyme activity. It is suggested that catalytic activity of the enzyme is associated with the absorption of water molecules on the protein's surface; however, in high water activity system, for instance, during vinyl *n*-octanoate hydrolysis, the enzyme gathered easily and resulted in the hydrolysis in the presence of water [54].

When the whole cells of *Rhizopus oryzae* were used in a transesterification reaction, there was a reduction in the yield of biodiesel when the water content rises from 5 to 10%. The reaction rate was at its maximum when 10% of water content was set, whereby at higher than 10% water content, lower rate was achieved with immobilized *Burkholderia* lipase. Thus, it can be resolved that the leverage effect between the unwanted side reactions and lipase activity maintenance with water content, requires being well-thought-out, particularly in lipase-catalyzed biodiesel production [61].

We may conclude that pure ILs act similarly as a polar organic solvent in which they devour essential water from the enzyme. Therefore, it is suggested that the appropriate level of water is critical to the catalytic activity of the enzyme in the IL system; to improve the enzymatic reactions over non-polar solvents.

5.6 Temperature and thermal stability

Temperature influences the activity of lipases significantly. IL-enzyme systems is affected by the temperature in a manner it mostly influences the enzyme's activity rather than the system as a whole. Nonetheless, the temperature has a significant effect on the IL viscosity and thus on the enzyme dissolution, in which it influences the activity.

The decomposition temperature attained from the thermo-gravimetric analysis (TGA) experimentally under vacuum, is used to assess the thermal stability of IL. Moderate but noteworthy decomposition of IL was denoted often by mass spectrometry. Also, weight losses were demoted at constant temperatures as different as 200 °C cooler than the "onset" TGA temperature [69].

In relevance to the temperature variation, a study on the enzyme's activity was reported by Yang et al. [68]. They have applied *Penicillium expansum* lipase (PEL) in transesterification and hydrolytic processes in the IL [BMIM][PF_6]. Optimal temperature of 70 °C in the IL was recorded. The temperature of the reaction of lipases was in the range of 35−60 °C [61]. The IL, [C_6mim][Tf_2N] was evaluated with CALB. The initial reaction rate (v_0) of vinyl n-octanoate conversion increased remarkably at elevated temperatures. The enzyme was deactivated at elevated temperatures (above 60 °C), resulting in reduced enantioselectivity and activity. The maximum conversion was achieved at 40 °C, while the highest enantioselectivity (> 400) was attained at lower temperatures (30–40 °C). The difference between *n*-hexane and IL at high temperature may be elucidated by the strong protective effect of IL against the enzyme's thermal inactivation [54]. Operating at higher temperatures minimizes the microbial contamination of the product and offers better solubility and homogenized mixtures of viscous plant oils and fats which are the natural substrates of lipase [70]. Apart from maintaining the lipases activity, reaction temperature affects the methanol loading as a result of evaporation which impacts the mass and heat transfer. During lipase-catalyzed reaction at high temperatures, the evaporation of alcohol donors shall also be considered [61].

We may also conclude that the reaction temperatures have a vital role in enzymatic reactions, by enhancing the reaction rate while at the same time, maintaining the enzyme activity.

5.7   Selectivity

Two conformations of lipases may be present: "open" (active) and "closed" (inactive). The two shapes exist because of structural rearrangements similar to those persuaded by substrates, depending on the medium of the reaction. Many lipases are enhanced in the non-polar solvents due to conformational modification [70].

ILs are designer solvents and could be adapted to fit in the desired reactions while enhancing the selectivity of the lipase towards the substrate. For instance, Lozano et al. [71]showed that [EMIM][BF$_4$] stabilized CALB with double of the half-life to 8.3 h at 50 °C. Continuous operations of CALB in [BMIM][PF$_6$] resulted in 2300 times greater half-life time and selectivity higher than 90%, than that recorded when the incubating the enzyme in the absence of substrate (3.2 h) [71]. Enantioselectivity is the selectivity of a reaction towards one of a pair of enantiomers. Enzymatic reactions performed in ILs have shown good enantioselectivity, high stability, better activity, and improved reaction rates. The enantioselectivity value was lesser in toluene than that in ILs, with [C$_6$mim][Tf$_2$N] being the best reaction medium. The substrate structural difference justifies the enantioselectivity of enzyme-aided acylation in ILs. The enantioselectivity enhancement in IL may result from the high thermostability of enzymes in ILs [54]. The enzyme's enantioselectivity might be influenced by several factors, such as the structure of the substrate, the nature of the medium and the enzyme's thermostability. Therefore, no satisfactory explanations are available. Thus, more exploration shall be addressed to understand the selectivity mechanism of enzymes in IL.

5.8   Product purification

Being designer solvents, ILs have the tendency to reduce the formation of by-products and the occurrence of the processes side reactions. Therefore, these are preferred in synthesis and hydrolysis involving the use of enzymes. Product separation or downstream processing is a challenge of the industrial application. Researchers observed that crude biodiesel post-treatment through conventional methods resulted in separation complications, as multiple steps are essential to eliminate the glycerides, catalyst, alcohol, glycerol, and soap [67]. Moreover, in lipase-catalyzed reactions, the free lipase has a homogeneous nature, which leads to technical complications including non-reusability of biocatalysts and the difficulty in separation processes. For instance, the extraction of 1-butanol from aqueous solution is a measure of the potential of "bis[(trifluoromethyl)sulfonyl] amide-based" ILs as selective separating agents. A comparison between imidazolium-based and ammonium-based ILs resulted in the conclusion that few of them affect the distribution ratio and butan-1-ol/water selectivity [72]. In this context, and to facilitate the product purification, free lipase CALB was immobilized in magnetic nanoparticles functionalized with the IL, [BMIN]BF$_4$. The products were collected after each reaction round via magnetic separation and then a new methanol and rapeseed oil were introduced to the reactor with the recycled immobilized lipase [48]. As a practical benefit, the efficient and facile separation of the solid catalyst was simply achieved by an external field of the magnet.

In industrial practice, ILs have demonstrated their merits through a few chemical manufacturing processes, including alkoxyphenolphospines production, which is a basic precursor for photoinitiator by BASF (BASIL^{TM} process). In BASIL^{TM} process, the reaction time was reduced significantly leading to a higher capacity. Moreover, the product separation from the IL and the catalyst was also easily achieved [73].

Lozano et al. [74] used Novozym® 435, lipase coupled with the hydrophobic IL,"N-octadecyl-N', N'', N'''-trimethylammonium bis (trifluoromethylsulfonyl) imide ([C$_{18}$tma][Tf$_2$N])" in the production of biodiesel from triolein. Enzyme's half-life was promoted up to 1370 days at 60 °C in the IL and the mixture of reaction was separated into pure biodiesel, solid IL and glycerol. Application of IL-based catalyst minimized the required steps in the preparation of biodiesel and separation which were the obstacles to develop an economically feasible process.

6.   Lipases and ILs in lipids reactions

Lipases are major players in enzyme industries which include but not limited to synthesis of surfactants, biodiesel preparation, catalysts in hydrolyses, ester synthesis, transesterification and inter esterification processes, foods, pharmaceutical, and cosmetics. Lipases strive in diverse environments and therefore advanced as an auspicious catalytic agent for biotransformation reactions in ILs, deep eutectic solvents (DESs) and in organic solvents [75].

6.1 IL-lipase-catalyzed biodiesel synthesis

Biodiesel, considered as an eco-friendly substitute to current petroleum fuel, is the transesterification product of animal fats or vegetable oils and is regarded as a biodegradable and renewable biofuel and considered as an eco-friendly substitute to current petroleum fuel. The most commonly used catalysts for transesterification include acids, alkalis, chemical catalysts and lipases [40]. Other approaches include thermal cracking (pyrolysis), blending and microemulsion. Transesterification is the reaction of alcohol and carboxylic acid esters in the presence of a catalyst to produce FAMEs and water. This process is influenced by several factors, mainly, temperature, alcohol/oil molar ratio, water content, free fatty acids, dosage and catalysts' type. The process is a three-step reversible process. During the transformation of triglycerides, diglycerides, monoglyceride, and then glycerol in the last phase, to generate one mole of ester in each step [17].

The use of ILs has unlocked new prospects to produce biodiesel via lipase-catalyzed process as the IL does not only serve as a solvent but also as a catalyst. Short-chain ILs such as [BMIM][PF$_6$] or [BMIM][Tf$_2$N] are well-documented and carry the biocatalytic synthesis and have been used for biodiesel production in enormous instances. Acidic ILs based on "sulfonate-functionalized quaternary ammonium salts (N,N-dimethyl-N-(3-sulfopropyl) cyclohexylammonium hydrogen sulfate ([Ps-N-Ch(Me)$_2$][HSO$_4$]) and N,N-dimethyl-N-(3-sulfopropyl) cyclohexylammoniumtosylate ([Ps-N-Ch(Me)$_2$][p-TSA]))" have been applied as catalysts for the biodiesel synthesis through *tung* oil transesterification with methanol; through which "[Ps-N-Ch(Me)$_2$][p-TSA]" displayed the highest catalytic activity and yielded 98.98% of biodiesel, with five times reuse of the IL [76].

Hydrophobic imidazolium-ILs, are well-organized in the lipase-catalyzed synthesis of biodiesel. The competence of hydrophobic ILsis more than their corresponding hydrophilic peers owing to their protective action towards the enzyme from dissociation into the aqueous layer [77]. The most used lipase for biodiesel production is CALB which has been commercialized as (Novozym® 435) in an immobilized form. In general, both the hydrophobic and hydrophilic ILs have been tested in biodiesel production.. However, most of the studies reported that higher yield of FAME was attained in hydrophobic ILs in contrast with that obtained in organic solvents or solvent-free media.

The addition of $[C_1C_3OHPyr][Tf_2N]$ improved the biodiesel yield due to the miscibility of methanol with the IL, leading to the reduction in the inhibition level on the lipase, resulting in higher biodiesel yield [40]. Sunitha et al. [33] have reported biodiesel production from sunflower oil with the catalysis by Novozyme® 435 with two hydrophilic ILs [HMIM][BF$_4$] and [BMIM][BF$_4$] and two hydrophobic ILs [EMIM][PF$_6$] and [BMIM][PF$_6$]. The yield of FAMEs was higher in the hydrophobic IL (98%) in both [EMIM][PF$_6$] and [BMIM][PF$_6$] compared to 10% of in [HMIM][BF$_4$], and no yield was obtained in [BMIM][BF$_4$] (hydrophilic IL).

It was due to the fact that ILs are capable to provide a greater contact area between the lipase and the substrate (oil), which increases the enzymatic activity. In addition, the use of IL produces a biphasic system which facilitates the separation of the products at the the last stage of the reaction [78]. *Table 4* summarizes the uses of some ILs for biodiesel production by lipase-catalyzed systems.

Table 4. Examples of lipase-catalyzed biodiesel production in ionic liquids with the optimum conditions.

Ionic Liquid (IL)	Substrate	Solvent	Lipase source	Optimum conditions	Yield of biodiesel	Ref.
[HMIM][PF$_6$]	Chinese tallow kernel oil	Methanol	*Candida rugosa*	40 °C, 48 h, methanol/oil: 4:1	95.4%	[79]
[BSO$_3$HMIM][HSO$_4$]	Rapeseed oil	Methanol	-	130 °C, 5 h, Methanol/oil 12:1, 2% catalyst (IL)	Nearly 100%	[80]
[Ps-N-Ch(Me)$_2$][p-TSA]	*Tung* oil	Methanol	-	120 °C, 2 h, Methanol/oil 21:1, 5% catalyst (IL)	98.98%	[76]
poly (ionic liquid): P(VB-VS)HSO$_4$	*Soapberry* oil	Methanol	-	150 °C ,8 h Methanol/oil 29:1, 8.7 wt% catalyst (IL)	95.2%	[81]

6.2 Synthesis of esters and other products in ionic liquids

Organic solvents and ILs work efficiently in enzymatic synthesis, however, organic solvents became unfavorable owing to the toxicity issues, particularly in food, cosmetic or pharmaceuticals industries. The tunable physicochemical properties of ILs, allow them to serve as media in enormous enzymatic reactions. The IL's selection is important, as the incorrect IL may negatively impair the enzyme. As elaborated in previous sections, PF_6 and BF_4 ions do not interact strongly with water whereby optimum enzyme hydration can ensue inactivation can be prevented.

The fatty acid ester synthesis of L-ascorbic acid was enhanced by IL. The substance, L-ascorbic acid ester is a naturally occurring antioxidant with various applications in fats and oils [82]. The ascorbic acid esterification with oleic acid or palmitic acid using CALB was enhanced in [EMIM][BF_4] with a yield as 61-65% of 6-O-L-ascorbyl oleate. In this context, IL was able to dissolve polar substances, such as ascorbic acid, while preserving the activity of the lipase.

Findrik et al. [28] used [BMIM][PF_6] in the fatty acids ester of sugars (SFAE) synthesis from glucose and palmitic acid catalyzed by Novozym® 435. SFAE, characterized by insecticidal and antimicrobial activities has been utilized as surfactants in various industries. The process could go for four cycles without purification. Starch palmitate was synthesized using a mixture of [BMIM][BF_4] and [BMIM][OAc] and *Candida rugosa* lipase.

Using ILs, the problems related to low enzyme stability and starch' solubility limit in many reactions have been overcome, as many substrates which are insoluble in conventional solvents have better solubilities in ILs (*Table 5*).

Table 5. Lipase catalyzed-reactions and their main products, in Ionic liquids.

Ionic Liquid (IL)	Lipase source	Acyl acceptor	Acyl donor	Product	Yield [%]	Application	Ref.
[BMIM] [BF_4] with DMSO as co-solvent	Lipozyme RMIM	Galactose	Oleic acid	Galactose oleate	87%	Emulsifiers, surfactants in pharmaceutical, cosmetics and food, Detergent.	[84]
[EMIM] [$MeSO_3$]	*Burkholderia Contaminans* (DFS3) lipase	Solvent-free	Vinyl acetate	methyl 6-O-acetyl-α-D-glucopyranoside	70%	Esters of α-D-glucose production	[85]

[BMIM] [TFSI] [BMIM] [PF$_6$] [HMIM][PF$_6$] [OMIM][PF$_6$]	Novozyme 435	Hexanol	dihydrocaff eic acid	Methyl dihydrocaffeate, hexyl dihydrocaffeate, dodecyl Dihydrocaffeate, octadecyl dihydrocaffeate	68.7% 84.0% 81.2% 74.8%	Antioxidant properties	[86]
[BMIM][BF$_4$] [BMIM][MS]	*Candida rugosa* lipase	Isopropano l	Ketoprofen ethyl ester	Ketoprofen	45%	Anti-inflammatory drugs (NSAIDs) Toothpaste additive to prevent periodontal disease.	[42]

IL-lipase reactions have been useful in biopolymers and food industries, for instance, corn starch esterification with *Rhizopus oryzae* lipase in novel surfactants from imidazolium-based ILs has been attempted by Adak and Banerjee [83]. An enhanced hydrophobicity and thermoelectricity were displayed by the modified starch.

6.3 Ionic liquids: The large-scale promise

The continuous growth of ionic liquids research has unlocked many limitations of their applications. Promising outcomes have been obtained on using ILs at laboratory scale, and the accomplishments at these primary applications have accelerated the potential of ILs at a greater scale.

Several commercialized and pilot processes using ILs have been recently recognized. Although ILs have not been used widely in industrial applications for commercial use, some corporations have initiated the industrial applications and distribution of ILs. In 1998, the commercial use of ILs was approved by the French Petroleum Institute. It was used in the preparation of polybutene (Difasol process), an important intermediate for producing rubber and plastic [87]. The initial industrial use of IL was introduced by BASF and was called the BASIL^{TM} process (Biphasic Acid Scavenging utilizing Ionic Liquid), in Ludwigshafen, Germany, in 2002. This process formed *in situ* ionic liquid to capture the HCl generated during the alkoxyphenyl phosphines production from reacting chlorophenyl phosphines with alcohols. The established process was utilizing 1-methylimidazole to capture the acid, as an alternative to trimethylamine. The reaction formed 1-methyl-imidazolium chloride, a liquid salt at the reaction conditions. The yield in BASIL^{TM} process was enhanced 80,000 times in contrast to the using tertiary amine in the conventional process [88].

Several firms such as Degussa, Central Glass Co. Ltd., IFP (Institut Français du Pétrole), Linde, Pionics, IoLiTec (Ionic Liquids Technologies) and Eastman Chemical Company have also engaged with the challenge of ILs technology scale up from the laboratory to industrial scale. Some of the industrial applications of ILs at pilot or commercial scales are presented in *Table 6*.

Table 6. Some of the leading companies in ILs industrial synthesis and applications.

Company	Application	Ionic Liquids	Ref.
BASF	Azeotropes breaking Process to break azeotropes, eg: H_2O-EtOH and H_2O-THF	$[RMI][BF_4]$	[88]
Supelco	Analytical chemistry: ionic liquid GC columns, such as SLB-IL111 ionic liquid column.	Example: [1,9-di (3-vinyl-imidazolium) nonane][Tf_2N]	[89]
Proionic TECHNOLOGY - CBILS®	Synthesis Biomass - Electrolytes - High-temperature cooling	[BMIM], [EMIM] based ILs on kg, tons. Customized ILs synthesis	[90]
IoLiTec	Customized synthesis Batteries, solar cells, Electrolytes, Fuel Cells, Sensor Electrolytes, Supercap-Electrolytes, solvents	Examples: TEGO®-R -IL-P9 1-Alkyl-2-methylimidazoles, Ammonium-based ILs	[91]

7. Recyclability of ILs and lipase

Recycling of lipases utilized for bioprocessing of oils and other biomaterials reduces the cost of the process and also minimizes the wastewater release into the environment. The reusability of lipase has been reported by several studies with promising results. Moniruzzaman and his group 44] have shown the potential of IL-lipase system to be reused as a biocatalyst, whereby the produced IL-microencapsulated *Candida rugosa* lipase in surfactant aggregates formed in ILvinyl-3-ethylimidazolium bis(trifluoromethyl-sulfonyl) amide) ([veim][Tf_2N]) monomer. Lipase encapsulated within the IL polymer retained its activity and showed excellent stability in aqueous solution. Moreover, most of the activity was retained after five cycles, with a complete recovery of the biopolymers the reaction mixture by centrifugation [44].

The feasible industrial method of membrane filtration could be overpriced way beyond using new IL in the process [26]. Therefore, the separation of hydrophobic ILs using the freezing process is very practical to release the IL from the reaction. With advanced

Industrial Applications of Green Solvents I   Materials Research Forum LLC
Materials Research Foundations **50** (2019) 21-60   doi: https://doi.org/10.21741/9781644900239-2

technology, more promising recyclability options have been introduced recently. For instance, a novel solid acidic IL polymer (PIL) has been used to catalyze the synthesis of biodiesel from waste oil. Results demonstrated that the PIL was very effective for both esterification of free fatty acids and the transesterification of triglycerides (99.0% yield). The recovery of the solid catalyst could be achieved by filtration. After the six cycles, 99% yield was obtained [92]. The hydroxyl-functionalized IL, $[C_1C_3OHPyr][Tf_2N]$ was utilized as a medium for the lipase (*Candida rugosa*)-catalyzed transesterification reaction. At the end of the reaction, the enzyme and IL-phase were recycled by washing with water followed by acetone. High biodiesel yield was obtained by extending the reaction time from two hours (fresh) to six hours (recovered) due to the reduction of the catalytic performance of the retrieved enzyme which resulted from lipase's active sites blockage by impurities [40]. Combination of SO_3H- functioned zwitterion and phosphotungstic acid (PW) formed heteropoly anion-based Brønsted acidic ILs (HPA-ILs). The activity of the reused catalyst was examined via the esterification of methanol with oleic acid. Upon each run, the reaction mixture was cooled, and the catalyst was collected at the bottom of the reactor as solid and filtered out for collection. There was no notable reduction in yield after four cycles, though 15% loss of the catalyst' weight was recorded [93].

Recently, magnetic nanocomposites have been obtained by combining the covalently modified chitosan using imidazole-based ILs with various functional groups and magnetic nanoparticles Fe_3O_4 to be a support matrix for porcine pancreatic lipase (PPL) immobilization.PPL immobilized on nanocomposites (modified with IL) displayed higher reusability and retained 91% (residual activity) after 10 repeated cycles. More importantly, using a magnetic field, the immobilized lipase could be readily retrieved [43]. Similarly, graphene oxide Fe_3O_4 nanocomposites were produced by Xie et al. [94] and the nanoparticles were used as magnetic carriers for *Candida rugosa* lipase, to serve as a biocatalyst for the production of biodiesel with 92% yield. The catalyst was reused for five cycles without activity loss. Similar attempts were also done by other groups using the idea of magnetic-IL and immobilized lipase for easy recovery [47].

It can be concluded that lipase-catalyzed reactions are currently promising and have enormous horizons. Due to the development of nanotechnology, immobilization and encapsulation of lipases on solid support and nanocomposites, as well as functionalization with ILs, are motivating advancement to carry on the exploration of more possibilities and opportunities of lipase and ILs processes.

8.   Kinetics parameters of lipases in ILs

Despite many proofs on enhancing the catalytic activity of lipases in ILs, the effects of IL on the behavior of the biocatalyst are still in the exploration stage. In fact, kinetic properties of the enzyme could disclose the mechanisms and the catalytic activity of the reaction, in which the enzyme acts. Enzymes operate to enhance the rate of reaction therefore, the rates of the reactions catalyzed by enzymes with different substrates and under different conditions have been examined. The main platform used to understand enzymatic reaction has been the Michaelis-Menten equation (Eq.1) which has been modified depending upon the enzyme and the inhibition.

$$v = \frac{V_{max}S}{K_m + S} \ldots\ldots\ldots\ldots\ldots\ldots\ldots\ldots\ldots\ldots\ldots\ldots\ldots\ldots\ldots\ldots\ldots\ldots (1)$$

where the maximal velocity is V_{max}, the rate of the enzyme is v, the substrate's concentration is S, and the constant for Michaelis-Menten's is K_m, which expresses the substrate concentration at $V_{max}/2$. K_m is also an indicator of the concentration of the substrate needed for a significant catalysis to take place [95].

The reaction rate of an enzyme fluctuates in a hyperbolic manner according to the substrate concentration. At V_{max}, enzyme molecules (active sites) are all occupied with the substrate. At a definite enzyme concentration, the maximum velocity (V_{max}) discloses the turnover number, or the catalytic constant, k_{cat}. k_{cat} is the maximum number of substrate molecules converted into a product by the enzyme, at the saturation of substrate on the enzyme per unit time. On the other hand, the specificity constant, k_{cat}/k_m, is the rate constant that describes the relationship between enzyme and substrate concentrations, to express the enzyme efficiency. The catalytic perfection is achieved when enzymes have k_{cat}/k_m ratio in the range of 10^8-10^9 M^{-1}s^{-1}[26].

ILs have been reported to enhance the rate of reaction, stabilize the lipase and prevent inhibition in some events. For instance, *Candida antarctic* lipase B immobilized on multiwall carbon nanotubes (MWNTs) modified by imidazolium-based ILs with different functional groups, showed a high affinity towards the substrate and high reaction rate. We can conclude that the IL modification contributes to improving the affinity between the carrier and the enzyme. The IL in use can lower the activation energy (E_a) required for the enzyme reaction and stabilize the enzyme-substrate complex [E-S] which improves the enzymatic activity [96].

The kinetic performance of IL-catalyzed transesterification of n-butanol and of methyl acetate revealed two models: (i) the ideal homogeneous (IH) model and (ii) the non-ideal homogeneous (NIH) model. The two models were suggested to interpret the kinetic data. Compared to conventional catalysts such as Amberlyst 15, sulfuric acid and ion-exchange resin catalysts, ILs were found more effective than conventional catalysts [98].

The enzymatic transesterification of ethyl ferulate (EF) with castor oil was recently reported in imidazolium with BF_4^-, TF_2N^- and PF_6^-, whereby [EMIM][PF$_6$] displayed the highest conversion (~100%) with excellent selectivity for the formation of feruloylated mono- and di-acylglycerols. Moreover, a significant shielding effect was observed against the thermal deactivation of lipase (Novozym 435®). The Arrhenius equation was ln $v° = 13.17 - 64.37/RT$, and E_a (activation energy) was 64.37±3.00 kJ. mol^{-1}. The initial reaction rate expression was in agreement with the Ping-Pong Bi-Bi kinetic model. The Ping- Pong Bi-Bi model follows the equation (Eq.2):

$$v = \frac{V_{max} [A][B]}{(K_{mB}[A] + K_{mA}[B]) + [A][B])} \dots\dots\dots\dots\dots\dots\dots\dots\dots\dots\dots\dots\dots\dots\dots\dots (2)$$

The E_a, V_{max}, K_{mA}, and K_{mB} were 64.37 kJ. mol^{-1}, 1.07×10^{-4} mol. L^{-1} min^{-1}, 0.11 mol. L^{-1}, and 8 mol. L^{-1}, respectively. The Ping-pong mechanism is characterized by changing of the enzyme into an intermediate form when the first substrate to product reaction occurs. At the end of the reaction, the enzyme does not undergo any alteration. Moreover, one product formed and released before the second substrate binds [99].

PEG-modified lipase-catalyzed the alcoholysis of 2-phenyl-1-propanol and vinyl acetate in n-hexane and [BMIM][PF$_6$]. The results showed some variations in the kinetics of the reaction between the organic solvent and the IL. The V_{max} was relatively lower in n-hexane (47 mmol. g^{-1} h^{-1}) than in [BMIM][PF$_6$] (55 mmol. g^{-1} h^{-1}). The value of K_m in [BMIM][PF$_6$] was about half of that in n-hexane. The results suggested that the stabilization seemed to be well-established in the IL in contrast with n-hexane and the complex of [E-S] was well-prepared to be formed in [BMIM][PF$_6$].

From our point of view, the kinetics of IL-catalyzed reactions is yet to be examined and established. With a number of 10^5 and more possibilities of ILs formulations, more simulation, molecular dynamics, and kinetics models are required to fit in the wide range of ILs in use. There is more to be explored on the mechanisms of activation or inhibition of the enzyme in ILs to best describe their behavior and optimize the reaction conditions.

Industrial Applications of Green Solvents I   Materials Research Forum LLC
Materials Research Foundations **50** (2019) 21-60   doi: https://doi.org/10.21741/9781644900239-2

9.   Limitations and concerns over ILs

At present, the physical and chemical characteristics of hundreds of ILs are still unexposed. Therefore, ILs toxicity is still unidentified, and hence, not all ILs could be considered green. There is no consistency in ILs toxic effect. Some ILs display low toxicity, while others show significant inhibition in several biological systems. It is presumed that the higher water solubility of ILs correlate with their high impact on the environment, as they could easily diffuse to the ecosystems. The IL's purity is also critical as impurities may impair the extraction processes or the enzyme activity [21].

The viscosity of some ILs is also one of the concerns in selected applications. The elongation of the cations' alkyl chain for a fixed anion generates stronger van der Waals interactions, leading to a negative impact on the stability and activity of the enzyme. The rate of biocatalytic reactions is influenced by IL's viscosity due to the mass transfer limitation, in case of vicious IL and fast reactions [51].

On the other hand, the recycling process of ILs requires volatile solvents or water. Therefore, establishing a separation technique is one of the solutions to resolve this issue. In this context, efforts have been made to immobilize enzymes and ILs on solid supports, which facilitate the recovery and the separation as mentioned in previous sections. Even so, many of the applications offered by ILs such as membrane separation processes and supercritical extraction technologies must be addressed. To date, the use of ILs in several industrial processes has provided the additional benefit of recyclability and enhanced yields. Despite the extensive research, ILs have not been generally assessed as eco-friendly solvents for the industrial applications in the food sector [87]. Moreover, ILs' biodegradability is a priority concern. The chemical components of the IL should decompose into harmless and easily degradable components.

10.   The outlook for ILs in industrial applications

ILs are rapidly emerging as important tools in the formulation of innovative and complicated materials and have been contributing remarkably towards advancing of new sustainable, efficient and eco-friendly processes. We have highlighted the benefits, advantages, and potential of ILs in the lipase-catalyzed reactions. To date, the challenges and concerns are directed to finding renewable resources with low energy consumption while minimizing costs and increasing efficiency towards industrial-scale processing using ILs (*Figure 3*).

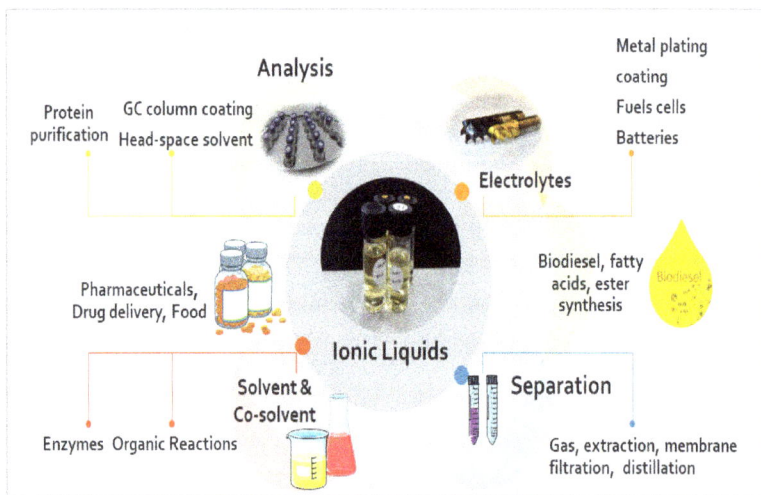

Figure 3. Summary of ionic liquid applications in the industry

For the ILs-catalyzed reactions, we need to highlight a few main points:

- When designing an IL, the researcher should define the potential in applications rather than inventing solvents. This is due to the indefinite numbers of possibilities of ILs to be formed (by altering the cations and anions). Thus, a full investigation of the existing ILs is the first priority.

- The publication trend is showing environmental concern. One of the most challenging tasks is waste management. Recycling and reuse of the ILs and the catalysts should be the direction of developing ILs processing and applications.

- ILs are expected to gain a significant impact in the food industry, considering the current flow of research and investigation.

- It is suggested to plan synthesis of the IL and run a cost-benefit analysis, including seizing the recovery technique and the purification method to be used.

- ILs field is expanding. So far, this field is branched to enormous applications including drug delivery, biofuels, biomass processing, pharmaceutical, metal plating, batteries and more. Enhancing the scale-up of production is a great future mark towards sustainable technology, which has been proven succeeded by new-born companies.

References

[1]   M. Moniruzzaman, K. Nakashima, N. Kamiya, M. Goto, Recent advances of enzymatic reactions in ionic liquids, Biochem. Eng. J. 48 (2010) 295–314. https://doi.org/10.1016/j.bej.2009.10.002

[2]   A.A. Elgharbawy, M.Z. Alam, M. Moniruzzaman, M. Goto, Ionic liquid pretreatment as emerging approaches for enhanced enzymatic hydrolysis of lignocellulosic biomass, Biochem. Eng. J. 109 (2016) 252–267. https://doi.org/10.1016/j.bej.2016.01.021

[3]   M. Moniruzzaman, T. Ono, Separation and characterization of cellulose fibers from cypress wood treated with ionic liquid prior to laccase treatment, Bioresour. Technol. 127 (2013) 132–137. https://doi.org/10.1016/j.biortech.2012.09.113

[4]   J. Gräsvik, D.G. Raut, J.-P.P. Mikkola, Challenges and perspectives of ionic liquids vs traditional solvents for cellulose processing, in: M. Jihoon, S. Haeum (Eds.), Handbook of ionic liquids. Properties, applications and hazards, Nova Science Publishers, Inc, New York, 2012, pp. 1-34.

[5]   M. Sivapragasam, M. Moniruzzaman, M. Goto, Recent advances in exploiting ionic liquids for biomolecules: Solubility, stability and applications, Biotechnol. J. 11 (2016) 1000–1013. https://doi.org/10.1002/biot.201500603

[6]   J. Kulhavy, R. Andrade, S. Barros, J. Serra, M. Iglesias, Influence of temperature on thermodynamics of protic ionic liquid 2-hydroxy diethylammonium lactate (2-HDEAL) + short hydroxylic solvents, J. Mol. Liq. 213 (2016) 92–106. https://doi.org/10.1016/j.molliq.2015.10.061

[7]   M. Cvjetko, J. Vorkapić-Furač, P. Žnidaršič-Plazl, Isoamyl acetate synthesis in imidazolium-based ionic liquids using packed bed enzyme microreactor, Process Biochem. 47 (2012) 1344–1350. https://doi.org/10.1016/j.procbio.2012.04.028

[8]   H. Olivier-Bourbigou, L. Magna, D. Morvan, Ionic liquids and catalysis: Recent progress from knowledge to applications, Appl. Catal. A Gen. 373 (2010) 1–56. https://doi.org/10.1016/j.apcata.2009.10.008

[9]   B. Dabirmanesh, K. Khajeh, J. Akbari, H. Falahati, S. Daneshjoo, A. Heydari, Mesophilic alcohol dehydrogenase behavior in imidazolium based ionic liquids, J. Mol. Liq. 161 (2011) 139–143. https://doi.org/10.1016/j.molliq.2011.05.007

[10]   M. da Graça Nascimento, J.M.R. da Silva, J.C. da Silva, M.M. Alves, The use of organic solvents/ionic liquids mixtures in reactions catalyzed by lipase from

Burkholderia cepacia immobilized in different supports, J. Mol. Catal. B Enzym. 112 (2015) 1–8. https://doi.org/10.1016/j.molcatb.2014.11.013

[11]   B. Zou, C. Song, X. Xu, J. Xia, S. Huo, F. Cui, Enhancing stabilities of lipase by enzyme aggregate coating immobilized onto ionic liquid modified mesoporous materials, Appl. Surf. Sci. 311 (2014) 62–67. https://doi.org/10.1016/j.apsusc.2014.04.210

[12]   R.A. Sheldon, S. van Pelt, Enzyme immobilisation in biocatalysis: why, what and how, Chem. Soc. Rev. 42 (2013) 6223–6235. https://doi.org/10.1039/c3cs60075k

[13]   L.B. Ramos Sanchez, Fungal lipase production by solid-state fermentation, J. Bioprocess. Biotech. 5 (2015) 1-9. https://doi.org/10.4172/2155-9821.1000203

[14]   C. Aouf, E. Durand, J. Lecomte, M.-C. Figueroa-Espinoza, E. Dubreucq, H. Fulcrand, P. Villeneuve, The use of lipases as biocatalysts for the epoxidation of fatty acids and phenolic compounds, Green Chem. 16 (2014) 17406–1754. https://doi.org/10.1039/c3gc42143k

[15]   F. Hasan, A.A. Shah, A. Hameed, Industrial applications of microbial lipases, Enzyme Microb. Technol. 39 (2006) 235–251. https://doi.org/10.1016/j.enzmictec.2005.10.016

[16]   R. Sharma, Y. Chisti, U.C. Banerjee, Production, purification, characterization, and applications of lipases, Biotechnol. Adv. 19 (2001) 627–662. https://doi.org/10.1016/s0734-9750(01)00086-6

[17]   T. De Diego, A. Manjon, P. Lozano, M. Vaultier, J.L. Iborra, An efficient activity ionic liquid-enzyme system for biodiesel production, Green Chem. 13 (2011) 444–451. https://doi.org/10.1039/c0gc00230e

[18]   B. Dabirmanesh, K. Khajeh, B. Ranjbar, F. Ghazi, A. Heydari, Inhibition mediated stabilization effect of imidazolium based ionic liquids on alcohol dehydrogenase, J. Mol. Liq. 170 (2012) 66–71. https://doi.org/10.1016/j.molliq.2012.03.004

[19]   Earle Martyn J, S.K. R, Ionic liquids. Green solvents for the future, Pure Appl. Chem. 72 (2000) 1391-1398. https://doi.org/10.1351/pac200072071391

[20]   C.Z. Liu, F. Wang, A.R. Stiles, C. Guo, Ionic liquids for biofuel production: Opportunities and challenges, Appl. Energy. 92 (2012) 406–414. https://doi.org/10.1016/j.apenergy.2011.11.031

[21]   K.S. Egorova, E.G. Gordeev, V.P. Ananikov, Biological activity of ionic liquids and their application in pharmaceutics and medicine, Chem. Rev. 117 (2017)7132-7189. https://doi.org/10.1021/acs.chemrev.6b00562

[22]   Y. Shi, Y. Wu, X. Lu, Y. Ren, Q. Wang, C. Zhu, D. Yu, H. Wang, Lipase-catalyzed esterification of ferulic acid with lauryl alcohol in ionic liquids and antibacterial properties in vitro against three food-related bacteria, Food Chem. 220 (2017) 249–256. https://doi.org/10.1016/j.foodchem.2016.09.187

[23]   Y. Fan, X. Wang, J. Li, L. Zhang, L. Yang, P. Gao, Z. Zhou, Kinetic study of the inhibition of ionic liquids on the trypsin activity, J. Mol. Liq. 252 (2018) 392–398. https://doi.org/10.1016/j.molliq.2018.01.014

[24]   W.Y. Lou, L. Chen, B.-B. Zhang, T.J. Smith, M.-H. Zong, Using a water-immiscible ionic liquid to improve asymmetric reduction of 4-(trimethylsilyl)-3-butyn-2-one catalyzed by immobilized *Candida parapsilosis* CCTCC M203011 cells, BMC Biotechnol. 9 (2009) 90-102. https://doi.org/10.1186/1472-6750-9-90

[25]   C. E. Paul, V. G. Fernández, Biocatalysis and biotransformation, in: X. Xu, Z. Guo, L.-Z. Cheong (Eds.), Ionic liquids in lipid processing and analysis opportunities and challenges, AOCS Press, London, 2017, pp. 11-58. https://doi.org/10.1016/b978-1-63067-047-4.00002-7

[26]   A.A. Elgharbawy, F.A. Riyadi, M.Z. Alam, M. Moniruzzaman, Ionic liquids as a potential solvent for lipase-catalysed reactions: A review, J. Mol. Liq. 251 (2018) 150-166. https://doi.org/10.1016/j.molliq.2017.12.050

[27]   M. Moniruzzaman, M. Goto, Ionic liquids: future solvents and reagents for pharmaceuticals, J. Chem. Eng. Japan. 44 (2011) 370–381. https://doi.org/10.1252/jcej.11we015

[28]   Z. Findrik, G. Megyeri, L. Gubicza, K. Bélafi-Bakó, N. Nemestóthy, M. Sudar, Lipase catalyzed synthesis of glucose palmitate in ionic liquid, J. Clean. Prod. 112 (2016) 1106–1111. https://doi.org/10.1016/j.jclepro.2015.07.098

[29]   M. Naushad, Z.A. ALOthman, A.B. Khan, M. Ali, Effect of ionic liquid on activity, stability, and structure of enzymes: A review, Int. J. Biol. Macromol. 51 (2012) 555–560.

[30]   H. Zhao, Protein stabilization and enzyme activation in ionic liquids: specific ion effects, J. Chem. Technol. Biotechnol. 91 (2016) 25–50. https://doi.org/10.1002/jctb.4837

[31]   S.P. Ã, R.J.K. Ãyz, S. Park, R.J. Kazlauskas, Biocatalysis in ionic liquids–advantages beyond green technology, Curr. Opin. Biotechnol. 14 (2003) 432–437. https://doi.org/10.1016/s0958-1669(03)00100-9

[32] H. Zhao, Methods for stabilizing and activating enzymes in ionic liquids-a review, J. Chem. Technol. Biotechnol. 85 (2010) 891–907. https://doi.org/10.1002/jctb.2375

[33] S. Sunitha, S. Kanjilal, P.S. Reddy, R.B.N. Prasad, Ionic liquids as a reaction medium for lipase-catalyzed methanolysis of sunflower oil, Biotechnol. Lett. 29 (2007) 1881–1885. https://doi.org/10.1007/s10529-007-9471-x

[34] M. Guncheva, D. Yancheva, P. Ossowicz, E. Janus, Structural basis for the inactivation of Candida rugosa lipase in the presence of amino acid ionic liquids, Bulg. Chem. Commun. 49 (2017) 132–136.

[35] O. Ulbert, K. Belafi-Bako, K. Tonova, L. Gubicza, Thermal stability enhancement of Candida rugosa lipase using ionic liquids, Biocatal. Biotransformation, 23 (2005) 177-183. https://doi.org/10.1080/10242420500192940

[36] H. Zhao, Z. Song, O. Olubajo, J. V Cowins, New ether-functionalized ionic liquids for lipase-catalyzed synthesis of biodiesel, Appl. Biochem. Biotechnol. 162 (2010) 13–23. https://doi.org/10.1007/s12010-009-8717-6

[37] S.H. Schöfer, N. Kaftzik, P. Wasserscheid, U. Kragl, Enzyme catalysis in ionic liquids: lipase catalysed kinetic resolution of 1-phenylethanol with improved enantioselectivity, Chem. Commun.0 (2001) 425–426. https://doi.org/10.1039/b009389k

[38] Y. Fan, X. Dong, X. Li, Y. Zhong, J. Kong, S. Hua, J. Miao, Y. Li, Spectroscopic studies on the inhibitory effects of ionic liquids on lipase activity, Spectrochim. Acta Part A Mol. Biomol. Spectrosc. 159 (2016) 128–133. https://doi.org/10.1016/j.saa.2016.01.047

[39] M.B.A. Rahman, K. Jumbri, N.A.M.A. Hanafiah, E. Abdulmalek, B.A. Tejo, M. Basri, A.B. Salleh, Enzymatic esterification of fatty acid esters by tetraethylammonium amino acid ionic liquids-coated *Candida rugosa* lipase, J. Mol. Catal. B Enzym. 79 (2012) 61–65. https://doi.org/10.1016/j.molcatb.2012.03.003

[40] Y. Fan, X. Wang, L. Zhang, J. Li, L. Yang, P. Gao, Z. Zhou, Lipase-catalyzed synthesis of biodiesel in a hydroxyl-functionalized ionic liquid, Chem. Eng. Res. Des. 2 (2018) 199–207. https://doi.org/10.1016/j.cherd.2018.01.020

[41] M.C. Lisboa, C.A. Rodrigues, A.S. Barbosa, S. Mattedi, L.S. Freitas, A.A. Mendes, C. Dariva, E. Franceschi, Á.S. Lima, C.M.F. Soares, New perspectives on the modification of silica aerogel particles with ionic liquid applied in lipase immobilization with platform in ethyl esters production, Process Biochem. (2018) 157-165. https://doi.org/10.1016/j.procbio.2018.09.015

[42] S. Park, T.T.N. Doan, Y.M. Koo, K.K. Oh, S.H. Lee, Ionic liquids as cosolvents for the lipase-catalyzed kinetic resolution of ketoprofen, Mol. Catal. 459 (2018) 113–118. https://doi.org/10.1016/j.mcat.2018.09.001

[43] H. Suo, L. Xu, C. Xu, H. Chen, D. Yu, Z. Gao, H. Huang, Y. Hu, L. Xu, C. Xu, H. Chen, D. Yu, Z. Gao, H. Huang, Y. Hu, Enhancement of catalytic performance of porcine pancreatic lipase immobilized on functional ionic liquid modified Fe3O4-Chitosan nanocomposites, Int. J. Biol. Macromol. 119 (2018) 624–632. https://doi.org/10.1016/j.ijbiomac.2018.07.187

[44] M. Moniruzzaman, K. Ino, N. Kamiya, M. Goto, Lipase incorporated ionic liquid polymers as active, stable and reusable biocatalysts, Org. Biomol. Chem. 10 (2012) 7707–7713. https://doi.org/10.1039/c2ob25529d

[45] R.L. de Souza, E.L.P. de Faria, R.T. Figueiredo, L. dos Santos Freitas, M. Iglesias, S. Mattedi, G.M. Zanin, O.A.A. dos Santos, J.A.P. Coutinho, Á.S. Lima, Protic ionic liquid as additive on lipase immobilization using silica sol-gel, Enzyme Microb. Technol. 52 (2013) 141–150. https://doi.org/10.1016/j.enzmictec.2012.12.007

[46] J.K. Lee, M.J. Kim, Ionic liquid co-lyophilized enzyme for biocatalysis in organic solvent: Remarkably enhanced activity and enantioselectivity, J. Mol. Catal. B Enzym. 68 (2011) 275–278. https://doi.org/10.1016/j.molcatb.2010.11.017

[47] D.H. Zhang, H.X. Xu, N. Chen, W.C. Che, The application of ionic liquids in enzyme immobilization and enzyme modification, Austin J. Biotechnol. Bioeng. 3 (2016) 1060–1064.

[48] C. Miao, L. Yang, Z. Wang, W. Luo, H. Li, Lipase immobilization on amino-silane modified superparamagnetic Fe_3O_4 nanoparticles as biocatalyst for biodiesel production, Fuel, 224 (2018) 774–782. https://doi.org/10.1016/j.fuel.2018.02.149

[49] M. Guncheva, K. Paunova, D. Yancheva, I. Svinyarov, M. Bogdanov, Effect of two series ionic liquids based on non-nutritive sweeteners on catalytic activity and stability of the industrially important lipases from *Candida rugosa* and *Rhizopus delemar*, J. Mol. Catal. B Enzym. 117 (2015) 62–68. https://doi.org/10.1016/j.molcatb.2015.04.009

[50] M. Solhtalab, H.R. Karbalaei-Heidari, G. Absalan, Tuning of hydrophilic ionic liquids concentration: A way to prevent enzyme instability, J. Mol. Catal. B Enzym. 122 (2015) 125–130. https://doi.org/10.1016/j.molcatb.2015.09.002

[51] T. Wang, L. Wang, Y. Jin, P. Chen, W. Xu, L. Yu, I. V. Voroshylova, S.R. Smaga, E. V. Lukinova, V. V. Chaban, O.N. Kalugin, E.G. Lemraski, Z. Pouyanfar, W.W.

Gao, F.-X. Zhang, G.-X. Zhang, C.H. Zhou, Key factors affecting the activity and stability of enzymes in ionic liquids and novel applications in biocatalysis, Biochem. Eng. J. 99 (2015) 67–84. https://doi.org/10.1016/j.bej.2015.03.005

[52]   S.H. Ha, S.H. Lee, D.T. Dang, M.S. Kwon, W.-J. Chang, Y.J. Yu, I.S. Byun, Y.-M. Koo, Enhanced stability of *Candida antarctica* lipase B in ionic liquids, Korean J. Chem. Eng. 25 (2008) 291–294. https://doi.org/10.1007/s11814-008-0051-0

[53]   S.H. Lee, T.T.N. Doan, S.H. Ha, W.-J. Chang, Y.-M. Koo, Influence of ionic liquids as additives on sol-gel immobilized lipase, J. Mol. Catal. B Enzym. 47 (2007) 129–134. https://doi.org/10.1016/j.molcatb.2007.05.002

[54]   B. Wang, C. Zhang, Q. He, H. Qin, G. Liang, W. Liu, Efficient resolution of (R,S)-1-(1-naphthyl)ethylamine by *Candida antarctica* lipase B in ionic liquids, Mol. Catal. 448 (2018) 116–121. https://doi.org/10.1016/j.mcat.2018.01.026

[55]   C. Reichardt, T. Welton, Solvents and solvent effects in organic chemistry, Wiley-VCH, Weinheim, 2011.

[56]   V. Strehmel, R. Lungwitz, H. Rexhausen, S. Spange, Relationship between hyperfine coupling constants of spin probes and empirical polarity parameters of some ionic liquids, New J. Chem. 34 (2010) 2125–2131. https://doi.org/10.1039/c0nj00253d

[57]   H. Zhao, G.A. Baker, Z. Song, O. Olubajo, L. Zanders, S.M. Campbell, Effect of ionic liquid properties on lipase stabilization under microwave irradiation, J. Mol. Catal. B Enzym. 57 (2009) 149-157. https://doi.org/10.1016/j.molcatb.2008.08.006

[58]   S. Park, R.J. Kazlauskas, Improved preparation and use of room-temperature ionic liquids in lipase-catalyzed enantio-and regioselective acylations, J. Org. Chem. 66 (2001) 8395–8401. https://doi.org/10.1021/jo015761e

[59]   J.-M. Lee, J.M. Prausnitz, Polarity and hydrogen-bond-donor strength for some ionic liquids: effect of alkyl chain length on the pyrrolidinium cation, Chem. Phys. Lett. 492 (2010) 55–59. https://doi.org/10.1016/j.cplett.2010.03.086

[60]   J. Dupont, On the solid, liquid and solution structural organization of imidazolium ionic liquids, J. Braz. Chem. Soc. 15 (2004) 341–350. https://doi.org/10.1590/s0103-50532004000300002

[61]   X. Chen, J. Li, L. Deng, Biodiesel production using lipases, in: T. Bornscheuer U, red. I (Eds), Lipid modification by enzymes and engineered microbes, Elsevier Inc., UK, 2018, pp. 203-238. https://doi.org/10.1016/b978-0-12-813167-1.00010-4

[62]   S.H. Ha, M.N. Lan, S.H. Lee, S.M. Hwang, Y.M. Koo, Lipase-catalyzed biodiesel production from soybean oil in ionic liquids, Enzyme Microb. Technol. 41 (2007) 480–483. https://doi.org/10.1016/j.enzmictec.2007.03.017

[63]   P. Antonia, F. van Rantwijk, R.A. Sheldon, Effective resolution of 1-phenyl ethanol by *Candida antarctica* lipase B catalysed acylation with vinyl acetate in protic ionic liquids (PILs), Green Chem. 14 (2012) 1584-1588. https://doi.org/10.1039/c2gc35196j

[64]   N.L. Mai, Y.M. Koo, Compatibility of ionic liquids with enzymes, in: Z. Fang, R.L. Smith (Jr.), X. Qi (Eds.), Production of Biofuels and Chemicals with Ionic Liquids, Netherlands Springer, Dordrecht, 2014, pp. 257–273. https://doi.org/10.1007/978-94-007-7711-8_10

[65]   R.Y. Cabrera-Padilla, E.B. Melo, M.M. Pereira, R.T. Figueiredo, A.T. Fricks, E. Franceschi, Á.S. Lima, D.P. Silva, C.M.F. Soares, Use of ionic liquids as additives for the immobilization of lipase from *Bacillus sp.*, J. Chem. Technol. Biotechnol. 90 (2014) 1308–1316. https://doi.org/10.1002/jctb.4438

[66]   B. Réjasse, T. Besson, M.-D. Legoy, S. Lamare, Influence of microwave radiation on free *Candida antarctica* lipase B activity and stability, Org. Biomol. Chem. 4 (2006) 3703–3707. https://doi.org/10.1039/b610265d

[67]   I.M. Atadashi, M.K. Aroua, A.R.A. Aziz, N.M.N. Sulaiman, Refining technologies for the purification of crude biodiesel, Appl. Energy. 88 (2011) 4239–4251. https://doi.org/10.1016/j.apenergy.2011.05.029

[68]   Z. Yang, K.-P. Zhang, Y. Huang, Z. Wang, Both hydrolytic and transesterification activities of *Penicillium expansum* lipase are significantly enhanced in ionic liquid [BMIm][PF6], J. Mol. Catal. B Enzym. 63 (2010) 23–30. https://doi.org/10.1016/j.molcatb.2009.11.014

[69]   A. Berthod, M.J. Ruiz-Ángel, S. Carda-Broch, Recent advances on ionic liquid uses in separation techniques, J. Chrom. 1559 (2018) 2-16. https://doi.org/10.1016/j.chroma.2017.09.044

[70]   M. Guncheva, K. Paunova, M. Dimitrov, D. Yancheva, Stabilization of *Candida rugosa* lipase on nanosized zirconia-based materials, J. Mol. Catal. B Enzym. 108 (2014) 43–50. https://doi.org/10.1016/j.molcatb.2014.06.012

[71]   P. Lozano, T. De Diego, D. Carrie, M. Vaultier, J.L. Iborra, Over-stabilization of *Candida antarctica* lipase B by ionic liquids in ester synthesis, Biotechnol. Lett. 23 (2001) 1529–1533. https://doi.org/10.1023/a:1011697609756

[72]   M. Wlazło, K. Paduszyn, U. Doman, U. Domańska, M. Wlazło, K. Paduszyński, Extraction of butan-1-ol from aqueous solution using ionic liquids: An effect of cation revealed by experiments and thermodynamic models, Sep. Purif. Technol. 196 (2018) 71–81. https://doi.org/10.1016/j.seppur.2017.05.056

[73]   H. Xing, T. Wang, Z. Zhou, Y. Dai, The sulfonic acid-functionalized ionic liquids with pyridinium cations: Acidities and their acidity–catalytic activity relationships, J. Mol. Catal. A Chem. 264 (2007) 53–59. https://doi.org/10.1016/j.molcata.2006.08.080

[74]   P. Lozano, J.M. Bernal, G. Sánchez-Gómez, G. López-López, M. Vaultier, How to produce biodiesel easily using a green biocatalytic approach in sponge-like ionic liquids, Energy Environ. Sci. 6 (2013) 1328–1338. https://doi.org/10.1039/c3ee24429f

[75]   M. Cvjetko Bubalo, A. Jurinjak Tušek, M. Vinković, K. Radošević, V. Gaurina Srček, I. Radojčić Redovniković, Cholinium-based deep eutectic solvents and ionic liquids for lipase-catalyzed synthesis of butyl acetate, J. Mol. Catal. B Enzym. 122 (2015) 188–198. https://doi.org/10.1016/j.molcatb.2015.09.005

[76]   J. Yang, Y. Feng, T. Zeng, X. Guo, L. Li, R. Hong, T. Qiu, Synthesis of biodiesel via transesterification of tung oil catalyzed by new Brönsted acidic ionic liquid, Chem. Eng. Res. Des. 117 (2017) 584–592. https://doi.org/10.1016/j.cherd.2016.09.038

[77]   N.I. Ruzich, A.S. Bassi, Investigation of enzymatic biodiesel production using ionic liquid as a co-solvent, Can. J. Chem. Eng. 88 (2010) 277–282. https://doi.org/10.1002/cjce.20263

[78]   S.J. Nara, J.R. Harjani, M.M. Salunkhe, Lipase-catalysed transesterification in ionic liquids and organic solvents: a comparative study, Tetrahedron Lett. 43 (2002) 2979–2982. https://doi.org/10.1016/s0040-4039(02)00420-3

[79]   F. Su, C. Peng, G.-L. Li, L. Xu, Y.-J. Yan, Biodiesel production from woody oil catalyzed by Candida rugosa lipase in ionic liquid, Renew. Energy. 90 (2016) 329–335. https://doi.org/10.1016/j.renene.2016.01.029

[80]   P. Fan, S. Xing, J. Wang, J. Fu, L. Yang, G. Yang, Sulfonated imidazolium ionic liquid-catalyzed transesterification for biodiesel synthesis, Fuel, 188 (2017) 483–488. https://doi.org/10.1016/j.fuel.2016.10.068

[81]   Y. Feng, L. Li, X. Wang, J. Yang, T. Qiu, Stable poly (ionic liquid) with unique crosslinked microsphere structure as efficient catalyst for transesterification of soapberry oil to biodiesel, Energy Convers. Manag. 153 (2017) 649–658. https://doi.org/10.1016/j.enconman.2017.10.018

[82]   S. Park, F. Viklund, K. Hult, R.J. Kazlauskas, Vacuum-driven lipase-catalysed direct condensation of l-ascorbic acid and fatty acids in ionic liquids: synthesis of a natural surface active antioxidant, Green Chem. 5 (2003) 715–719. https://doi.org/10.1039/b307715b

[83]   S. Adak, R. Banerjee, A green approach for starch modification: Esterification by lipase and novel imidazolium surfactant, Carbohydr. Polym. 150 (2016) 359–368. https://doi.org/10.1016/j.carbpol.2016.05.038

[84]   E. Abdulmalek, H.S.M. Saupi, B.A. Tejo, M. Basri, A.B. Salleh, R.N.Z.R.A. Rahman, M.B.A. Rahman, Improved enzymatic galactose oleate ester synthesis in ionic liquids, J. Mol. Catal. B Enzym. 76 (2012) 37–43. https://doi.org/10.1016/j.molcatb.2011.12.004

[85]   M.C. Villalobos, A.G. Gonçalves, M.D. Noseda, D.A. Mitchell, N. Krieger, A novel enzymatic method for the synthesis of methyl 6-O-acetyl-α-D-glucopyranoside using a fermented solid containing lipases produced by *Burkholderia contaminans* LTEB11, Process Biochem. 73 (2018) 0–1. https://doi.org/10.1016/j.procbio.2018.07.023

[86]   S. Gholivand, O. Lasekan, C. Ping, F. Abas, L. Sze, Comparative study of the antioxidant activities of some lipase-catalyzed alkyl dihydrocaffeates synthesized in ionic liquid, Food Chem. 224 (2017) 365–371. https://doi.org/10.1016/j.foodchem.2016.12.075

[87]   P.L.G. Martins, A.R. Braga, V.V. de Rosso, Can ionic liquid solvents be applied in the food industry?, Trends Food Sci. Technol. 66 (2017) 117–124. https://doi.org/10.1016/j.tifs.2017.06.002

[88]   M.K. Potdar, G.F. Kelso, L. Schwarz, C. Zhang, M.T.W. Hearn, Recent developments in chemical synthesis with biocatalysts in ionic liquids, Molecules. 20 (2015) 16788–16816. https://doi.org/10.3390/molecules200916788

[89]   A.I. Siriwardana, Industrial applications of ionic liquids, in: A. J. Torriero (Ed.), Electrochemistry in ionic liquids, Springer International Publishing, New York, 2015, pp. 563-603. https://doi.org/10.1007/978-3-319-15132-8_20

[90]   Proionic, Ionic Liquids par Excellence, https://www.proionic.com/ionic-liquids/services-ionic-liquid-production.php, 2018 (accessed October 8, 2018).

[91]   Io.li.tec, Ionic Liquids Technologies,https://iolitec.de/index.php/en, 2018 (accessed October 8, 2018).

[92]   X. Liang, Novel acidic ionic liquid polymer for biodiesel synthesis from waste oils, Appl. Catal. A Gen. 455 (2013) 206–210. https://doi.org/10.1016/j.apcata.2013.01.036

[93]   E. Ra, F. Mirnezami, Temperature regulated Brønsted acidic ionic liquid-catalyze esterification of oleic acid for biodiesel application, J. Mol. Struct. 1130 (2017) 296–302. https://doi.org/10.1016/j.molstruc.2016.10.049

[94]   W. Xie, M. Huang, Immobilization of *Candida rugosa* lipase onto graphene oxide Fe3O4 nanocomposite : Characterization and application for biodiesel production, Energy Convers. Manag. 159 (2018) 42–53. https://doi.org/10.1016/j.enconman.2018.01.021

[95]   J.M. Berg, J.L. Tymoczko, L. Stryer, The Michaelis-Menten model accounts for the kinetic properties of many enzymes, in: Biochemistry,WH Freeman and Company, New York, 2002.

[96]   Z. Yang, Y.-J. Yue, W.-C. Huang, X.-M. Zhuang, Z.-T. Chen, M. Xing, Importance of the ionic nature of ionic liquids in affecting enzyme performance, J. Biochem. 145 (2008) 355–364. https://doi.org/10.1093/jb/mvn173

[97]   X. Wan, X. Xiang, S. Tang, D. Yu, H. Huang, Y. Hu, Immobilization of *Candida antarctic* lipase B on MWNTs modified by ionic liquids with different functional groups, Colloids Surfaces B Biointerfaces. 160 (2017) 416–422. https://doi.org/10.1016/j.colsurfb.2017.09.037

[98]   X. Cui, J. Cai, Y. Zhang, R. Li, T. Feng, Kinetics of Transesterification of Methyl Acetate and n-Butanol Catalyzed by Ionic Liquid, Ind. Eng. Chem. Res. 50 (2011) 11521–11527. https://doi.org/10.1021/ie2000715

[99]   S. Sun, S. Zhu, Enzymatic preparation of castor oil-based feruloylated lipids using ionic liquids as reaction medium and kinetic model, Ind. Crops Prod. 73 (2015) 127–133. https://doi.org/10.1016/j.indcrop.2015.04.019

Industrial Applications of Green Solvents I   Materials Research Forum LLC
Materials Research Foundations 50 (2019) 61-106   doi: https://doi.org/10.21741/9781644900239-3

Chapter 3

Water in Organic Synthesis as a Green Solvent

Muhammad Faisal*

Department of Chemistry, Quaid-i-Azam University-45320, Islamabad, Pakistan

*mfaisal4646@gmail.com, mfaisal@chem.qau.edu.pk

Abstract

The growing demand for more sustainable approaches in synthetic chemistry has led to increasing attention in the application of H_2O as a solvent in the last decade. H_2O as a solvent is not only environmentally benign and inexpensive but also offers promising advantages, for instance, improves rates and yields; enhances chemo-, enantio-, regio-, stereo-selectivities; simplifies the process of reaction handling and workup; enables the recovering and reusing of the catalysts; avoids protection-deprotection steps; and allows milder reaction conditions. This book chapter focuses on the potential application of H_2O as a solvent for organic synthesis, highlighting benefits and the spectrum of important organic reactions that can be performed in H_2O with a green chemistry perspective.

Keywords

Green Chemistry, Water, Reactivity and Selectivity, Protecting-Group Free Synthesis, Recycling of Catalysts

Abbreviations

Expanded term	Acronym
2,2'-Bis(diphenylphosphino)-1,1'-binaphthyl	BINAP
9-Fluorenylmethoxycarbonyl	Fmoc
Acetonitrile	CH_3CN
Also known as	*aka.*
Ammonium chloride	NH_4Cl
Atom-transfer radical cyclisation	ATRC
Baylis-Hillman	BH
Chloroform	$CHCl_3$
Critical micelle concentration	CMC
Cross-metathesis	CM
Cu(I)-catalyzed alkyne-azide [3+2] cycloaddition	CuAAC
Diels-Alder	DA
Dimethyl sulfoxide	DMSO

Diphenylborinic acid	Ph_2BOH
Environmental protection agency	EPA
Ethanol	EtOH
Green chemistry institute	GCI
Guanidine hydrochloride	$C(NH_2)_3{}^+Cl^-$
Hour	H
Isooctane	IOA
Lithium chloride	ClLi
Methanol	MeOH
Microwave	MW
Polyethylene glycol	PEG
Polyethyleneglycolubiquinolsebacate	PQS
Presidential green chemistry challenge award	PGCCA
Ring-closing metathesis	RCM
Rxempli gratia	*e.g.*
Scandium(III) triflate	$Sc(OTf)_3$
Siloxy	R_3SiO-
sodium dodecyl sulfate	SDS
Solid-supported evan's oxazolidin-2-one	IRORI Kan
Sulfonated polystyrene	PS
Tetrahydrofuran	THF
Tetra-*n*-butylammonium bromide	TBAB
Thermolysin	TML
Trifluoroethanol	TFE
Water	H_2O
Zirconium dioxide	ZrO_2

Contents

1.   Introduction

1.1   The development of green chemistry

Chemistry dates back to alchemy in the seventeenth and eighteen centuries, and it has been constantly progressing human life and economic prosperity ever since through makes better materials, effective drugs, safer food, and enhanced health. There is a growing demand for the indirect or direct employment of products currently mediated on petrochemicals, such as personal care and products, agrochemicals, paints, pharmaceuticals, and coatings, as well as more advanced materials for the advancement of society [1]. Despite the evident advantages of chemistry to a human being, chemicals have badly affected the environment and human health. It is a fact that all these chemicals, including the more toxic ones, are released into the atmosphere during their production, application or disposal. For our planet, this will cause an inevitable eco-toxicological hazard. As awareness of this is growing, non-governmental and governments organizations are forcing industries to decrease toxic waste during the manufacture of their products and to mitigate their effects owing to the consequent risk for the environment and human health [2,3].

At the end of the twentieth century, the term "green chemistry *aka.* sustainable chemistry" was first used by P.T Anastas as in a special program organized by the United States environmental protection agency (EPA) in order to support advanced technologies that reduce or eliminate the impact of dangerous chemicals in the manufacture, design and use of chemical products, and to encourage sustainable development of chemical technology and chemistry by industry, academia and government [4]. The United States presidential green chemistry challenge award (PGCCA) was published in 1995 and was followed by similar awards in European countries. The green chemistry institute (GCI) was developed to create contacts between governmental agencies and industrial corporations with universities and research centers to implement and design green, innovative and operative technologies. In 1997, the first conference on the theme of green chemistry was held in Washington followed by subsequent conferences all around the world [5].

1.2   Green chemistry and its requirements in organic synthesis

The interdisciplinary cooperation of scientists, governmental agencies, research institutes, universities, and industries, which have their own plans and approaches to decrease the pollution, has resulted in the creation of the "sustainable" or "green" concept. In adhering to the ethos of "green chemistry", all procedures, for example, preparation, processing, and application of chemical materials, must be performed in such a way as to decrease impact on the environment and human health. Green chemistry *aka* sustainable chemistry is defined as "environmentally benign chemical synthesis" [5,6]. Under green chemistry, the schemes for the synthesis of compounds are planned in such a manner that there is no or minimum toxic waste to the atmosphere. To carry out the green organic reactions, the following basic principles of "green chemistry" formulated by P.T Anastas as must be maintained: (i) minimization or prevention of hazardous byproducts and waste products,(ii) minimum energy requirement for any synthesis, (iii) choosing the appropriate catalyst, (iv) selecting the most appropriate solvent, (v) maximum incorporation of the starting materials into the product, (vi) selecting the appropriate starting materials, and (vii) following the green metric correlations. The synthesis that does not follow these basic principles is not "green". It is essential to carefully select the catalysts and the solvents for carrying out such reactions [7,8].

1.3   Green and alternative solvents in organic synthesis

Green chemistry is highly depended on the solvent. Solvents used by chemists are generally utilized as a reaction media, in separation or purification methods. The appropriateness and quality appropriateness of reactions and chemical processes are highly dependent on the nature of the solvent utilized. Most of the common solvents (such as dichloromethane, toluene, and benzene, *etc*) that are usually used are harmful. Benzene is popular for causing cancer, toluene causes brain, liver and kidney problems, used halogenated solvents carbon tetrachloride, chloroform, methylene chloride have been recognized as human carcinogens. Volatility is also one of the major problems with various organic solvents that may damage the environment and health of human [9,10].

Solvents employed in organic reactions as well as in other areas have had a growing impact on the awareness of researchers attributed to having effects on the environment and human health. The chosen solvent should carefully be considered for the effect it has on the environment and human health. There are huge consumptions, an almost fifteen billion kilograms of organic halogenated solvents, worth more than $5 billion, are used worldwide per year [11]. As per the fifth principle of green chemistry, expressed by J.C. Warner and P.T. Anastas, "The use of auxiliary substances (*e.g.* separation agents, solvents, *etc.*) must be made unnecessary wherever possible and, innocuous when used". We should initially consider how to maximize the yield and minimize the application of solvents in order to eliminate waste disposal after usage. Solvent usage should also include some of the following considerations [12]: (i) knowledge of its environmental fate, (ii) the toxicity effect on the environment, and (iii) volatility. All of these effects have made researchers to find alternative solvents that can be the replacement for classical petrochemical solvents. There are various solvents which have been examined and employed in the organic reactions to move to more environmentally benign and innocuous conditions for human health and nature [13,14]. Following are the most employed and researched solvents in attendance as well as the solvent-free conditions: (i) water; (ii) supercritical fluids (widely carbon dioxide, water); (iii) ethanol, aqueous surfactant micelles and polymers [15]; (iv) fluorous solvents; and (v) ionic liquids.

1.4   Water as a green solvent in organic synthesis

For a long time, H_2O has been regarded as contamination when it comes to organic synthesis. But investigations have shown that water can have striking effects on some organic reactions. Use of H_2O as a solvent in organic reactions is desirable from a green chemistry perspective since it is cheap, readily available, non-flammable and non-toxic. In the 1980s, Breslow discovered that H_2O had a promising effect on the selectivity and rate of reaction in Diels-Alder (DA) cycloaddition between cyclopenta-1,3-diene **1** and but-3-en-2-one **2** to furnish 1-(bicyclo[2.2.1]hept-5-en-2-yl)ethanone **3** (Table 1). This has contributed a lot to the interest in aqueous synthesis [16,17]. H_2O as solvent exhibited the formation of hydrophobic aggregates in order to eliminate the contact surface between H_2O and the organic phase. In order to keep the complex net of hydrogen bonding interactions, H_2O developed networks around the aggregates; therefore, served as internal pressure. The resulting effect is the enhancement of reactions with the negative volume of activation, like the DA cycloaddition [18]. These findings were considered an important development in the area of organic chemistry in aqueous media.

H_2O is an exclusively valuable solvent for high temperature and high-pressure reactions, where the polarity is inferior, making it a better solvent for organic reactions, and where the ionic product of H_2O is high, making it a stronger base and acid. Furthermore, H_2O is a renewable resource, unlike numerous organic solvents which rely on non-renewable petroleum resources. H_2O has desirable benefits but it also has some shortcomings. Its high energy capacity results in demand for high energy in order to regulate its

Industrial Applications of Green Solvents I   Materials Research Forum LLC
Materials Research Foundations **50** (2019) 61-106   doi: https://doi.org/10.21741/9781644900239-3

temperature when implemented on an industrial scale compared to other solvents like MeOH. The high energy demand also becomes a problem when H_2O is to be isolated through rotary evaporation [19].

Table 1. Diels-Alder reaction in water by Breslow

Solvent	Additive	$K_2 \times 10^5$ $(M^{-1}s^{-1})$
IOA		5.94 ± 0.3
MeOH		75.5
H_2O		4400 ± 70
H_2O	ClLi (4.86 M)	10800
H_2O	$C(NH_2)_3{}^+Cl^-$ (4.86 M)	4300
H_2O	α-Cyclodextrin (10 nM)	10900
H_2O	β-Cyclodextrin (10 nM)	2610

Isooctane= IOA; methnanol= MeOH; lithium chloride=ClLi; Guanidine hydrochloride $=C(NH_2)_3{}^+Cl^-$

2. On water reactions and concept of micellar catalysis in organic synthesis

2.1 On water reactions

Traditional organic reactions are frequently carried out in organic solvents. On account of concern about the shortcoming of using an organic solvent which causes a large amount of waste, H_2O became the most interesting alternative solvent for organic synthesis. An early example discovered by Sharpless in 2005 displayed the benefit of using H_2O over organic solvents in term of rate acceleration in Claisen rearrangements form **4** to **5** (Table 2) [20,21]. When performing a reaction in H_2O, organic substrates are insoluble and usually form emulsions. This phenomenon actually shows a faster rate of the reaction compared to organic solvents owing to the highly concentrated conditions. This heterogeneous emulsion is developed by an aggregation of hydrophobic organic molecules into very small droplets. The reaction then takes place inside these isolated droplets, reacting at a faster rate. This kind of reaction is called an "on water" reaction which is the original idea for the subsequent further advancement of micellar catalysis in organic synthesis [22].

Table 2. An early example of "on water" Claisen rearrangements by Sharpless

Solvent	Yield (%)
Toluene	16
Dimethylformamide	21
Acetonitrile	27
Methanol	56
Neat	73
Water	100

2.2   Micellar catalysis

Micellar catalysis is the technology which offers a more striking possibility for organic synthesis to take place in H_2O. By using the idea of self-assembling of surfactant molecules into micelles, numerous classical reactions take place efficiently in H_2O. The surfactant is a substance comprising of two main components: lipophilic and hydrophilic portions. When a surfactant is dissolved in H_2O above its CMC (critical micelle concentration), it self-aggregated into micelles with a lipophilic internal and hydrophilic external in contact with H_2O. The lipophilic inner cores of micelles then act as the organic solvent pockets for the organic reaction to take place [21,22]. Although several of the commercially available surfactants have been extensively used in industries (Fig. 1), they still cannot facilitate many reactions in H_2O, especially transition metal-catalyzed reactions. In 2008, the Lipshutz group [23] started a program of green chemistry by disclosing PTS-600 as the first generation designer surfactant. Particle size, shape, and concentration in H_2O are crucial to the performance of organic reactions in the medium of a surfactant. The vitamin E mediated surfactant was formed to have better characteristics. It was later launched as the second generation TPGS-750-M [24]. Both are commercially available and extensively used in industry and academia (Fig.2).

The first two generations of designer surfactants had been disclosed to permit transition metal- catalyzed reactions in H_2O. Both showed outstanding performance in many of the name reactions as previously reported [25]. The structures of these two surfactants are similar, consisting of three main parts: polyethylene glycol (PEG) as a hydrophilic section, a diacid linker, and vitamin E or α-tocopherol as the lipophilic part. By systematically studying, different lengths of linker and PEG, to affect the particle shape, size and performance of the surfactants in chemical reactions could be determined [25].

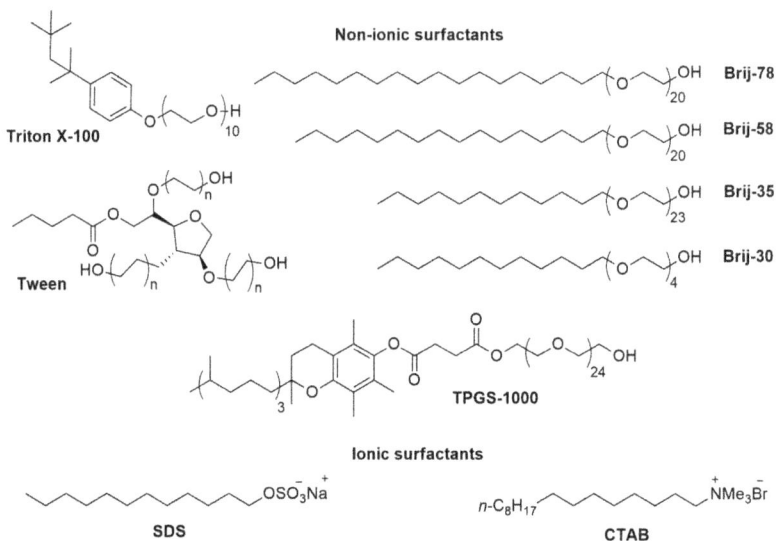

Figure 1. Structure of some commercially available surfactants.

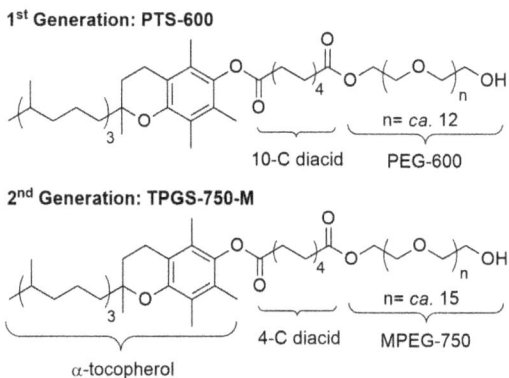

Figure 2. The first two-generation surfactants developed by the Lipshutz group.

Although micellar catalysis can solve the problem of incompatibility between water and organic substances, there are some more challenging problems which limit the usage of H₂O as a solvent for some reactions. An example of H₂O-sensitive reaction includes the employment of very sensitive substrates or catalysts. Exclusively, H₂O-sensitive organometallic compounds are the main concern when performing a reaction in H_2O. By optimizing conditions and modifying the reaction process, these H_2O-sensitive substances can be produced *in-situ* in water. Hence, it is possible that the reaction can take place in H_2O. An obvious example is the Negishi reaction which involves H_2O-sensitive organo-zinc reagents (Scheme 1) [24]. The optimized conditions involve *in-situ* generation of organo-zinc in a micellar environment that can facilitate the cross-coupling reaction of **6** with **7** to afford **8** successfully. Likewise, an example of Kumada-type reactions between *in-situ* generated organo-magnesium and aryl halides **9** can be performed in Likewise as well to produce **10** (Scheme 2) [26]. As mentioned, micellar catalysis offers an opportunity to use H_2O-sensitive materials in H_2O condition which broadens the scope of applications of H_2O as a replacement for organic solvents.

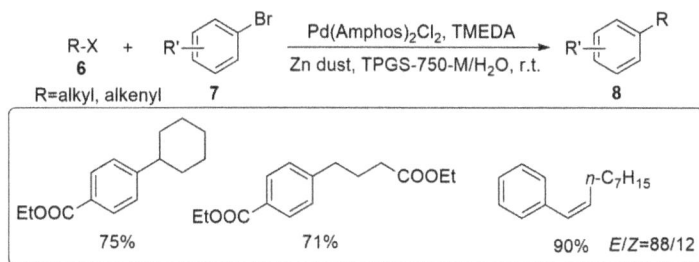

Scheme 1. A Negishi-like reaction in water.

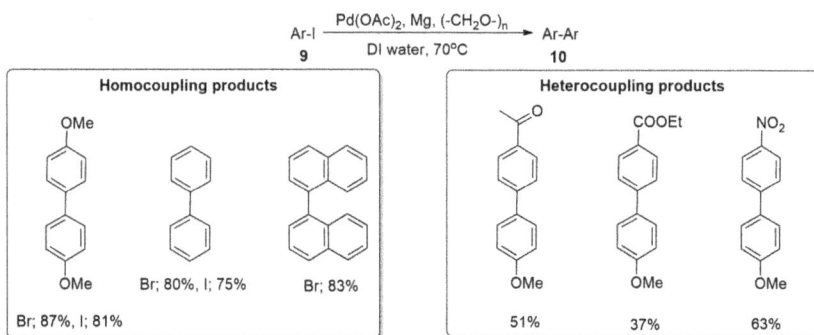

Scheme 2. Kumada-Grignard type biaryl coupling in water.

Industrial Applications of Green Solvents I   Materials Research Forum LLC
Materials Research Foundations **50** (2019) 61-106   doi: https://doi.org/10.21741/9781644900239-3

In green chemistry, the development of micellar catalysis is ongoing in academia and extending to the industry as well. A new generation of surfactants, as well as a new area of reactions, is being investigated to expand the scope of this application. It would be remarkable to witness all organic reactions in H_2O as the only medium. This will simplify the process of waste management, purification, and solvent selection [21].

3.   Enhancement in rate and yield of organic reactions

Increasing the reactivity of a reaction affects prominently its sustainability since it may permit easier purification, lower catalytic loadings, lower temperature, better yields and shorter reaction time. In the 1980s, Breslow discovered that H_2O over organic solvents had a promising effect on the reactivity of reaction in DA cycloaddition between cyclopenta-1,3-diene **1** and but-3-en-2-one **2** to furnish 1-(bicyclo[2.2.1]hept-5-en-2-yl)ethanone **3** (Table 1) [17]. The results of DA cycloaddition were explained on the basis of hydrophobic effect [27]. This characteristic of H_2O arises from the repulsive forces between H_2O and hydrophobic species, which results in the development of hydrophobic aggregates that permit plummeting the contact surface among them. In order to keep the complex net of hydrogen bonding interactions, H_2O wraps itself around these hydrophobic aggregates; therefore, serving as an internal pressure [18]. The resulting effect is that reactions with the negative volume of activation, like the DA cycloaddition, will be enhanced. Occasionally, the enhancements in the rate of organic reaction owe to interfacial forces among some free hydroxyl moieties of H_2O and the organic compounds (particularly the transition states) [28].

In 2005, the time for completion of the reaction between quadricyclane **11** and dimethyl diazene-1,2-dicarboxylate **12** to furnish **13** was observed under a range of conditions by K. Barry and co-workers [29]. The investigation revealed that polar protic solvents enhanced the rate of reaction. The following order was observed for the rates of reaction: water > methanol > dimethyl sulfoxide> acetonitrile ≈ dichloromethane >ethyl acetate ≈ toluene (Table 3). This sequence reveals that dipolar effects, charge stabilization and hydrogen bonding may each be significant for acceleration of rate. It was interesting noted that due to the isotope effect, the cycloaddition rate was significantly decreased when deuterium oxide (heavy water) was employed in place of H_2O. Presumably, it happened due to a decrease in the hydrophobic effects.

Million-time rate acceleration of a DA cycloaddition owing to micellar catalysis and combined Lewis acid in H_2O was reported by Otto et al. [30]. They reported that the cycloaddition between **1** and **14** performed in H_2O as a reaction medium to afford **15** was 287-times faster as compared to the same addition reaction in CH_3CN (Table 4). Furthermore, they observed that the cycloaddition in H_2O in the presence of micellar catalysis and Lewis acid was 1.8×10^6-fold faster as compared to the same addition reaction in CH_3CN.

Table 3. Cycloaddition of dimethyl azodicarboxylate and quadricyclane under diverse conditions[a]

Solvent	Conc. [M]	Time to completion
Toluene	2	> 5 days
Ethyl acetate	2	> 5 days
Acetonitrile	2	3.5 days
Dichloromethane	2	3 days
Dimethyl sulfoxide	2	1.5 days
Methanol	2	18 h
Neat	4.53	2 days
Heavy water	4.53	45 min
Perfluorohexane	4.53	1.5 days
Water	4.53	10 min

[a]Compound **13** was the only observed product in each case.

Table 4. Joint effect of micellar catalysis and H_2O on rate of DA cycloaddition

Conditions	Reaction kinetics
Acetonitrile	1
Water	287
Cu(OSO$_3$C$_{12}$H$_{25}$)$_2$ 2.4 mM, Water	1.8×10^6

The rate of Claisen rearrangement involving a [3,3]-sigmatropic rearrangement with negative volume changes of activation when compared with DA cycloaddition was also enhanced in H_2O. The enhancement in yield and rate of Claisen rearrangements when carried out in water relative to organic media or neat conditions was noticed by K. Barry and co-workers [29]. The rearrangement of 1-chloro-4-((2-methylbut-3-en-2-yl)oxy)naphthalene **4** to 4-chloro-2-(3-methylbut-2-en-1-yl)naphthalen-1-ol **5** in water at 23°C was accomplished within 120 h whereas, it was found to be considerably slower in organic media showing yield in the range of 16-56% (Table 2). Compared to the reaction in water, the yield and rate of rearrangement reaction were also observed to be slower in neat conditions.

In literature, the aqueous solutions have been extensively used to accelerate yield and rate of pericyclic reactions in the total synthesis of a diverse library of bioactive drugs and natural products. Total synthesis of gambogin 21 by Xu and co-workers [31], a bioactive natural product that exhibits cytotoxic properties against the HEL and Hela cell lines, involves two important steps of a Claisen/DA cascade cycloaddition and Claisen rearrangement (Scheme 3). In aqueous solutions, the Claisen/DA cascade cycloaddition from **17** to **18** displayed dramatic improvements in yield and rate of reaction. In aqueous solutions, promising improvement in rate and yield of Claisen rearrangement from **19** to **20** were also noticed (Table 5) [31].

Table 5. Improvement in rate and yield of the DA cycloaddition and Claisen rearrangement by aqueous solutions

DA cycloaddition from **17** to **18**				Claisen rearrangement from **19** to **20**			
Solvent	T (°C)	t (h)	Conversion (%)	Solvent	T (°C)	t (h)	Conversion (%)
MeOH	65	4	0	MeOH	50	4.5	50
TFE	65	4	0	TFE	25	4	0
EtOH	65	4	0	EtOH	25	4	0
MeOH/H_2O (1:1)	65	4	100	MeOH/H_2O (1:1)	50	2.5	100
TFE/H_2O (1:1)	65	4	100	TFE/H_2O (1:1)	25	75	100
EtOH/H_2O (1:1)	65	4	100	EtOH/H_2O (1:1)	25	72	100

Trifluoroethanol=TFE

Various investigations have confirmed the hydrophobic effect as well as the hydrogen bonding contribution to the enhancements in the rate of pericyclic reactions [32]. For instance, Butler and co-workers have reported that the rate of the DA cycloaddition reaction of dicyano (pyridazin-1-ium-1-yl)methanide **22** with the pent-1-en-3-one **23** to produce **24** was improved by slowly rising the mole fraction of H_2O in the organic media (from 0 to 1) of cyanomethane, propanone, MeOH, EtOH, and *t*-butanol at 37°C (Scheme 4). In each case, exponential improvement in rate was noticed as the mole fraction of H_2O surpassed *ca.* 0.9. No triggering effect was observed when MeOH was used in the place of H_2O. These experimental data support the dominating effects of hydrophobicity and hydrogen bonding on the rate enhancement [32].

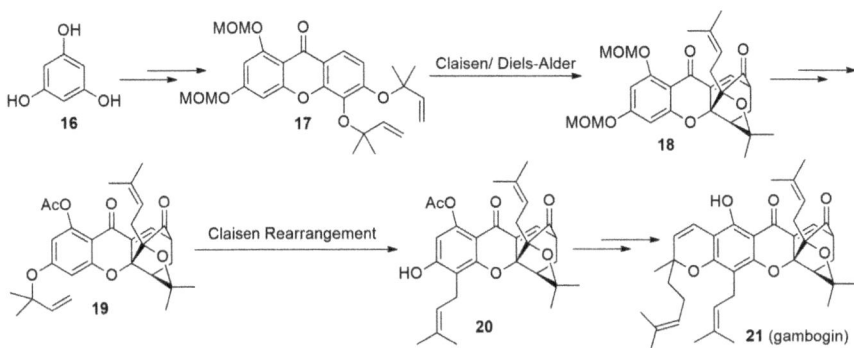

Scheme 3. Pericyclic reactions for formation of gambogin.

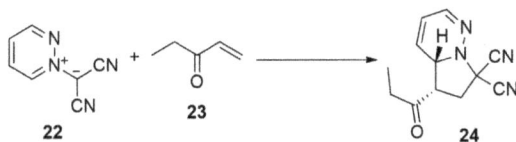

Scheme 4. Synthesis of 5-propionyl-5,6-dihydropyrrolo[1,2]pyridazine-7,7(4aH)-dicarbonitrile.

Tandon and Maurya [33] reported the effect of water on nucleophilic addition and substitution reactions of 1, 4-benzoquinones. The reaction of 2,3-dichloronaphthalene-1,4-dione **25** with phenylamine **26** was performed in water at 50°C to furnish the respective target **27** in quantitative yield. This approach was observed to be more superior than other reported techniques (Table 6) [34,35]. A diverse range of thiols, amides, hydrazines, amine, and amino acids effectively underwent nucleophilic addition and substitution reactions (Fig. 3).

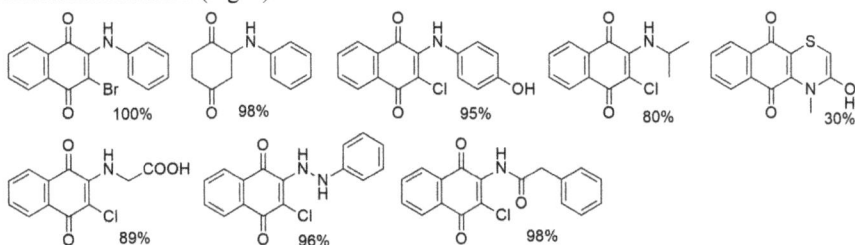

Figure 3. Structure of synthesized products by nucleophilic substitution and addition reactions.

Triethylboron-mediated atom-transfer radical cyclisation (ATRC) of iodoacetates and iodoacetals in H_2O has been reported by Koichiro and co-worker [36]. Radical cyclisation reaction of iodoacetal efficiently proceeded in water (Table 7). ATRC of prop-2-en-1-yl iodoacetate **28** to deliver 4-(iodomethyl)oxolan-2-one **29** was much more effective in H_2O as compared to in hexane, acetonitrile, alcohols, dimethylformamide, dimethyl sulfoxide, benzene, dichloromethane or tetrahydrofuran. The noteworthy effect of H_2O on yield was noticed in this cyclisation reaction.

Table 6. Catalyst-free nucleophilic addition and substitution reaction in water

Solvent	Temperature (°C)	Time (h)	Yield (%)
Benzene	50-60	0.5	81
MeOH	-	-	73
EtOH	r.t.	1	90
Water	r.t.	0.83	100
Water	50	0.25	100

Table 7. Effect of water on the triethylboron-mediated ATRC

Solvent	Yield (%)
Dichloromethane	<1
Tetrahydrofuran	<1
Benzene	<1
Hexane	<1
Acetonitrile	13
Methanol	6
Ethanol	3
Dimethylformamide	13
Dimethyl sulfoxide	37
Water	78

4. Improvement in chemo-, enantio-, regio- and stereoselectivity

Besides improvements in the rate and yield, the improvements in chemo-, enantio- and regio-, stereoselectivities of reactions under water conditions have been noticed not only in Lewis acid catalyzed reactions but also in non-catalyzed DA cycloaddition. For example, according to the report by Otto and Engberts [37], in comparison with organic media, the enantioselectivity of Cu-catalyzed DA cycloaddition of (E)-3-phenyl-1-(pyridin-2-yl)prop-2-en-1-one **30** with cyclopenta-1,3-diene**1** to produce **31** was highly increased by the application of H$_2$O as the reaction solvent (Table 8) [37].

Table 8. The improvement of enantioselectivity of DA cycloaddition by water

Solvent	ee (%)
CH$_3$CN	17
THF	24
EtOH	39
CHCl$_3$	44
H$_2$O	74

Chakraborti et al. [38] have described the formation of aryl alkyl/2-alkyl benzothiazolines and styryl/ heteroaryl/2-aryl benzothiazoles under water conditions (Scheme 5) [38]. Diverse types of aldehydes **32** (heteroaryl, aryl, and alkyl) were treated with 2-aminothiophenols **33** in water to obtain the respective substituted benzo[d]thiazoles **34**. The reactions were chemoselective without debenzoylation/O-dealkylation, thia-Michael addition, reduction of the α,β-unsaturated carbonyl or nitro groups, and substitution of the nitro group or the halogen atom.

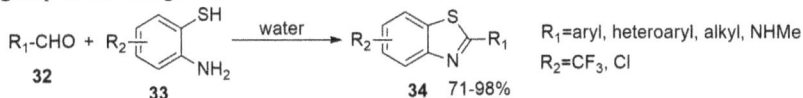

Scheme 5. Catalyst-free reaction of 2-aminothiophenols with aldehydes in water.

In organic synthesis, the Wittig olefination [39,40] is a significant reaction as it produces olefins with a high level of stereoselectivity [41]. Dambacher et al. [42] described diverse examples of Wittig olefination in H$_2$O and organic media by using numerous aryl aldehydes **35** and ylides **36** (Scheme 6) [42]. In terms of selectivity and yield, H$_2$O was observed to be the most effective solvent for the formation of olefins **37** as compared to organic media such as dichloromethane, benzene, and methanol. The investigation also revealed that the solubility of reactants in H_2O was not a significant parameter to accomplish appropriate E/Z-ratios and chemical yields as exposed in the successful Wittig olefinations of protected, aliphatic and heterocyclic aldehydes.

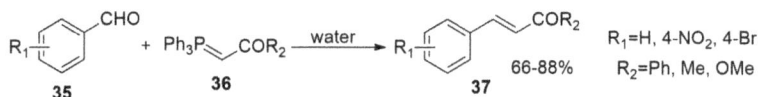

Scheme 6. Catalyst-free Wittig olefinations of ylides and aromatic aldehydes in water.

Formation of **40** and **41** through aminolysis of a diverse range of substituted epoxides **39**througharomatic and aliphatic amines **39** in H_2O has described by Azizi and Saidi [43]. α-Amino alcohols in this economical and practical method were developed under milder conditions with high regio- and stereoselectivity (except styrene oxide) and in excellent yields (Scheme 7). This water-based approach also offers operational simplicity, simple purification procedure, and environmentally friendly conditions.

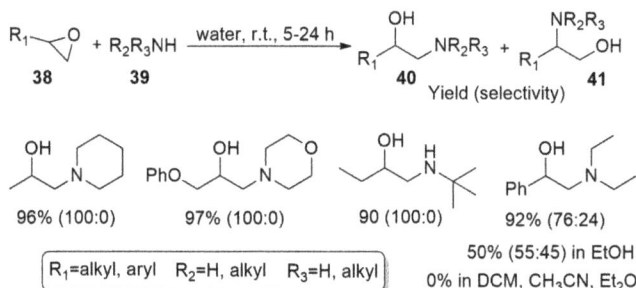

Scheme 7. Development of β-aminoalcohols through aminolysis in H_2O.

Azoulay et al. [44] reported a highly efficient approach for enantioselective addition of aniline **26** to $(2R,3S)$-2,3-diphenyloxirane **42** by using a bipyridine-scandium organocatalytic system (1.2 mol % of a chiral bipyridine ligand **45** and 1 mol % of $Sc(OSO_3C1_2H_{25})_3$ **44**) in pure H_2O and other organic solvents. In water as a solvent, chiral α-amino alcohol **43** was synthesized in excellent yield (89%) as well as in high enantioselectivity (91%) (Table 9). Noteworthy, in this environmentally benign and

Materials Research Forum LLC
doi: https://doi.org/10.21741/9781644900239-3

simple methodology for the organocatalytic asymmetric ring opening reaction, the employment of H_2O as a medium provided excellent enantio-selectivity and yield than that of DCM.

A highly stereoselective and facile approach for the development of highly substituted chiral tetrahydro naphthalene scaffolds attached with an oxazolidine entity **49** from the reaction of **46** with **47** in the presence of organocatalytic **48** has been developed by Zhong and co-workers [45]. The course included catalytic tandem Michael addition/nitrone development/intramolecular [3+2] olefin-nitrone cycloaddition. During the reaction conditions optimization, the authors examined various reaction media and H_2O was found to be the best solvent in terms of selectivity when compared to organic solvents (dichloromethane or hexane) (Table 10). The H_2O employed in the scheme not only constitutes an eco-benign medium but also assists to enhance the stereoselectivity and reactivity.

Table 9. Effect of the solvent on the organocatalyticenantioselective addition reaction

Solvent	Sc(III)	Yield (%)	ee (%)
DCM	Sc(OTf)$_3$	85	74
THF/H$_2$O (9:1)	Sc(OTf)$_3$	<5	71
H$_2$O	Sc(OTf)$_3$	15	85
H$_2$O	Sc(OSO$_3$C$_{12}$H$_{25}$)$_3$	89	91

Table 10. One-pot synthesis of functionalized tetrahydro naphthalene in various solvents

Solvent	Additive	Yield (%)	d.r.	ee (%)
DCM	AcOH	71	68:32	>99
Hexane	none	35	72:28	>99
H$_2$O	AcOH	56	92:8	>99
H$_2$O	PhCO$_2$H	67	92:8	>99

Industrial Applications of Green Solvents I   Materials Research Forum LLC
Materials Research Foundations **50** (2019) 61-106   doi: https://doi.org/10.21741/9781644900239-3

Mori et al. [46] generated boron enolates by using a diphenylborinic acid (in catalytic amount) and used them for stereoselective aldol reactions of carbocation equivalents in H_2O. Treatment of (Z)-trimethyl((1-phenylprop-1-en-1-yl)oxy)silane **50** with benzaldehyde **51** in the presence of sodium dodecyl sulfate (SDS), benzoic acid and a boron source (Ph_2BOH) at room temperature for 24h under aqueous condition afforded aldol adduct **52** in high diastereoselectivity and yield (Table 11). On the other hand, conventional organo-boron-catalyzed reactions requiring extremely anhydrous conditions were carried out at lower temperatures. Noteworthy, the employment of H_2O as reaction medium was important in this reaction. Nearly no respective aldol adduct was achieved in organic media, for instance, dichloromethane and ethyl ether, and poor yield of aldol adduct was realized under the neat reaction condition (Table 11).

Table 11. Organoboron-catalyzed organic synthesis in H_2O

Solvent	Yield (%)	Syn/anti
Neat	24%	90/10
DCM	Trace	-
Et₂O	Trace	-
Water	90%	92/8

In recent years, Kobayashi and co-workers [47] investigated the effectiveness of optically active ligands together with copper(II) towards stereoselective aldol reaction. Treatment of (Z)-trimethyl(pent-2-en-3-yloxy)silane **53** with benzaldehyde **51** in the presence of bis(oxazoline) chiral ligand **55** and a catalytic amount of copper(II) triflate in H_2O/THF at -15°C afforded aldol adduct **54** in excellent enantioselectivity (81%) and yield (81%) (Table 12). On the contrary, dry organic media; for example, dry DCM and dry EtOH, provided much lower selectivities and yields [47]. Silyl cation species, which developed from (Z)-trimethyl(pent-2-en-3-yloxy)silane **53** during aldol reaction, were presumably responsible for the reduction of enantioselectivity in stereoselective aldol reactions in anhydrous solvents because they are capable to catalyze the aldol reactions to deliver racemic products. In contrast, these silyl cation species are quickly hydrolyzed in aqueous media; therefore, the unrequired reaction route avoided.

Racemization is an inescapable issue when hydrolysis of thioester **56** is performed under basic conditions. PS-SO_3H-catalyzed acidic hydrolysis of achiral (R)-S-ethyl 2-

phenylpropane thioate **56** smoothly proceeded to give excellent enantioselectivity (93% ee) under the condition of refluxing H_2O for 72 h (Scheme 8) [48]. This is one of the benefits of the acidic hydrolysis of thioesters in H_2O.

Table 12. Acid-catalyzed stereoselective organic synthesis in aqueous solution

Solvent	Yield (%)	Syn/anti	ee (%)
H_2O/THF	81	78/22	81
Dry ethanol	10	70/30	41
Dry DCM	11	68/32	20

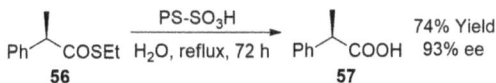

Scheme 8. Aqueous synthesis of (R)-2-phenylpropanoic acid usingPS-SO₃H as catalyst.

5. Towards milder reaction conditions and catalyst-free synthesis

Certainly, for a synthetic procedure to be eco-friendly, it is important and valuable that the reaction could be performed under mild conditions. In literature, there are numerous examples in which the employment of water has led reactions towards milder conditions [49–51].

Triazoles exhibit significant biological potencies, for instance, anti-bacterial, anti-

epileptic, anti-viral and anti-allergic behaviors and are important industrial compounds. Numerous approaches have been reported in the literature for the construction of triazole frameworks; nonetheless, most of these approaches are expensive and difficult to handle and desired accelerating ligands and harsh conditions. The Cu(I)-catalyzed alkyne-azide[3+2] cycloadditionreaction (CuAAC) reported by Rostovtsev et al. [52] has been an outstanding example for formation of triazoles under milder aqueous conditions. In this approach, reaction of acetylide, which is produced by the *in situ* reduction of a Cu(II) salt in H_2O without or with an assistance of organic co-solvent, with organic azides (for example, (*R*)-methyl 2-(2-azidoacetamido)-3-(benzylthio)propanoate **58**) afforded triazoles (for example, **60**) in nearly quantitative yields. This approach did not require accelerating ligands or the protection of reaction mixture from atmospheric oxygen (Scheme 9).

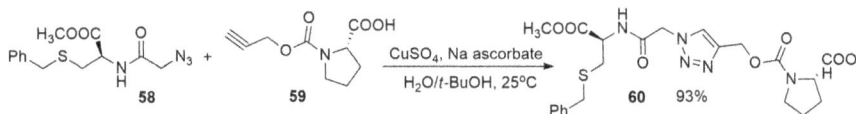

Scheme 9. CuAAC reaction example for formation of triazoles under milder aqueous conditions.

In organic synthesis, the Baeyer-Villiger oxidation (*aka* Baeyer-Villiger rearrangement) of ketones and aldehydes to the respective lactones or esters has been the broadly employed reaction [53]. Conventionally, it is carried out with an organic peroxy acid for example peroxyacetic acid. However, the utilization of a peroxy acid leads to the development of one equivalent of the respective carboxylic acid salt as unwanted impurity, which has to be disposed of or recycled. Additionally, organic peroxy acids are dangerous for health (on account of shock sensitivity) or/and expensive which minimizes their commercial utility. The storage and transport of peroxyacetic acid, for instance, have been strictly restricted, making its application prohibitive. Therefore, growing attention has been devoted on the formation of techniques deploying aqueous hydrogen peroxide (H_2O_2) as the major oxidizer, preferably in H_2O as reaction solvent in order to perform reaction under milder conditions [53,54]. The striking example of Baeyer-Villiger rearrangement in aqueous H_2O_2 was disclosed by Sheldon and co-workers [55], in which selenium-catalyzed reaction was performed in the presence of aqueous H_2O_2 to afford **62** from **61** (Scheme 10) [55].

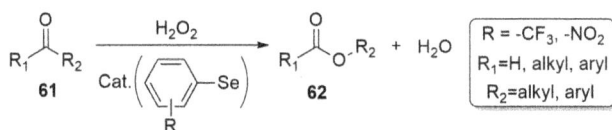

Scheme 10. Baeyer-Villiger rearrangement in aqueous H_2O_2.

Owing to numerous pharmacological activities of pyrimido[4,5-*d*]pyrimidines **66**, these compounds have been regarded as significant frameworks. The literature approaches for the formation of pyrimido[4,5-d]pyrimidines **66** involved multi-step synthetic steps and harsh conditions, *e. g.* the application of tosylateas catalyst, using phosphoryl chloride with dimethylformamide as a solvent [56]. In order to overcome these problems, an eco-friendly and practically useful methodology for one-pot formation of **66** from the reaction of barbituric acid **63**, aldehyde **32** and thiourea or urea **64** in the presence of water as reaction medium under microwave irradiation has been reported by Mazaahir et al. [56]. This technique completely circumvents the application of corrosive bases or acids, hazardous organic solvents, and catalysts (Scheme 11). Water-insoluble solid pyrimido[4,5-*d*]pyrimidines produced in short reaction time were in high purity and yield [57].

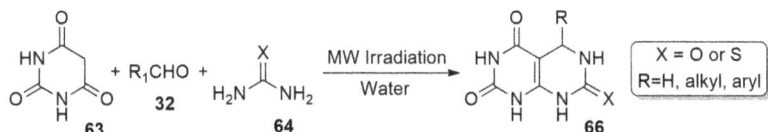

Scheme 11. Development of pyrimido[4,5-d]pyrimidine frameworks under water conditions.

Aminolysis is a significant process for the formation of different important compounds. A broad spectrum of procedures have been established for epoxide ring opening, for instance, the use of silica, alkali metal perchlorates, metal halides, metal triflates, metal alkoxides, metal amides, alumina, enzymes and montmorillonite clay as activators or catalysts. Nevertheless, most of these catalysts desired in stoichiometric amounts and are expensive, difficult to handle and moisture sensitive. Also, with these catalysts, some sterically hindered amines and deactivated aromatic amines fail to open up epoxide rings or still need a high pressure or temperature [58]. Recently, effective aminolysis of a library of epoxide-containing compounds **67** by aromatic or aliphatic amines (for example **68**) was carried out in H_2O by Azizi and Saidi [43] in order to overcome issues, which were reported in the literature (Scheme 12). The substituted 2-morpholino-1-phenoxyethanol **69** was attained under mild conditions in excellent yield and selectivity without any organic co-solvent or catalyst.

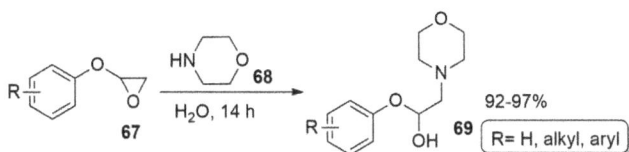

Scheme 12. Formation of 2-morpholino-1-phenoxyethanol in aqueous medium.

Organo-boron compounds are mostly useful reagents because of low toxicity, high stability, and ease of handling. Particularly, they have been extensively utilized in Suzuki-Miyaura cross-coupling reactions. Therefore, in this reaction, the application of H_2O as a reaction medium has been an interesting challenge [49]. In 2005, Anderson and Buchwald [59] reported a novel sulfonated ligand **72** which was used to develop an activated complex for conducting Suzuki–Miyaura cross-coupling reaction of boronic acids **71** with aryl chlorides **70** under aqueous conditions to afford coupling products **73** (Scheme 13) [59]. In most cases, low catalyst loadings (0.1–0.5 mol%) and room temperature conditions were used to accomplish coupling reaction. Fortunately, a wide range of boronic acids **71** or aryl chlorides **70** with a diverse range of functional groups coupled under these aqueous conditions without the demand of any protecting groups.

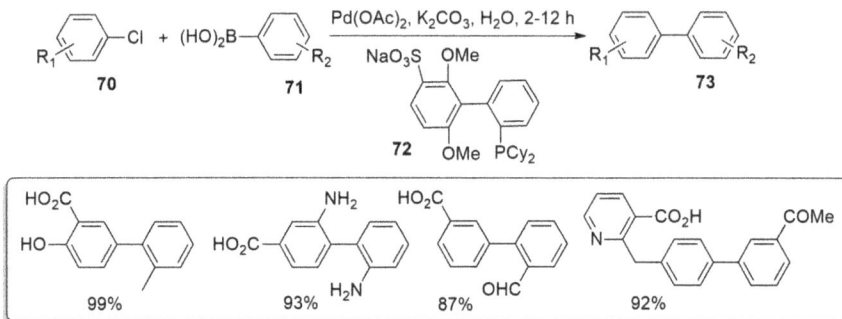

Scheme 13. Suzuki-Miyaura cross-coupling under aqueous conditions and sulfonated ligand.

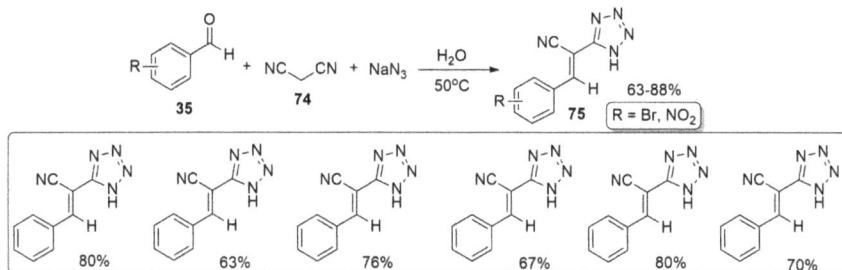

Scheme 14. 5-Substituted-tetrazoles synthesis under catalytic-free and milder aqueous conditions.

Tetrazoles have gained significant consideration on account of their broad spectrum of utilities in rocket propellants, explosives, photography, information recording systems, and pharmaceuticals. These are important ligands for numerous valuable functionalizations and also intermediates for diverse of nitrogen-containing heterocyclic compounds. Most techniques reported in literature for the formation of tetrazole frameworks have disadvantages; for example, tedious workup, difficulty in preparing and/or obtaining the starting materials, harsh reaction conditions, long reaction times, low yields, application of high boiling solvents and the use of expensive and toxic reagents are some to mention [60]. In order to overcome these problems, Tisseh et al. [61] developed the catalytic-free multi-component one-pot domino Knoevenagel condensation and 1,3-dipolar cycloaddition reaction of sodium azide, malonodinitrile**74** and carbonyl compounds **35** under milder aqueous conditions to synthesize respective 5-substituted-tetrazoles **75**. This green approach offered good to excellent yields (63–88%, Scheme 14) [61].

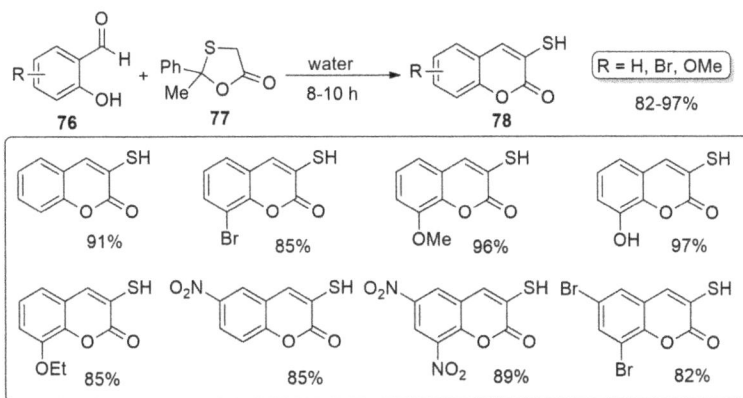

Scheme 15. Synthesis of 3-sulfanyl-2H-chromen-2-onesunder aqueous conditions.

Coumarins and 3-mercapto coumarins belong to the important family of heterocycles existed in numerous natural products which have found broad applications such as useful medicinal products, laser dyes, dispersed fluorescent, in the preparation of optical brighteners, cosmetics and pharmaceuticals, perfumes and additives in food [62]. Many traditional methods to form coumarin skeletons are well known, which include Wittig condensation, Reformatsky, Perkin, Knoevenagel and Pechmann reactions. To expedite these traditional methods for synthesis of coumarin frameworks, some variations in terms of reaction conditions and catalysts including cation-exchange resins, ionic liquids, W/ZrO_2 solid acid, Amberlyst-15, Nafion-H, clays, zeolite, sodium hydroxide in water, organo-palladium and solid-phase synthesis have been established. In contrast, only a few approaches have been reported on the formation of 3-mercapto coumarins **78** (*aka.*3-sulfanyl-2*H*-chromen-2-one). However, all of the approaches suffer from many shortcomings including low selectivity, tedious workup, low yields, long reaction times, use of expensive reagents, the requirement of high amounts of catalysts and the generation of large amounts of toxic waste [63]. In order to overcome these issues, Yadav and co-workers [63] reported the preparation of 3-sulfanyl-2*H*-chromen-2-ones **78** *via* mercaptoacetylative cyclization reaction of substituted 2-hydroxybenzaldehyde **76** with 2-methyl-2-phenyl-1,3-oxathiolan-5-one **77** under aqueous conditions. The 3-sulfanyl-2*H*-chromen-2-ones **78** were attained in high yields (82–97%) (Scheme 15) [63]. In this green,

Industrial Applications of Green Solvents I   Materials Research Forum LLC
Materials Research Foundations **50** (2019) 61-106   doi: https://doi.org/10.21741/9781644900239-3

efficient and convenient procedure, H_2O itself acted as catalyst during cyclization reaction by hydrogen bonding and thus circumvented the employment of any other catalyst.

Double bonds reduction has been an extensively employed approach for both interconversion of the functional group and the insertion of chiral centers in achiral entities. Nevertheless, this approach usually involved the employment of pressure reaction vessels and hazardous hydrogen gas and, hence recently, transfer hydrogenation has been developed as a safer, cheaper and more reliable alternative [64]. Noteworthy, in recent years, the application of H_2O as the reaction medium in transfer hydrogenation has received considerable attention [65]. For instance, Wu et al. [66] reported a chemoselective and efficient methodology for iridium-catalyzed reduction of aromatic and aliphatic aldehydes **32** to alcohols **79** under very mild aqueous conditions (Scheme 16) [66]. They reported that their method tolerated various functional groups such as nitro, olefins, and halogens; did not desire vacuum conditions; and operated with low catalyst loading.

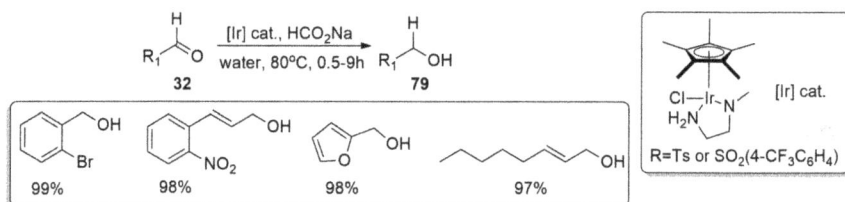

Scheme 16. Aqueous transfer hydrogenation of aldehydes.

Recently, Li and co-workers [67] described a highly effective alkynylation of aromatic and aliphatic aldehydes **32** with substituted alkene **80** by using phosphine/silver complexes as catalysts in H_2O to obtain **81** [67]. The synthetic approach was dually promoted by H_2O and electron-releasing phosphine ligand to provide good to excellent yields of 63–98%. A hydroxyl group-containing aldehyde *viz.* 4-(hydroxymethyl)benzaldehyde **82** was successfully alkynylated to 1-(4-(hydroxymethyl)phenyl)-3-phenylprop-2-yn-1-ol **83** without protecting the hydroxyl group (Scheme 17).

Another utility that has been rarely investigated is the selective functionalization of H_2O-soluble biopolymers such as oligosaccharides, oligonucleotides, and polypeptides. It has been also reported that H_2O has promising potential for carrying out carbon-carbon bond creating reactions with unprotected sugars under mild aqueous alkaline conditions. Also, in literature, various H_2O-soluble organometallic catalytic systems have been reported to shorten the long synthetic pathways by eliminating steps of multiple protection and deprotection [68–70].

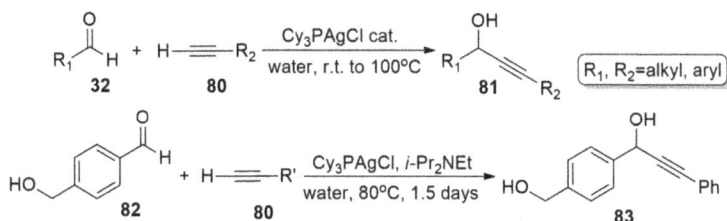

Scheme 17. Alkynylation of aldehyde derivatives by using phosphine/silver complexes in H$_2$O.

Branco and Gawande [71] have described an eco-friendly and novel 9-fluorenylmethoxycarbonyl (Fmoc) protection of a diverse range of amino phenols, amino alcohols, amino acids, and aromatic and aliphatic amines **84** in H$_2$O as reaction solvent under catalyst-free and mild conditions using 9-fluorenylmethyl chloroformate **85** (Scheme 18) [71]. The Fmoc protected compounds **86** were attained in high yields (75–92%).

Scheme 18. Fmoc protection of amines and amino acids in aqueous media.

The Baylis-Hillman (BH) reaction has gained great attention owing to the fascinating tandem Michael/aldol reaction sequence catalyzed through a Lewis acid or a Lewis base and the valuable potential of synthesizing multi-functional products [72]. In order to establish a green, catalyst-free and stereoselective protocol for formation of (*Z*)- and (*E*)-allyldithiocarbamate derivatives (**88 and 89**) from acetates of BH adducts **87**, Yadav et al. [73] have developed a three-component one-pot coupling reaction of acetated-BH adducts **87** with substituted amines **84** and carbon disulfide in the presence of water as reaction solvent at room temperate [73]. The allyl dithiocarbamate derivatives (**88 and 89**) were in excellent yields (80–94%) (Scheme 19). The reaction route involved the nucleophilic substitution of acetate groups by the dithiocarbamate ions using H$_2$O as a promoter and a solvent; hence, eliminating the application of toxic solvents and catalysts.

85

Industrial Applications of Green Solvents I   Materials Research Forum LLC
Materials Research Foundations **50** (2019) 61-106   doi: https://doi.org/10.21741/9781644900239-3

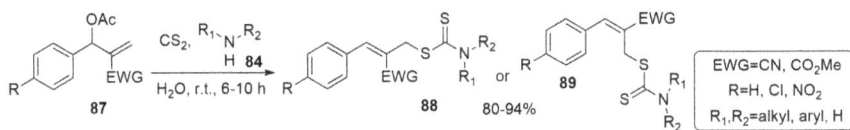

Scheme 19. Coupling reaction for formation allyldithiocarbamate derivatives in water.

Moberg and Rákos [74] documented the first pseudo-five-component one-pot catalytic-free formation of 1,2-dihydro[1,6]naphthyridine derivatives **91** from malononitrile **74**, amines **84** and methyl ketones **90** under mild aqueous conditions (Scheme 20) in high yields (79–93%) [74]. The **92-94** were the significant precursors in the formation of 1,2-dihydro[1,6]naphthyridines **91**.

The 1,2-dihydro[1,6]naphthyridine derivatives **91** have been observed in several natural marine products with bioactive affinities such as selective antagonistic potency for 5-HT$_4$ receptors, allosteric inhibition of Akt2 and Aktl and Akt2, HIV-1 integrase inhibition and anti-proliferative efficacy [51]. Classically, the preparation of [1,6]-naphthyridines was performed either using multistep synthetic route or expensive catalysts, but these eco-friendly procedures provided a valuable alternative to synthesize bioactive drugs. The formation of 1,2-dihydro[1,6]naphthyridines **91** using the above-mentioned approach can be performed under anhydrous conditions without the use of toxic organic solvents and expensive catalysts [75,76].

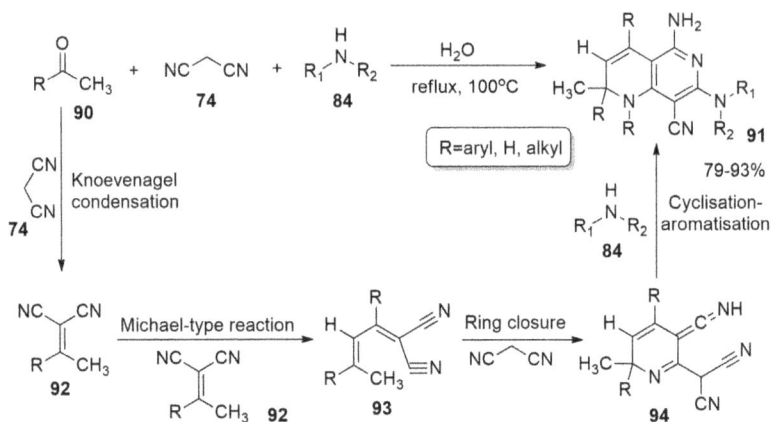

Scheme 20. One-pot catalytic-free formation of 1,2-dihydro[1,6]naphthyridine derivatives in water.

Wurz and Charette [77] reported the asymmetric and racemic cyclopropanation reactions of several alkanes **96** in H_2O catalyzed by transition metals. However, this approach towards cyclopropanation included the preparation and subsequent utilization of highly explosive diazoacetic ester **98**. To resolve this problem, the same two researchers documented conditions permitting the *in situ* formation of the diazoacetic ester **98**, starting reaction from ethyl glycinate hydrochloride **95** and introducing sulfuric acid and sodium nitrite, which then reacted with the styrene **96** and rhodium catalytic system to lead to the required cyclopropane **97** in moderate selectivity and high yield (Scheme 21). Furthermore, this reaction was accompanied effectively on a three grams scale, allowing a straightforward, secured and cheap approach to cyclopropane entities **97** [77].

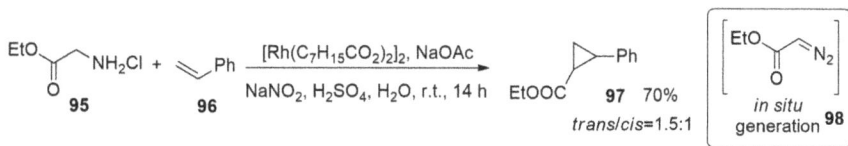

Scheme 21. Cyclopropane synthesis in H_2O using an in situ generated diazoacetic ester.

The domino-Knoevenagel-hetero-DA reaction introduced by Tietze [78] has been very popular for its promising sequential conversion. The reactions can be used for the construction of dihydropyran-containing compounds (**101**, **102**, **104** and **105**) and for the formation of several bioactive drugs [79]. In literature, amine-based catalytic systems and Lewis acid catalyzed reactions were reported for this reaction [80]. In order to eliminate the use of Lewis acid or catalyst, Ghandi et al. [81] described a one-pot methodology for the synthesis of tetrahydro-2H-pyrano[2,3-d]pyrimidines (**101** and **102**) and benzo-δ-sultones having hexahydro-2H-chromene (**104** and **105**) under aqueous conditions. A large range of 2-formyl-4-phenyl-(E)-2-phenylethenesulfonate derivatives **99** underwent a one-pot domino-Knoevenagel-hetero-DA reaction with 1,3-diethyl barbituric acid **100** and Meldrum's acid **103** in water (Schemes 22) [50].

The reaction between 2-formyl-4-phenyl-(E)-2-phenylethenesulfonate derivatives **99** and 1,3-dimethylbarbituric acid **100** in water produced respective adducts (**101** and **102**) in good to excellent yields (49–93%). Similarly, the reaction with Meldrum's acid in water afforded respective compounds (**104** and **105**) as mixtures of diastereomers in moderate yields (42–78%). The precursors **99** were synthesized by simple condensation of (E)-2-phenylethenesulfonyl chloride and 2-hydroxybenzaldehyde in propanone as solvent under basic conditions. The attractive features of this reaction were the avoidance of numerous sequential steps, short reaction time, use of water and simplicity [51].

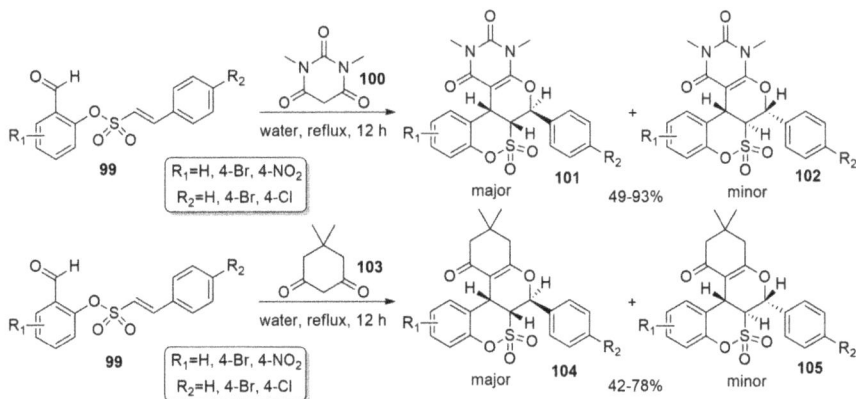

Scheme 22. Catalytic-free preparation of hexahydro-2H-chromene-annulated and tetrahydro-2H-pyrano[2,3-d]pyrimidine-annulated benzo-δ-sultones in aqueous media.

6. Simplification in the course of workup

In organic synthesis, the workup process, through chromatography purifications or extractions, for example, may be responsible for the utilization of a large amount of solvent. In order to make the process eco-friendly and economically feasible, it is essential to simplify the workup process by converting multi-step synthesis into one-pot synthesis, using water as solvent in the workup step(s) instead of organic solvents or adopting such synthetic approaches, which replace the tedious separations (extractions or chromatography purifications for instance) with simple filtration. Investigations have revealed the tendency of water to convert tedious workup into simple work-up in many cases.

Amongst heterocyclic compounds, nitrogen-containing heterocycles have been extensively used as core framework of heterocycles in natural science and other areas of science [82]. Varma and co-workers [83] have established a novel procedure for the preparation of nitrogen-containing heterocyclic compounds (108 and 111) from aromatic, aliphatic and cyclic amines 106 and diazane109, respectively, and alkyl dihalides (107 and 110) (Scheme 23). Using water as solvent and microwave (MW) irradiation, reaction duration was decreased to 20 minutes, side reactions were eliminated and yields were enhanced. This technique provided the benefit of easy purification since phase isolation of the products (**108** or **111**) from the water occurred after the reaction and only decantation or filtration is required [83–86].

$$R-NH_2 \; + \; X(CH_2)_nX \xrightarrow[\text{MW, 120°C, 20 min}]{K_2CO_3,\ H_2O} R-N\overset{\frown}{\underset{\smile}{}}(CH_2)_n$$

106   **107**   **108**

n= 3-6
R= alkyl, aryl, cyclic
X= Cl, Br, I, OTs

$$R_1-NHNH_2 \; + \; \underset{\textbf{110}}{X\overset{R_2\ R_3}{\diagup\diagdown}X} \xrightarrow[\text{MW, 120°C, 20 min}]{K_2CO_3,\ H_2O} \underset{\textbf{111}}{R_1-N-N}$$

109   60-81%

$R_1, R_2, R_3 =$ H, aryl, alkyl
X = Cl, Br, OTs

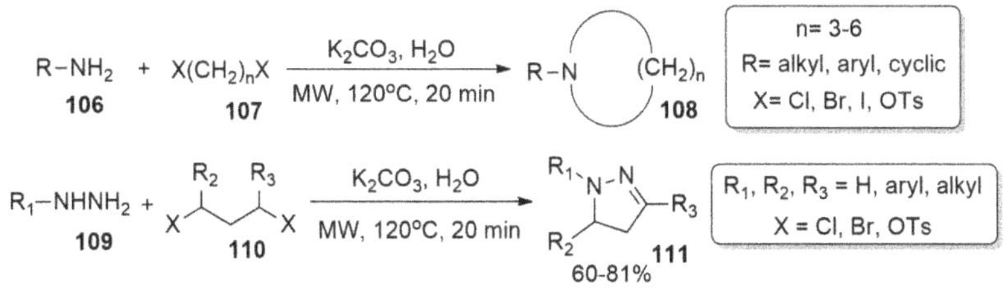

Scheme 23. Formation of nitrogen-containing heterocycles.

The enzyme-based highly enantioselective hydrolysis of amides has been broadly investigated. These organic reactions are catalyzed by lipases, amidases, and acylases. Aspartame **115**, a popular low-calorie non-saccharide intense sweetener has been prepared by thermolysin (TML), a thermostable neutral metalloproteinase enzyme (Scheme 24). In this synthetic process, the l-enantiomer of racemic methyl L-phenylalaninate **111** selectively reacted with the α-carbonylmoiety of N-protected (S)-2-aminosuccinic acid **112** in water to produce **113** and **114**. This procedure is superior from the point of view of yield and is more eco-friendly than the reported methods [87]. Also, this technique provides the benefit of easy purification.

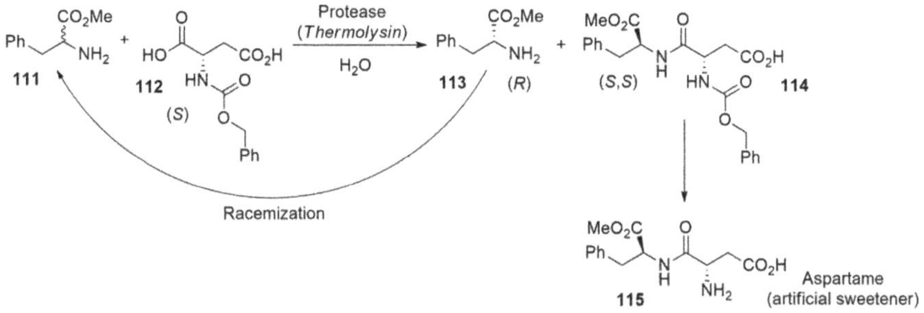

Scheme 24. Synthesis of aspartame: an artificial sweetener.

Hayashi and co-workers [88] disclosed an effective organocatalytic system **119**, containing both tetrazole and siloxy (R_3SiO-) moieties within a tetrahydropyrrole framework, for the asymmetric Mannich reaction of numerous cyclic and aliphatic ketones (for example **118**) with glyoxal 1-(dimethyl acetal) **116** and 4-methoxyaniline **117** under aqueous conditions to deliver **120** (Scheme 25) [88]. In particular, aqueous solutions of the aldehydes were employed and no additional volume of H_2O was essential to achieve high enantioselectivities and yields. This allows to directly charge the crude mixture on a column chromatography for purification; therefore, avoiding the step of extraction.

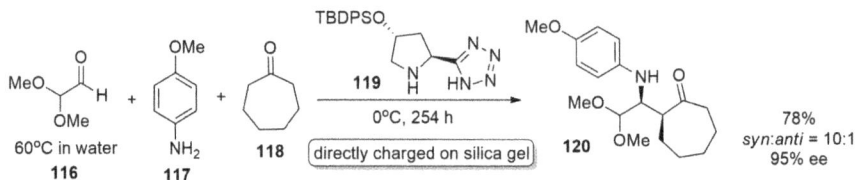

Scheme 25. Extraction-free Mannich reaction catalyzed by organocatalytic system.

The group of Cheng [89] described an asymmetric Michael reaction between oxocyclohexane **121** and nitroolefines **122** in water using a surfactant-kind asymmetric organocatalytic system **123** to obtain **124** (Scheme 26) [89]. The reaction provided adducts **124** in high stereoselectivities and yields and operated at room temperature without any additional promoters. Generally, no organic solvent required for the process of extraction since the separation of the crude adduct was carried out by phase separation or filtration.

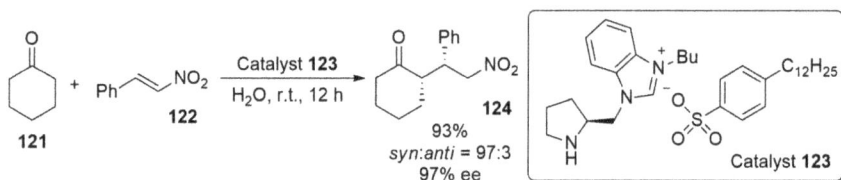

Scheme 26. Surfactant-type organocatalyzed Michael reaction in water.

Recently, Kacprzak [90] reported a straightforward two steps, a one-pot technique for the regioselective synthesis of 1,2,3-triazole derivatives **127**. Aqueous copper sulfate, sodium ascorbate, alkyne and water introduced into the dimethyl sulfoxide solution of aryl or alkyl azides **126** created *in situ* reaction of **125** with sodium azide in anhydrous dimethyl sulfoxide (Scheme 27). The 1,2,3-triazole derivatives **127**, produced in high yields, frequently precipitated and purified by simply filtering the aqueous mixture and thus eliminating the step of purification *via* column chromatography. When mono substituted olefins **128** were used only 1,4-regioisomers were detected.

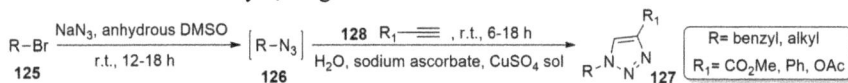

Scheme 27. Regioselective synthesis of 1,2,3-triazole derivatives by Kacprzak.

Industrial Applications of Green Solvents I   Materials Research Forum LLC
Materials Research Foundations **50** (2019) 61-106   doi: https://doi.org/10.21741/9781644900239-3

In the highly efficient and novel two-step protocol by Sharpless and co-workers, bistriazoles (**131** and **134**) were synthesized regioselectively by ring-opening reaction of isomeric diepoxides (**129** and **132**) with azide ion in the presences of NH_4Cl and H_2O at refluxing condition followed by treatment with 2-butynedioic acid in H_2O (Scheme 28). The crystalline solid bistriazoles (**131** and **134**) were separated by simply filtering the reaction mixture avoiding the chromatographic purification and extraction steps [91].

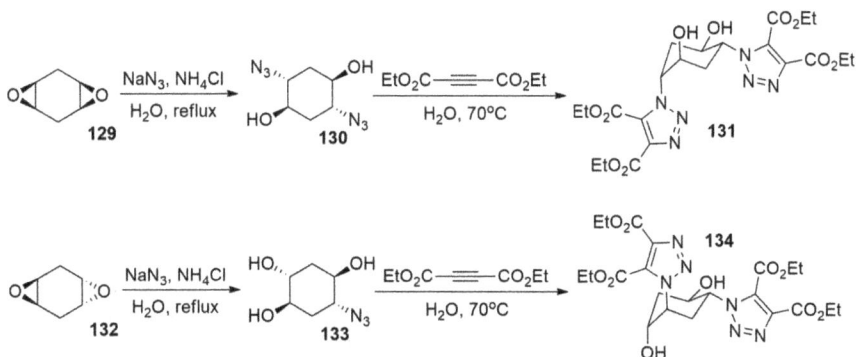

Scheme 28. Two-step development of bistriazole frameworks.

In 2007, Pang et al. [92] developed a novel method to synthesize the dihydropyrazoles **137** through the Bamford–Stevens reaction of N'-(2,3,4,5,6-pentafluorobenzyl)-2,4,6-tri(propan-2-yl)benzenesulfonohydrazide **135** with nitriles or methyl acrylate **136** in the presences of triethylamine and water or THF (Scheme 29) [92]. Although the reaction could be performed in THF, the on-water method provided almost quantitative yields. The solid insoluble products were simply separated by filtering the reaction mixture.

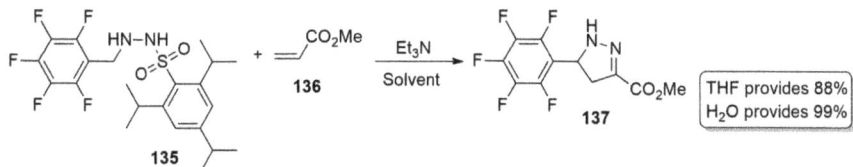

Scheme 29. Synthesis of pyrazolines through Bamford-Stevens reaction.

Tu et al. [93] reported an effective approach for synthesis of poly-functionalized indeno[1,2-*b*]quinolone derivatives **140** *via* a three-component reaction of functionalized enaminone derivatives **139**, (het)- alkyl or aryl aldehydes and 1,3-dioxoindan **138** in the presence of water and 4-methylbenzenesulfonic acid as catalyst under MW irradiation

Industrial Applications of Green Solvents I
Materials Research Forum LLC
Materials Research Foundations **50** (2019) 61-106
doi: https://doi.org/10.21741/9781644900239-3

(Scheme 30) [93]. Aromatic aldehydes electron donating groups reacted within 5-7 min, while aromatic aldehydes with electron deactivating moieties reacted within 2-3 min. Also, some three-component reactions were also carried out through conventional heating at the 150°C, delivering the derivatives **140** in longer reaction time (2 h) and relatively low yields. Owing to the aqueous reaction conditions, the separation of products from reaction was also simple, *i.e.*, only neutralization followed by simple filtration was required for solid products separation.

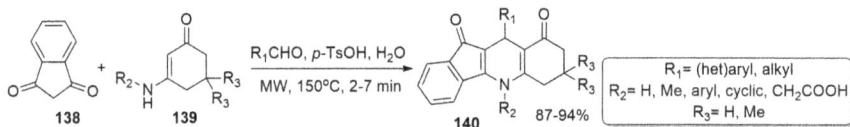

Scheme 30. Formation of poly- functionalized indeno[1,2-b]quinolones.

In order to investigate the structure-activity relationship for the binding of phenothiazines to HIV-1 TAR RNA, Mayer et al. [94] prepared a small series of 10*H*-phenothiazine derivatives**142** with unique substitutions around the scaffold (Scheme 31) [94]. The formation started by a mild iodo-catalyzed reaction of diarylamine derivatives**141** with sulphur in distilled H_2O at a temperature of 190°C for 20 min to afford 10*H*-phenothiazine products **142** in appropriated yields. Attributed to the hydrophobic nature of the synthesized products **142**, they easily precipitated out it upon cooling and separated by simply filtering the reaction mixture. Additionally, alkylation at the NH followed by amination (MW irradiation, 100 °C, 40 min) provided frameworks **143** which were screened for binding to HIV-1 TAR RNA.

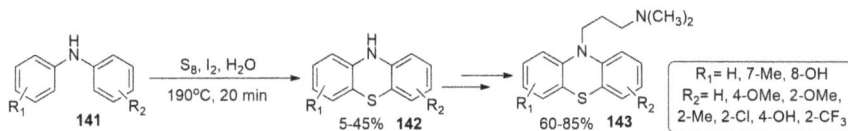

Scheme 31. Synthesis of 10H-phenothiazines and aliphatic amine functionalized phenothiazines.

In order to show the importance of H_2O as reaction medium for the high-throughput formation of compounds in a parallel synthesis technique, Pirrung and Sarma [95,96] accomplished the Ugimulti-component condensation between four aldehydes **32**, four acids **145** and isonitriles **144** to achieve a series of 32 *β*-lactams **146** (Scheme 32) [95,96]. Mostly, the *β*-lactams were precipitates and separated by simply filtering the reaction mixture.

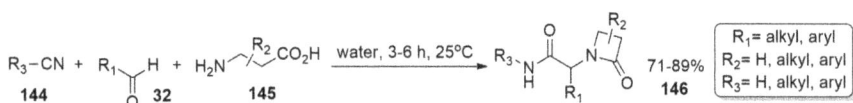

R₃–CN + R₁CHO + H₂N—CO₂H →(water, 3-6 h, 25°C)→ 146   71-89%

144   32   145

R₁= alkyl, aryl
R₂= H, alkyl, aryl
R₃= H, alkyl, aryl

Scheme 32. Conversion access to a series of 32 β-lactams with H$_2$O as reaction medium.

During the formation of an organo-catalyzed asymmetric DA cycloaddition of diene1 and cinnamaldehyde **147** using a chiral salt of diarylprolinolsilyl ethers **148** to give **149**, the Hayashi and co-workers [97] revealed that scaling up the reaction to a twenty millimoles scale can eliminate the employment of organic solvents (Scheme 33) [97]. Further, the H$_2$O phase can be easily removed by distillation and decantation to obtain the product with excellent yields.

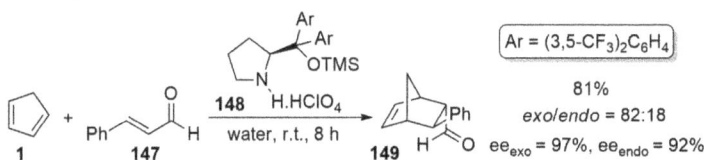

Ar = (3,5-CF$_3$)$_2$C$_6$H$_4$

81%
exo/endo = 82:18
ee$_{exo}$ = 97%, ee$_{endo}$ = 92%

Scheme 33. DA cycloaddition through asymmetric organocatalytic system.

7. Enhancement in recycling the catalyst

The mission to implement principles of sustainable chemistry is a driving strength towards the formation of regenerable and reusable catalytic systems. H$_2$O plays a significant role during the process of catalyst recycling by offering assistance in the isolation of catalyst from the reaction mixture.

Recently, 2-((diphenylphosphino)oxy)aniline **151** as a new ligand has been introduced by Firouzabadi and co-workers [98] for the Heck cross-coupling reactions of aromatic halides **125** with ethenylbenzene **150** in H$_2$O at 80-95°C in the presence of palladium acetate [98]. The 2-aminophenyl diphenylphosphinite **151**, which is an air and water stable ligand, was simply synthesized from the reaction of 2-aminophenol with chlorodiphenyl phosphine in excellent yield. After the reaction completion, the insoluble organocatalyst was simply isolated from the reaction mixture by centrifugation or filtration and recycled six-fold to give the product **152** in 79–83% yields (Scheme 34).

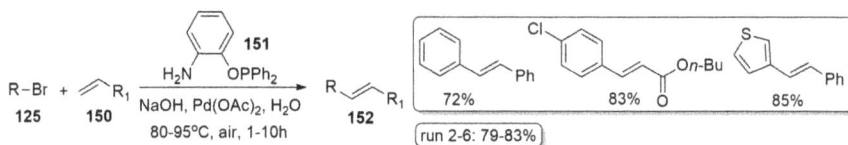

Scheme 34. Pd-phosphinite complex-catalyzed Heck cross-coupling reaction in water.

In organocatalysis, immobilization of the organocatalyst on a solid matrix to allow its recovery and reuse has been important. For instance, Font et al. [99] reported a Merrifield resin (chloromethylpolystyrene) supported 4-functionalized proline organocatalyst **154** synthesized *via* click chemistry for the aqueous asymmetric aldol condensation of cyclohexanone**121** with benzaldehyde **51** to deliver **152** (Scheme 35) [99]. This organocatalyst could be recovered and reused at least five-fold without apparent loss of selectivity and yield.

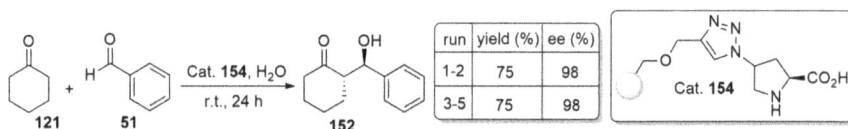

Scheme 35. Merrifield resin supported prolineorganocatalyst for asymmetric aldol condensation in water.

Berthod et al. [100] have reported the utility of ammonium moieties in the design of a H_2O-soluble derivative of BINAP **157** for the ruthenium-organocatalyzed asymmetric reduction of ethyl 3-oxobutanoate **155** to (*R*)-ethyl 3-hydroxybutanoate **156** under aqueous conditions (Scheme 36) [100]. The organocatalyst **157** in aqueous medium was recycled up to eight-fold with ee always over 97% and with no substantial loss of potency.

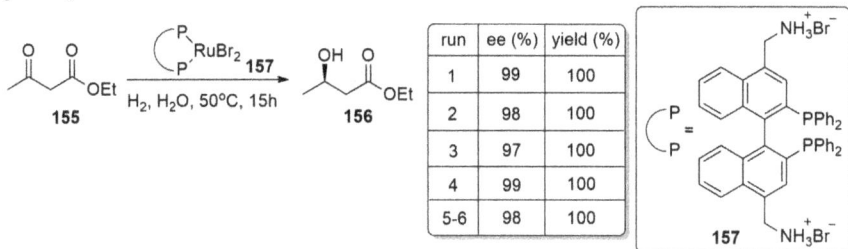

Scheme 36. Reusable H_2O soluble organocatalyst for the asymmetric reduction of β–ketoesaters.

Ghorai and Lipshutz [101] described the ring-closing metathesis (RCM) and the cross-metathesis (CM) of olefins using a unique, potent and promising catalytic system. In particular, they developed a novel ruthenium organocatalyst **160** with a PQS-attached Grubbs-Hoveyda-type complex (PQS = polyethyleneglycolubiquinolsebacate) comprising in the association of a covalently bound ruthenium carbine to catalyze the metathesis, a hydrophilic region for solubility in pure H_2O and a hydrophobic region for the solubilization of the substrate. The species freely dissolved in H_2O, creating nanomicelles in which RCM reactions withH_2O-insoluble dienic compounds (for example **158**) can take place in pure H_2O at room temperature. Extraction of the cyclic target compound **159** and subsequent introduction of fresh diene**158** in the aqueous phase permitted the organo-catalyst to be recycled nine-fold with still an appropriate catalytic capability (Scheme 37).

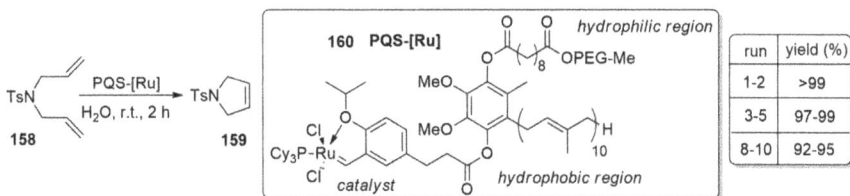

run	yield (%)
1-2	>99
3-5	97-99
8-10	92-95

Scheme 37. Metathesis of alkenes in water using a ruthenium organocatalyst.

In the recent years, a new metal- and organic solvent-free catalytic system **162** for the functionalization of cyclohexene**161** to the *trans*-1,2-cyclohexanediol **163**, using 30% hydrogen peroxide (Scheme 38), has been reported by Usui et al. [102]. The catalytic system **162** is a resin-supported sulphonic acid, for instance, Nafion^{TM}, Amberlyst-15 or the related silica-Nafion complexes, and could be regenerated by filtering the reaction mixture and reused five-fold without loss of efficacy.

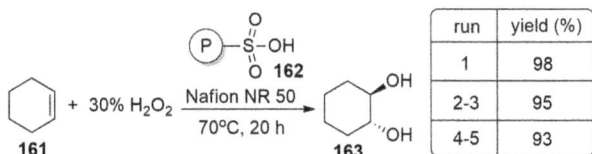

run	yield (%)
1	98
2-3	95
4-5	93

Scheme 38. Olefin dihydroxylation with H_2O_2 over Nafion resin.

A resin-supported scandium-based catalytic system, synthesized from sulfonated polystyrene (PS) resin **165**, was observed to be a highly operative organocatalyst for Mukaiyama aldol addition of **51** in H_2O [103]. The organo-catalyst was easily regenerated by simply filtering the reaction mixture and recycled without any important loss of catalytic potency (Scheme 39).

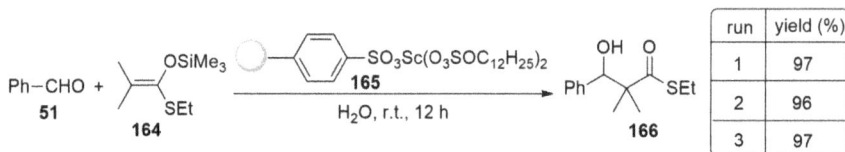

run	yield (%)
1	97
2	96
3	97

Scheme 39. Mukaiyamaaldol reactions in H$_2$O using supported scandium catalytic system.

The hydration of 2-propenenitrile **167** using Ru(OH)$_x$/Al$_2$O$_3$ catalyst produced acrylic amide **168** (Scheme 40). No side reactions, like polymerization of 2-propenenitriles and hydration of carbon-carbon double bonds, occurred [104]. The Ru(OH)$_x$/Al$_2$O$_3$catalytic system was simply isolated by filtering the reaction mixture. After isolation of the catalytic system, it can be recycled at least twice with no significant loss of catalytic affinity.

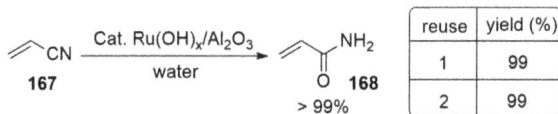

reuse	yield (%)
1	99
2	99

Scheme 40. Ruthenium-catalyzed hydration of 2-propenenitrile to acrylic amide in H$_2$O.

Apart from instances involving aqueous palladium-catalyzed Suzuki-Miyaura cross-couplings, there are increasing number of research papers describing the employment of immobilized, reusable Pd-catalytic systems under aqueous conditions. Solodenko et al. [105] used a heterogeneous palladium(II) precatalytic system **171** that is insoluble in organic solvents and H$_2$O [105]. In most cases, high yields (48-100%) were obtained for the cross-couplings of aromatic halides **169** and trifluoromethane sulfonate with a broad range of functionalized boronic acids **71** in the presence of TBAB, potassium carbonate and catalyst **171** at 120°C for 20 min (Scheme 41). Loaded into an IRORI Kan, the precatalytic system **171** could be regenerated and reemployed up to fourteen-fold without any substantial loss of catalytic potency.

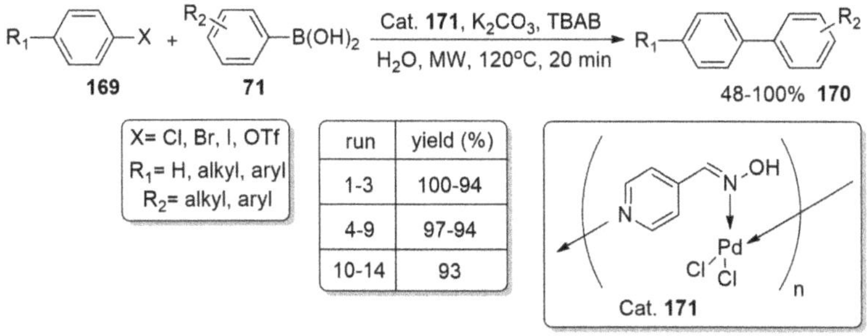

Scheme 41. Suzuki couplings with standard soluble palladium catalyst.

A unique method for the organocatalytic asymmetric Michael reaction of aliphatic or aromatic aldehydes **173** with nitroalkenes **172** on H_2O has been established by Zheng et al. [106]. The synthesis was performed in pure H_2O at room temperature using low loading of catalyst **174** and the produced Michael products**175** were in high enantio- and diastereo-selectivities (Scheme 42) [106]. The catalytic system **174** can be reused for more than six-fold without a considerable loss of stereo chemical control and catalytic ability. Additionally, the synthetic technique is environmentally benign and practically simple.

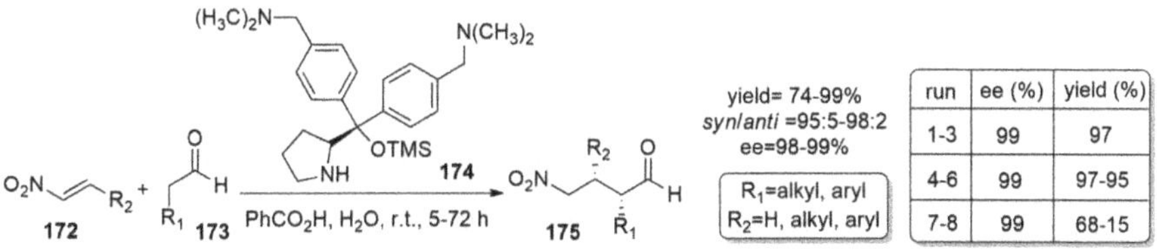

Scheme 42. Organocatalytic system for the Michael reaction in water.

Conclusion

This book chapter focuses on the potential of H_2O as a reaction solvent. In the past decade, amazing consideration has been devoted to organic reactions in H_2O and the research activities in this field are growing exponentially. Key advancements in the area have concentrated on resolving typical problems of organic reactions including low yields, long reaction times, insolubility of reagents/products and selectivity matters in non-classical solvents or in the absence of catalytic systems. H_2O holds many capabilities for employment in organic processes as a solvent owing to its availability and low cost. However, the major purpose to pursue H_2O as a solvent is its hydrophobic effect that leads to such outstanding new chemistry not otherwise possible. Several examples highlighting the benefits of H_2O as a solvent have been summarized in this book chapter, which hopefully will help researchers across the globe to create new, environmentally benign, clean and inexpensive approaches for organic synthesis in H_2O.

References

[1]   P.J. Dunn, The importance of green chemistry in process research and development, Chem. Soc. Rev. 41 (2012) 1452–1461. https://doi.org/10.1039/c1cs15041c

[2]   P.T. Anastas, Green chemistry and the role of analytical methodology development, Crit. Rev. Anal. Chem. 29 (1999) 167–175.

[3]   P. Anastas, N. Eghbali, Green chemistry: principles and practice, Chem. Soc. Rev. 39 (2010) 301–312. https://doi.org/10.1039/b918763b

[4]   S. Eminov, Catalytic conversion of fructose, glucose and cellulose to 5-(hydroxymethyl) furfural (HMF), (2017).

[5]   V.K. Ahluwalia, M. Kidwai, New trends in green chemistry, Springer Science & Business Media, 2004.

[6]   V.K. Ahluwalia, M. Kidwai, Basic principles of green chemistry, in: New Trends Green Chem., Springer, 2004: pp. 5–14. https://doi.org/10.1007/978-1-4020-3175-5_3

[7]   P.T. Anastas, M.M. Kirchhoff, Origins, current status, and future challenges of green chemistry, Acc. Chem. Res. 35 (2002) 686–694. https://doi.org/10.1021/ar010065m

[8]   S.L.Y. Tang, R.L. Smith, M. Poliakoff, Principles of green chemistry: Productively, Green Chem. 7 (2005) 761–762. https://doi.org/10.1039/b513020b

[9]   F.M. Kerton, R. Marriott, Alternative solvents for green chemistry, 2nd Ed., Royal Society of Chemistry, 2013.

[10]   E. Buncel, R.A. Stairs, H. Wilson, The role of the solvent in chemical reactions, Oxford University Press, Oxford, 2003.

[11]   J.H. Clark, S.J. Tavener, Alternative solvents: shades of green, Org. Process Res. Dev. 11 (2007) 149–155. https://doi.org/10.1021/op060160g

[12]   P.T. Anastas, J.C. Warner, Green chemistry: theory and practice, Oxford university press Oxford, 2000.

[13]   J.M. DeSimone, Practical approaches to green solvents, Science 297 (2002) 799–803. https://doi.org/10.1126/science.1069622

[14] J. Fraga-Dubreuil, G. Çomak, A.W. Taylor, M. Poliakoff, Rapid and clean synthesis of phthalimide derivatives in high-temperature, high-pressure H_2O/EtOH mixtures, Green Chem. 9 (2007) 1067–1072. https://doi.org/10.1039/b704405d

[15] N.D. Gullickson, J.F. Scamehorn, J.H. Harwell, Liquid-coacervate extraction, in: J.F. Scamehorn, J.H. Harwell (Eds.) Surfactant based separation process, Marcel Dekker, New York, 1989, pp. 139–152. https://doi.org/10.1080/03602548908050926

[16] R. Breslow, U. Maitra, On the origin of product selectivity in aqueous diels-alder reactions, Tetrahedron Lett. 25 (1984) 1239–1240. https://doi.org/10.1016/s0040-4039(01)80122-2

[17] D.C. Rideout, R. Breslow, Hydrophobic acceleration of Diels-Alder reactions, J. Am. Chem. Soc. 102 (1980) 7816–7817. https://doi.org/10.1021/ja00546a048

[18] A. Lubineau, J. Augé, Y. Queneau, Water-promoted organic reactions, Synthesis (1994) 741–760. https://doi.org/10.1055/s-1994-25562

[19] V. Strand, Implementation of water as solvent in aerobic oxidative NHC-catalysis, Chalmers University of Technology Gothenburg, Sweden 2018

[20] S. Narayan, J. Muldoon, M.G. Finn, V.V Fokin, H.C. Kolb, K.B. Sharpless, On water: Unique reactivity of organic compounds in aqueous suspension, Angew. Chem. Int. Ed. 44 (2005) 3157. https://doi.org/10.1002/anie.200590069

[21] P. Klumphu, Development of new methodology in organic synthesis enabled by aqueous micellar catalysis, University of California, Santa Barbara, 2016.

[22] A. Chanda, V.V Fokin, Organic synthesis "on water," Chem. Rev. 109 (2009) 725–748.

[23] B.H. Lipshutz, S. Ghorai, W.W.Y. Leong, B.R. Taft, D. V Krogstad, Manipulating micellar environments for enhancing transition metal-catalyzed cross-couplings in water at room temperature, J. Org. Chem. 76 (2011) 5061–5073. https://doi.org/10.1021/jo200746y

[24] B.H. Lipshutz, S. Ghorai, A.R. Abela, R. Moser, T. Nishikata, C. Duplais, A. Krasovskiy, R.D. Gaston, R.C. Gadwood, TPGS-750-M: a second-generation amphiphile for metal-catalyzed cross-couplings in water at room temperature, J. Org. Chem. 76 (2011) 4379–4391. https://doi.org/10.1021/jo101974u

[25] G. La Sorella, G. Strukul, A. Scarso, Recent advances in catalysis in micellar media, Green Chem. 17 (2015) 644–683. https://doi.org/10.1039/c4gc01368a

[26] A. Bhattacharjya, P. Klumphu, B.H. Lipshutz, Kumada–Grignard-type biaryl couplings on water, Nat. Comm. 6 (2015) 7401. https://doi.org/10.1038/ncomms8401

[27] R. Breslow, Hydrophobic effects on simple organic reactions in water, Acc. Chem. Res. 24 (1991) 159–164. https://doi.org/10.1021/ar00006a001

Industrial Applications of Green Solvents I   Materials Research Forum LLC
Materials Research Foundations **50** (2019) 61-106   doi: https://doi.org/10.21741/9781644900239-3

[28] Y. Jung, R.A. Marcus, On the theory of organic catalysis "on water," J. Am. Chem. Soc. 129 (2007) 5492–5502.

[29] S. Narayan, J. Muldoon, M.G. Finn, V.V Fokin, H.C. Kolb, K.B. Sharpless, "On water": Unique reactivity of organic compounds in aqueous suspension, Angew. Chem. Int. Ed. 44 (2005) 3275–3279. https://doi.org/10.1002/anie.200462883

[30] S. Otto, J.B.F.N. Engberts, J.C.T. Kwak, Million-fold acceleration of a Diels-Alder reaction due to combined Lewis acid and micellar catalysis in water, J. Am. Chem. Soc. 120 (1998) 9517–9525. https://doi.org/10.1021/ja9816537

[31] K.C. Nicolaou, H. Xu, M. Wartmann, Biomimetic total synthesis of gambogin and rate acceleration of pericyclic reactions in aqueous media, Angew. Chem. Int. Ed. 44 (2005) 756–761. https://doi.org/10.1002/anie.200462211

[32] R.N. Butler, W.J. Cunningham, A.G. Coyne, L.A. Burke, The influence of water on the rates of 1, 3-dipolar cycloaddition reactions: trigger points for exponential rate increases in water-organic solvent mixtures. water-super versus water-normal dipolarophiles, J. Am. Chem. Soc. 126 (2004) 11923–11929. https://doi.org/10.1021/ja040119y

[33] V.K. Tandon, H.K. Maurya, 'On water': unprecedented nucleophilic substitution and addition reactions with 1, 4-quinones in aqueous suspension, Tetrahedron Lett. 50 (2009) 5896–5902. https://doi.org/10.1016/j.tetlet.2009.07.149

[34] J.C. Lien, L.J. Huang, J.P. Wang, C.M. Teng, K.H. Lee, S.C. Kuo, Synthesis and antiplatelet, antiinflammatory, and antiallergic activities of 2-substituted 3-chloro-1, 4-naphthoquinone derivatives, Bioorg. Med. Chem. 5 (1997) 2111–2120. https://doi.org/10.1016/s0968-0896(97)00133-8

[35] A.El.Wareth.A.O. Sarhan, A.M.K. El-Dean, M.I. Abdel-Monem, Chemoselective reactions of 2, 3-dichloro-1, 4-naphthoquinone, Monatshefte Für Chemie 129 (1998) 205–212. https://doi.org/10.1007/pl00010156

[36] H. Yorimitsu, T. Nakamura, H. Shinokubo, K. Oshima, K. Omoto, H. Fujimoto, Powerful solvent effect of water in radical reaction: triethylborane-induced atom-transfer radical cyclization in water, J. Am. Chem. Soc. 122 (2000) 11041–11047. https://doi.org/10.1021/ja0014281

[37] S. Otto, J.B.F.N. Engberts, A systematic study of ligand effects on a lewis-acid-catalyzed Diels− Alder reaction in water. water-enhanced enantioselectivity, J. Am. Chem. Soc. 121 (1999) 6798–6806. https://doi.org/10.1021/ja984273u

[38] A.K. Chakraborti, S. Rudrawar, K.B. Jadhav, G. Kaur, S. V Chankeshwara, "On water" organic synthesis: a highly efficient and clean synthesis of 2-

aryl/heteroaryl/styryl benzothiazoles and 2-alkyl/aryl alkyl benzothiazolines, Green Chem. 9 (2007) 1335–1340. https://doi.org/10.1039/b710414f

[39]   G. Wittig, H. Laib, Zur Stevensschen Umlagerung von Oniumsalzen, Eu. J. Org. Chem. 580 (1953) 57–68. https://doi.org/10.1002/jlac.19535800108

[40]   G. Wittig, G. Geissler, Zur Reaktionsweise des Pentaphenyl-phosphors und einiger Derivate, Eu. J. Org. Chem. 580 (1953) 44–57. https://doi.org/10.1002/jlac.19535800107

[41]   B.E. Maryanoff, A.B. Reitz, The Wittig olefination reaction and modifications involving phosphoryl-stabilized carbanions. Stereochemistry, mechanism, and selected synthetic aspects, Chem. Rev. 89 (1989) 863–927. https://doi.org/10.1021/cr00094a007

[42]   J. Dambacher, W. Zhao, A. El-Batta, R. Anness, C. Jiang, M. Bergdahl, Water is an efficient medium for Wittig reactions employing stabilized ylides and aldehydes, Tetrahedron Lett. 46 (2005) 4473–4477. https://doi.org/10.1016/j.tetlet.2005.04.105

[43]   N. Azizi, M.R. Saidi, Highly chemoselective addition of amines to epoxides in water, Org. Lett. 7 (2005) 3649–3651. https://doi.org/10.1021/ol051220q

[44]   S. Azoulay, K. Manabe, S. Kobayashi, Catalytic asymmetric ring opening of meso-epoxides with aromatic amines in water, Org. Lett. 7 (2005) 4593–4595. https://doi.org/10.1021/ol051546z

[45]   B. Tan, D. Zhu, L. Zhang, P.J. Chua, X. Zeng, G. Zhong, Water-more than just a green solvent: a stereoselective one-pot access to all-chiral tetrahydronaphthalenes in aqueous media, Chem. Eur. J. 16 (2010) 3842–3848. https://doi.org/10.1002/chem.200902932

[46]   Y. Mori, K. Manabe, S. Kobayashi, Catalytic use of a boron source for boron enolate mediated stereoselective aldol reactions in water, Angew. Chem. Int. Ed. 40 (2001) 2815–2818. https://doi.org/10.1002/1521-3773(20010803)40:15%3C2815::aid-anie2815%3E3.0.co;2-f

[47]   S. Kobayashi, K. Manabe, Development of novel Lewis acid catalysts for selective organic reactions in aqueous media, Acc. Chem. Res. 35 (2002) 209–217. https://doi.org/10.1021/ar000145a

[48]   U.M. Lindstrom, (Ed.) Organic reactions in water: principles, strategies and applications, Wiley-Blackwell, 2008.

[49]   M.O. Simon, C.J. Li, Green chemistry oriented organic synthesis in water, Chem. Soc. Rev. 41 (2012) 1415-1427. https://doi.org/10.1039/c1cs15222j

[50]   D. Dallinger, C.O. Kappe, Microwave-assisted synthesis in water as solvent, Chem. Rev. 107 (2007) 2563-2591. https://doi.org/10.1021/cr0509410

[51]   M.B. Gawande, V.D.B. Bonifácio, R. Luque, P.S. Branco, R.S. Varma, Benign by design: catalyst-free in-water, on-water green chemical methodologies in organic synthesis, Chem. Soc. Rev. 42 (2013) 5522–5551. https://doi.org/10.1039/c3cs60025d

[52]   V.V Rostovtsev, L.G. Green, V.V Fokin, K.B. Sharpless, A stepwise huisgen cycloaddition process: copper (I)-catalyzed regioselective "ligation" of azides and terminal alkynes, Angew. Chem. Int. Ed. 114 (2002) 2708–2711. https://doi.org/10.1002/chin.200243045

[53]   G.J. ten Brink, I.W.C.E. Arends, R.A. Sheldon, The Baeyer-Villiger reaction: New developments toward greener procedures, Chem. Rev. 104 (2004) 4105–4124. https://doi.org/10.1021/cr0300111

[54]   G.R. Krow, The Baeyer-Villiger oxidation of ketones and aldehydes, Org. React. 43 (2004) 251–798.

[55]   G.J. ten Brink, J.M. Vis, I.W.C.E. Arends, R.A. Sheldon, Selenium-catalyzed oxidations with aqueous hydrogen peroxide.2.Baeyer-Villiger reactions in homogeneous solution[1], J. Org. Chem. 66 (2001) 2429–2433. https://doi.org/10.1021/jo0057710

[56]   M. Kidwai, K. Singhal, S. Kukreja, One-pot green synthesis for pyrimido [4,5-d] pyrimidine derivatives, Zeitschrift Für Naturforsch. B. 62 (2007) 732–736. https://doi.org/10.1515/znb-2007-0518

[57]   H.C. Hailes, Reaction solvent selection: The potential of water as a solvent for organic transformations, Org. Process Res. Dev. 11 (2007) 114–120. https://doi.org/10.1021/op060157x

[58]   S.V Malhotra, R.P. Andal, V. Kumar, Aminolysis of epoxides in ionic liquid 1-ethylpyridinium trifluoroacetate as green and efficient reaction medium, Synth. Comm. 38 (2008) 4160–4169. https://doi.org/10.1080/00397910802323056

[59]   K.W. Anderson, S.L. Buchwald, General catalysts for the Suzuki–Miyaura and Sonogashira coupling reactions of aryl chlorides and for the coupling of challenging substrate combinations in water, Angew. Chem. Int. Ed. 117 (2005) 6329–6333. https://doi.org/10.1002/ange.200502017

[60]   D. Habibi, M. Nasrollahzadeh, T.A. Kamali, Green synthesis of the 1-substituted 1H-1,2,3,4-tetrazoles by application of the Natrolite zeolite as a new and reusable heterogeneous catalyst, Green Chem. 13 (2011) 3499–3504. https://doi.org/10.1039/c1gc15245a

[61]   Z.N. Tisseh, M. Dabiri, M. Nobahar, A.A. Soorki, A. Bazgir, Catalyst-free synthesis of N-rich heterocycles via multi-component reactions, Tetrahedron, 68 (2012) 3351–3356. https://doi.org/10.1016/j.tet.2012.02.051

[62] S.H. Bairagi, P.P. Salaskar, S.D. Loke, N.N. Surve, D. V Tandel, M.D. Dusara, Medicinal significance of coumarins: A review, Int. J. Pharm. Res. 4 (2012) 16–19.

[63] L.D.S. Yadav, S. Singh, V.K. Rai, Catalyst-free, step and pot economic, efficient mercaptoacetylative cyclisation in H_2O: synthesis of 3-mercaptocoumarins, Green Chem. 11 (2009) 878–882. https://doi.org/10.1039/b904655k

[64] T. Ikariya, K. Murata, R. Noyori, Bifunctional transition metal-based molecular catalysts for asymmetric syntheses, Org. Biomol. Chem. 4 (2006) 393–406. https://doi.org/10.1039/b513564h

[65] X. Wu, J. Xiao, Aqueous-phase asymmetric transfer hydrogenation of ketones–a greener approach to chiral alcohols, Chem. Commun. (2007) 2449–2466. https://doi.org/10.1039/b618340a

[66] X. Wu, J. Liu, X. Li, A. Zanotti-Gerosa, F. Hancock, D. Vinci, J. Ruan, J. Xiao, On water and in air: fast and highly chemoselective transfer hydrogenation of aldehydes with iridium catalysts, Angew. Chem. Int. Ed. 45 (2006) 6718–6722. https://doi.org/10.1002/anie.200602122

[67] X. Yao, C.J. Li, Phosphine-triggered complete chemo-switch: from efficient aldehyde– alkyne– amine coupling to efficient aldehyde– alkyne coupling in water, Org. Lett. 7 (2005) 4395–4398. https://doi.org/10.1021/ol051575+

[68] A.L. Casalnuovo, J.C. Calabrese, Palladium-catalyzed alkylations in aqueous media, J. Am. Chem. Soc. 112 (1990) 4324–4330. https://doi.org/10.1021/ja00167a032

[69] F. Rodrigues, Y. Canac, A. Lubineau, A convenient, one-step, synthesis of β-C-glycosidic ketones in aqueous media, Chem. Commun. (2000) 2049–2050. https://doi.org/10.1039/b006642g

[70] S. Peters, F.W. Lichtenthaler, H.J. Lindner, A 2-C-fructosyl-propanone locked in a 2,7-dioxabicyclo [3.2.1] octane framework, Tetrahedron: Asymmetry, 14 (2003) 2475–2479. https://doi.org/10.1016/s0957-4166(03)00501-9

[71] M.B. Gawande, P.S. Branco, An efficient and expeditious Fmoc protection of amines and amino acids in aqueous media, Green Chem. 13 (2011) 3355–3359. https://doi.org/10.1039/c1gc15868f

[72] D. Basavaiah, A.J. Rao, T. Satyanarayana, Recent advances in the Baylis-Hillman reaction and applications, Chem. Rev. 103 (2003) 811–892. https://doi.org/10.1021/cr010043d

[73] L.D.S. Yadav, R. Patel, V.P. Srivastava, An easy access to functionalized allyl dithiocarbamates from Baylis-Hillman adducts in water, Tetrahedron Lett. 50 (2009) 1335–1339. https://doi.org/10.1016/j.tetlet.2009.01.023

[74]   C. Moberg, L. Rákos, Preparation of chiral polymer-supported epoxides-Application in the synthesis of chelating ligands, React. Polym. 15 (1991) 25–35. https://doi.org/10.1016/0923-1137(91)90144-d

[75]   F. Bossert, W. Vater, 1,4-Dihydropyridines-a basis for developing new drugs, Med. Res. Rev. 9 (1989) 291–324. https://doi.org/10.1002/med.2610090304

[76]   V.P. Litvinov, Advances in the Chemistry of Naphthyridines, Adv. Heterocycl. Chem. 91 (2006) 189–300.

[77]   R.P. Wurz, A.B. Charette, Transition metal-catalyzed cyclopropanation of alkenes in water: Catalyst efficiency and in situ generation of the diazo reagent, Org. Lett. 4 (2002) 4531–4533. https://doi.org/10.1021/ol0270879

[78]   L.F. Tietze, Domino reactions in organic synthesis, Chem. Rev. 96 (1996) 115–136.

[79]   J.S. Yadav, B.V.S. Reddy, D. Narsimhaswamy, P.N. Lakshmi, K. Narsimulu, G. Srinivasulu, A.C. Kunwar, Domino Knoevenagel hetero-Diels–Alder reactions: a stereoselective synthesis of sugar fused furo [3,2-b] pyrano [4,3-d] pyran derivatives, Tetrahedron Lett. 45 (2004) 3493–3497. https://doi.org/10.1016/j.tetlet.2004.02.149

[80]   M. Kiamehr, F.M. Moghaddam, An efficient ZnO-catalyzed synthesis of novel indole-annulated thiopyrano-chromene derivatives via Domino Knoevenagel-hetero-Diels–Alder reaction, Tetrahedron Lett. 50 (2009) 6723–6727. https://doi.org/10.1016/j.tetlet.2009.09.106

[81]   M. Ghandi, E. Mohammadimehr, M. Sadeghzadeh, A.H. Bozcheloei, Efficient access to novel hexahydro-chromene and tetrahydro-pyrano [2,3-d] pyrimidine-annulated benzo-δ-sultones via a domino Knöevenagel-hetero-Diels–Alder reaction in water, Tetrahedron, 67 (2011) 8484–8491. https://doi.org/10.1016/j.tet.2011.09.010

[82]   A. Deiters, S.F. Martin, Synthesis of oxygen-and nitrogen-containing heterocycles by ring-closing metathesis, Chem. Rev. 104 (2004) 2199–2238. https://doi.org/10.1021/cr0200872

[83]   Y. Ju, R.S. Varma, Aqueous N-alkylation of amines using alkyl halides: direct generation of tertiary amines under microwave irradiation, Green Chem. 6 (2004) 219–221. https://doi.org/10.1039/b401620c

[84]   Y. Ju, R.S. Varma, Aqueous N-heterocyclization of primary amines and hydrazines with dihalides: microwave-assisted syntheses of N-azacycloalkanes, isoindole, pyrazole, pyrazolidine, and phthalazine derivatives, J. Org. Chem. 71 (2006) 135–141. https://doi.org/10.1002/chin.200620093

[85]   Y. Ju, R.S. Varma, Microwave-assisted cyclocondensation of hydrazine derivatives with alkyl dihalides or ditosylates in aqueous media: syntheses of pyrazole,

pyrazolidine and phthalazine derivatives, Tetrahedron Lett. 46 (2005) 6011–6014. https://doi.org/10.1016/j.tetlet.2005.07.018

[86] Y. Ju, R.S. Varma, An efficient and simple aqueous N-heterocyclization of aniline derivatives: microwave-assisted synthesis of N-aryl azacycloalkanes, Org. Lett. 7 (2005) 2409–2411. https://doi.org/10.1021/ol050683t

[87] S.M. Roberts, N.J. Turner, A.J. Willetts, M.K. Turner, Introduction to biocatalysis using enzymes and microorganisms, Cambridge University Press, 1995.

[88] Y. Hayashi, T. Urushima, S. Aratake, T. Okano, K. Obi, Organic solvent-free, enantio-and diastereoselective, direct Mannich reaction in the presence of water, Org. Lett. 10 (2008) 21–24. https://doi.org/10.1021/ol702489k

[89] S. Luo, X. Mi, S. Liu, H. Xu, J.-P. Cheng, Surfactant-type asymmetric organocatalyst: organocatalytic asymmetric Michael addition to nitrostyrenes in water, Chem. Commun. (2006) 3687–3689. https://doi.org/10.1039/b607846j

[90] K. Kacprzak, Efficient one-pot synthesis of 1, 2, 3-triazoles from benzyl and alkyl halides, Synlett. 2005 (2005) 943–946. https://doi.org/10.1055/s-2005-864809

[91] H.C. Kolb, M.G. Finn, K.B. Sharpless, Click chemistry: diverse chemical function from a few good reactions, Angew. Chem. Int. Ed. 40 (2001) 2004–2021. https://doi.org/10.1002/1521-3773(20010601)40:11%3C2004::aid-anie2004%3E3.0.co;2-5

[92] W. Pang, S. Zhu, H. Jiang, S. Zhu, A novel synthesis of 5-perfluorophenyl 4,5-dihydro-1H-pyrazoles in THF or water, Journal of Fluorine Chemistry, 128 (2007) 1379–1384. https://doi.org/10.1016/j.jfluchem.2007.06.010

[93] S.-J. Tu, B. Jiang, J.-Y. Zhang, R.-H. Jia, Y. Zhang, C.-S. Yao, Efficient and direct synthesis of poly-substituted indeno [1,2-b] quinolines assisted by p-toluene sulfonic acid using high-temperature water and microwave heating via one-pot, three-component reaction, Org. Biomol. Chem. 4 (2006) 3980–3985. https://doi.org/10.1039/b611462h

[94] M. Mayer, P.T. Lang, S. Gerber, P.B. Madrid, I.G. Pinto, R.K. Guy, T.L. James, Synthesis and testing of a focused phenothiazine library for binding to HIV-1 TAR RNA, Chem. Biol. 13 (2006) 993–1000. https://doi.org/10.1016/j.chembiol.2006.07.009

[95] M.C. Pirrung, K. Das Sarma, Aqueous medium effects on multi-component reactions, Tetrahedron, 61 (2005) 11456-11472. https://doi.org/10.1016/j.tet.2005.08.068

[96] M.C. Pirrung, K. Das Sarma, Multicomponent reactions are accelerated in water, J. Am. Chem. Soc. 126 (2004) 444–445. https://doi.org/10.1021/ja038583a

Industrial Applications of Green Solvents I   Materials Research Forum LLC
Materials Research Foundations **50** (2019) 61-106   doi: https://doi.org/10.21741/9781644900239-3

[97]   Y. Hayashi, S. Samanta, H. Gotoh, H. Ishikawa, Asymmetric Diels-Alder reactions of α, β-unsaturated aldehydes catalyzed by a diarylprolinol silyl ether salt in the presence of water, Angew. Chem. Int. Ed. 120 (2008) 6736–6739. https://doi.org/10.1002/ange.200801408

[98]   H. Firouzabadi, N. Iranpoor, M. Gholinejad, 2-Aminophenyl diphenylphosphinite as a new ligand for heterogeneous palladium-catalyzed Heck–Mizoroki reactions in water in the absence of any organic co-solvent, Tetrahedron, 65 (2009) 7079–7084. https://doi.org/10.1016/j.tet.2009.06.081

[99]   D. Font, S. Sayalero, A. Bastero, C. Jimeno, M.A. Pericas, Toward an artificial aldolase, Org. Lett. 10 (2008) 337-340. https://doi.org/10.1021/ol702901z

[100] M. Berthod, C. Saluzzo, G. Mignani, M. Lemaire, 4,4′ and 5,5′-DiamBINAP as a hydrosoluble chiral ligand: syntheses and use in Ru (II) asymmetric biphasic catalytic hydrogenation, Tetrahedron: Asymmetry, 15 (2004) 573-575. https://doi.org/10.1016/j.tetasy.2003.12.033

[101] B.H. Lipshutz, S. Ghorai, PQS: A new platform for micellar catalysis. RCM reactions in water, with catalyst recycling, Org. Lett. 11 (2009) 705–708. https://doi.org/10.1021/ol8027829

[102] Y. Usui, K. Sato, M. Tanaka, catalytic dihydroxylation of olefins with hydrogen peroxide: An organic-solvent-and metal-free system, Angew. Chem. Int. Ed. 42 (2003) 5623–5625. https://doi.org/10.1002/anie.200352568

[103] S. Iimura, K. Manabe, S. Kobayashi, Hydrophobic polymer-supported scandium catalyst for carbon–carbon bond-forming reactions in water, Tetrahedron, 60 (2004) 7673–7678. https://doi.org/10.1016/j.tet.2004.06.083

[104] K. Yamaguchi, M. Matsushita, N. Mizuno, Efficient hydration of nitriles to amides in water, catalyzed by ruthenium hydroxide supported on alumina, Angew. Chem. Int. Ed. 43 (2004) 1576–1580. https://doi.org/10.1002/anie.200353461

[105] W. Solodenko, U. Schoen, J. Messinger, A. Glinschert, A. Kirschning, Microwave-assisted Suzuki-Miyaura reactions with an insoluble pyridine-aldoxime Pd-catalyst, Synlett. 2004 (2004) 1699–1702. https://doi.org/10.1055/s-2004-829546

[106] Z. Zheng, B.L. Perkins, B. Ni, Diarylprolinol silyl ether salts as new, efficient, water-soluble, and recyclable organocatalysts for the asymmetric Michael addition on water, J. Am. Chem. Soc. 132 (2009) 50–51. https://doi.org/10.1002/chin.201028025

Industrial Applications of Green Solvents I   Materials Research Forum LLC
Materials Research Foundations **50** (2019) 107-124   doi: https://doi.org/10.21741/9781644900239-4

Chapter 4

Industrial Application of Ionic Liquids in the Paint Industry

Muzammil Kuddushi, Monika Jain, Naved I Malek*

Applied Chemistry Department, S.V.National Institute of Technology, Surat- 395 007, Gujarat, India

navedmalek@chem.svnit.ac.in, navedmalek@gmail.com

Abstract

Paints, used to decorate or protect the substrate from external forces are composed of pigments, binders, and thinner. The paint industry uses various volatile organic solvents (VOCs) as one of the integral components of paints despite their detrimental effect on the environment. Among the tested strategies to develop environmentally benign alternatives to the VOCs, ionic liquids (ILs) have emerged as potential candidates. In this context, the main objective of this chapter is to introduce the new types of low viscous, highly efficient ILs that can be used in the paint industries. The strategy entails not only an exploration of the influence of the nature of the ILs but also to develop the much needed fundamental, molecular-level view of the heterogeneity of such systems.

Keywords

Paints, Volatile Organic Compounds, Ionic Liquids, Varnish Removal, Paint Removal

Contents

1. Introduction

Scientific inventions have reached new heights in recent time, benefiting to the mankind in all aspects, though destructive in some cases. So, it is the role of science itself to find the possible solution to these destructive inventions that are detrimental to the universe and also to see that further development should not carry the same effects or cause even more damage [1-4]. In this context, 'Environmentally-Benign' inventions have emerged with possible alternative approaches [5-9].

Among the spectrum of inventions, paints that are applied to any object of interest either to protect it from the detrimental effect of the environment or to enhance the interior and exterior properties of the surfaces. The major components of paint include: pigments, binders, and thinner [10-12]. The role of the binder is to bind the pigments to the object of interest after evaporation of the thinner, once the paint is applied. Thinner are volatile organic compounds (VOCs) which evaporate from paints during drying and enter into the environment causing harmful effect to human beings. In the present chapter, we are reporting alternatives to these VOCs for the paint industry. The new age 'Green' alternative; recently rediscovered are neoteric solvents, Ionic Liquids (ILs). As per our knowledge, there are only a few reports pertaining to the use of ILs in the paint industry (SCOPUS search with Ionic Liquid and Paint search criteria on 30.12.2018, only seven articles). This chapter focuses on the future development of the ILs to be used in the paint industry in order to encourage academic and industrial researchers to design 'environmentally benign' solvents for the paint industries.

2. A short history of paints

The first report on paint by Antediluvian date back on paints in caves found in Southern France (Font-de-Gaume, Niaux, Lascaux), Spain (Altamira), and South Africa [10,11]. The paints used over there were constituted of animal fats that were mixed with ocher, manganese and iron dioxide and chalk. The proof of paints was also found during various civilizations; during 1st and 2nd millennium B.C. from an Egyptian mummy, Greek temples, and shields of warring European tribes. Rock paintings found in Sahara region of North Africa are of 5th and 7th millennium B.C. The first painted utensils were found from China in around 200 B.C. where milky juice from the *Rhusvernicifera*, the lacquer tree was used as the lacquer and minerals including gold were used to colour them,

Industrial Applications of Green Solvents I   Materials Research Forum LLC
Materials Research Foundations **50** (2019) 107-124   doi: https://doi.org/10.21741/9781644900239-4

though the oldest recipe for lacquer was found in 100 A.D. Paints were used as the preservatives or to decorate the objects in the periods of 600 to 400 B.C. by the Greeks and Romans, who introduced the varnishes with drying oils. The varnishes were composed of a mixture of resins and several vegetable and animal oils (fats) and showed poor stability with time [10-13].

With increased attention due to their lucrative nature, the demand for paints had increased many folds during the late eighteenth century, which led to the emergence of the paint business. The first paint mill was established by Bostonian Thomas Child in the US before 1700. The total spending has crossed $10 billion annually on paints across the US only. As a major player of the chemical industry, paint technology is a multidisciplinary field as it utilizes the knowledge of chemistry, physics, and engineering through overlapping various fields including inks, plastics, adhesives, and rubber. Before World War I, nitrocellulose was used as the rapid-drying binder. After nitro-cellulose, phenolic resins and alkyd resins were the first synthetic binders used in the paints. Presently, numerous synthetic binders and resins are available but these are mostly based on petrochemical products [10, 11].

Vegetable oils, ethanol, and even water were used as the liquid components in the paints that did not allow the binder to dry quickly. This problem was solved at the beginning of the 20th century by the use of organic solvents that allowed faster drying with manageable properties of the paint. These organic solvents were an integral part of the paint industry since long but the environmental hazards, damaging nature to the eco-system and strict environmental regulations are now the major drawbacks that have to be dealt with cautiously.

3. Traditional solvents in the paint industry

Solvents are an integral part of many chemical reactions and are useful for the preparations of countless chemical substances, as catalysis, in separation, extraction, in the electrochemistry etc. Further, solvents decide the chemical reaction rates as well as the end products of these chemical reactions which are dependent on various physicochemical properties of the solvents including polarity, density, viscosity (including shear stress and strain), dipole moment, refractive index and boiling point among others. Selection of solvent for any specific application (synthesis or chemical kinetics) is of utmost importance before preceding any chemical reaction [13-17].

As far as the paint industry is concerned, except water, low molecular weight hydrocarbons, chlorinated and oxygenated compounds (ethers, ketones, esters, alcohols) and nitro-paraffin are used to maintain the consistency of the paint by reducing the viscosity. In most cases, a mixture of solvents is used instead of a single solvent depending on the viscosity, boiling point, evaporation rate, chemical nature, toxicity, solvency, cost and also the nature of the binder used along with other ingredients of the paint. Our research group is currently engaged in designing various thermo-physical properties of the mixture of solvents by judiciously selecting the mixture of two solvents, called binary mixtures [18-25]. Based on the application and curing conditions, the solvents of proper compositions for paints are selected, as the rate of evaporation depends on the composition of the solvent. The rate of solvent evaporation from the thin film of paint governs the propensity of the adherence of the paint and hence it is advisable to use mixed solvents. Evaporation of solvent causes a rapid increase in viscosity of the paint. This phenomenon takes place in two stages: initially, the vapour pressure of the solvent decides the evaporation of the solvent, this process is independent of the paint composition. After the primary removal of the solvent phase, the remaining solvent within the paint film is lost by a diffusion-controlled process, which is arelatively slow process [10-13].

Among the solvents, most organic solvents are volatile, toxic, and hazardous which have detrimental ecological effects on the environment and are considered harmful to humans and other ecosystems. The effect of solvent on human, animal, and plant organisms is concentration and exposure dependent. Some of the solvents cause acute damage to the body when exposed for a shorter time and minimal dosage. However, some of the solvents when absorbed by the humans in trace amounts cause chronic damage and sensitization. Further, the disposing of difficulties without affecting the environment and ecosystem are of major concerns. These detrimental effects are the root cause of several legislations, including the Montreal Protocol that suggests the minimum use of VOCs or even bans on several of these. Copenhagen climate summit was held in 2009 during a meeting in United Nations climate change conferences (UNCCC) to discuss the socio-economic impact of the climate change and to find a solution [26-30]. Yet strict regulations and policies are expected looking at the current status of the environmental damages and to save the ecosystems from the harmful effect of these VOCs emissions.

Working on the basic principle, "Prevention is Better Than Cure", the search for a more sustainable and eco-friendly replacement of the VOCs is emerging as one of the most profound and vital technological challenges of the 21st century. Academic and industrial researchers are working on various probable strategies as alternatives to VOCs and have developed several environmentally-benign systems that cause minimum or even no harm to the ecosystem and are a step towards the 'green approach' as far as their impact on the global environment is concerned. Various alternatives to VOCs are water, supercritical fluids, and recently rediscovered ionic liquids (ILs) and deep eutectic solvents [31-33].

Among these, water could be the best in several cases but not all due to its inability to dissolve many organic solutes and to be miscible with several organic solvents [34, 35]. Aqueous mixtures that are contaminated with several toxic organic solutes are difficult and expensive to dispose of. The other alternative, supercritical fluids need high pressure and strict temperature conditions have limited solubilizing capacity for many solutes and are also not cost-effective, though are applied in various synthesis and extractions as well as catalytic reactions as a catalyst [36-40]. Carbon dioxide is the most widely used supercritical fluid at present due to its nontoxic and non-flammable nature as well as low cost and high solvating power for many nonpolar organic compounds. But still, the tunability of the properties is not achieved through supercritical carbon dioxide [36-40]. Out of these alternatives, recently rediscovered 'green solvent' ionic liquids, defined as organic salts with a melting point below 100˚C; and made up of ions only are attracting the attention of the scientific world. Unlike the molecular solvents, ILs form a different class of solvent system comprising of crystalline state made entirely of ions in a liquid state. Among the bunch of unique properties, negligible vapour pressure and tunable nature are in favour of making them the best alternative to VOCs [41-57]. Their unique properties make them one of the best candidates for several industrial applications including paint and pharmaceutical industries and several others [42-44]. We had investigated the application of ILs in various fields such as surface active agents, coacervates, drug delivery vehicles, ionogels to name a few [58-68]. Among the spectrum of applications, herein we aim to discuss the application of ILs in the paint industry. As 30.12.2018 (SCOPUS), only seven reports are available on the use of ionic liquid in the paint industry. The results indicate the space available for researchers to explore this developing field [69-75].

4.   Alternatives to the traditional solvents, ionic liquid (IL)

The traditional convention says, "salt with a melting point below 100°C is 'ionic liquid' (IL)." The ionic liquids' (ILs) are also known as low-temperature molten salts, room temperature molten salts, ionic fluids, liquid organic salts, fused salts and neoteric solvents [42-44]. ILs are the combinations of organic cations and/or organic or inorganic anions, which decide the properties of ILs based on their shape, size, and geometry. The difference in the symmetry of the constituting cations and anions lower the lattice energy and hence the melting point of the ILs. For the inorganic salts, e.g. NaCl, the identical size of the cations and anions leads to the closer packing, higher lattice energy and as a result higher melting point [44].

Historically, "red oil," a liquid salt phase that was separated during the Friedel-Crafts reaction and determined recently by NMR spectroscopy is considered the first IL as per the current definition [76, 77]. The first mention of IL was found in 1914 when Paul Walden published his famous seminal paper "*Über die Moleculargröße und electrische Leitfähigkeit einiger geschmolzener Salze*", where the electrical properties of ethylammonium nitrate with a melting point of 13–14 °C were discussed [44]. The first generation room temperature IL, chloroaluminate was synthesized by Osteryoung and Wilkes in the 1970s [41]. Seddon et al. [77] have used ILs as the solvents in the 1980s, where they have used alkylpyridiniumtetrahalidoaluminate, [Rpy][AlCl₃X], to study transition metal complexes. The first report on IL as the solvent media was by Chauvin et. al. and Wilkes et. al. who used ILs as the solvent media for homogeneous transition metal catalysis [78-85]. The hygroscopic nature of these haloaluminates based ILs and their high reactivity with water, unfortunately, limits their utility. Among the arbitrarily selected generations of ILs, the first generation ILs were hygroscopic and air-sensitive which include aluminium trichloride based and haloaluminates based ILs. These first generation ILs have limited industrial applications. The issue was solved by the introduction of comparatively air- and moisture-stable ILs. The second generation ILs were air- and water-stable and have common anions like halides, [PF₆]⁻, [BF₄]⁻ and [CF₃CO₂]⁻ and were first reported by Wilkes and Zaworotko [80]. The properties of the second generation ILs were adjusted through changing the structural entity of the cations and anions. For example, imidazolium-based ILs with shorter chain length were used as solvents, reagents, catalysts, and materials and were most investigated in various industrial applications [83-86].

During the last two decades, ILs with highly asymmetric structured organic cations and bulky inorganic anions exhibiting low lattice energy have constituted an important class of alternatives to the VOCs. The most popular ILs belonging to this class are tetra-alkylammonium and phosphonium salts, tri-alkylimidazolium salts, di-alkylpyridinium and pyrrolidinium salts [84]. Except for some scattered studies, the breakthrough research on ILs may be credited to the Seddon's group from QUILL. Their group had used chloro-aluminate ILs for the exploration of transition metal complexes, for various electrochemical, spectroscopic and other complex chemical investigations [85-88]. ILs have been voted as best Scientific Innovation to impact the most in the 21st century and have the ability to set the basic rules of chemistry forever [87].

ILs can be designed according to their applications through judicious selection of the cations and anions, which make them interesting materials for many different applications. The inter as well as intra-molecular interactions, such as hydrogen bonding, van der Waals, cation–π and π – π interactions could be adjusted by changing the constituting ions. Such novel properties make them bright alternative materials in chemical synthesis, homogeneous catalysis, batteries, and surfactants sciences to name a few [88]. Various cations and anions studied in the literature have been outlined hereafter.

4.1 Cations for the ILs

The cationic centers most often used in the preparation of ILs involve a positively charged nitrogen or phosphorus containing organic structure with lower symmetry and are: (a) five-membered heterocyclic cations e.g. imidazolium, oxazolium, and triazolium, (b) six-membered and benzo-fused heterocyclic cations e.g. pyridinium, benzotriazolium, and isoquinolinium, (c) ammonium, phosphonium and sulphonium based and recently studied (d) functionalized (mostly within the alkyl chain) as well as chiral cations. Various structures of the above-mentioned cations are illustrated in Scheme 1.

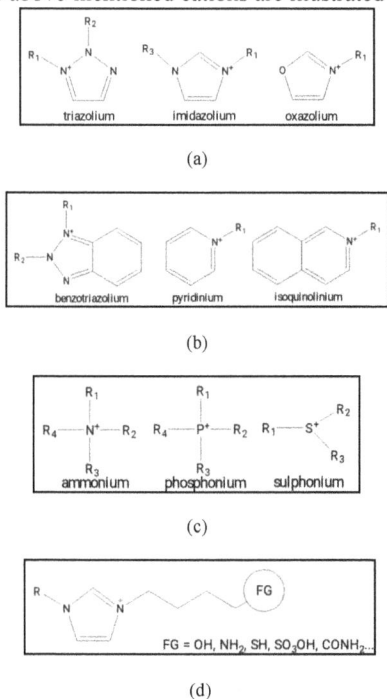

(a)

(b)

(c)

(d)

Scheme 1: *Representation of (a) five-membered, (b) six-membered, benzo-fused, (c) ammonium, phosphonium, sulphonium and (d) functionalized imidazolium-based cations.*

The properties of the ILs that are influenced by the nature of cations include melting point, toxicity, viscosity, and miscibility with other solvents [41-44]. For example, the toxicity of the 1-ethyl-3-methylimidazolium chloride ([EMIM][Cl]) increases just by increasing the alkyl chain length from ethyl to butyl, i.e. 1-butyl-3-methylimidazolium chloride ([BMIM][Cl]) is more toxic.

4.2   Anions for the ILs

The anions used as the constituting component of ILs are weakly basic inorganic or organic compounds with diffuse or protected negative charge. These are basically of two types: (a)   fluorous   anions   such   as   hexafluorophosphate, bis(trifluoromethanesulfonyl)amide,   tris(trifluoromethanesulfonyl)   methanide, tetrafluoroborate, fluoroacetoxyborate and (b) non-fluorous anions such as tetrachloroaluminate, nitrate, halides,   organic salts, and bis(oxalato)borate. The structures of anions are shown in Scheme 2 [41-44, 89].

The properties of ILs that are influenced the most are hydrophobicity, viscosity, density and solvation through changing the anions. e.g. 1-n-butyl-3-methylimidazolium hexafluorophosphate ([Bmim][PF$_6$]) is immiscible with water, whereas IL with the same cation and BF$_4$$^-$ anion is water soluble [89].

There are about 10^{18} ILs available theoretically, mostly unexplored yet. The physicochemical properties of ILs could be modulated by merely changing the constituting cations and/or anions. The key physicochemical properties include density, viscosity, vapour pressure, solvation, melting point, surface tension, heat capacity, and thermal conductivity among others. Among all these properties, the negligible vapor pressure is most important as far as the environmental protection concern from the VOCs. The paint industry uses VOCs to effectively clean the varnishes and also the paint despite their deteriorated properties for the environment. The most common features of the ILs to be used in the paint industry include low volatility, adequate viscosity, and proper tunability. Low volatility protects the environment from the harmful effect of the VOCs. If highly viscous ILs are to be used in place of the low viscous VOCs, they will not get penetrated into the object onto which the paint is applied and could not interact with the object to influence its originality. The tunable nature of IL could help in designing the IL in such way that it could be miscible with water or even with lesser toxic solvents that may reduce the harmful effect of the organic solvents. Further, the binary and ternary mixtures of ILs with molecular organic solvents or VOCs could be helpful in designing the thermo-physical properties of these industrially relevant solvents. The altered thermo-physical properties of pure ILs could be advantageous in removing of the ILs in the paint industry. In the present chapter, we aim to focus on these properties of ILs as a perspective to use these neoteric solvents in the paint industry, i.e. tenability, viscosity, and negligible vapour pressure.

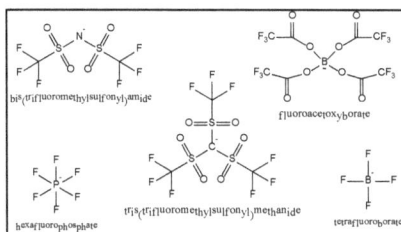

(a)

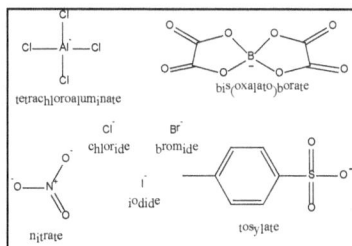

(b)

Scheme 2 *Representation of (a) non-fluorous anions and (b) fluorous anions.*

5.   Ionic liquids in the paint industry

Recently Sarwono et al. [71] investigated the role of ILs in removing the alkyd paint and demonstrated the difference between the ILs and VOCs through comparison. They have used five imidazolium-based ILs with 1-butyl-3-methylimidazolium as cation and dicyanamide [DCA], bis (trifluoromethanesulfonyl) imide [NTf₂], hydrogen sulphate [HSO₄], acetate [OAc] and chloride [Cl] as the anions. They have also compared the results obtained with three organic solvents (toluene, acetonitrile, and ethanol). Alkyd paint is usually applied onto wood, metal, plastic, composite, and other substrates to coat them. Due to various obvious reasons, it is necessary to remove the coating from the substrates. The wooden sticks used in the study were coated with alkyd paints through a dip-coating method to get coating thickness of 85±16.5μm that was measured by Digital Micrometer (Digmatic Mitatyo). The paint removal was then conducted through immersing the painted wooden sticks in the representative five ILs and three organic solvents. The sticks were periodically tested for the removal of paints through measuring the percentage uptake of the ILs or solvents and by capturing the microscopic images. Due to higher viscosity and bulky anion (for NTf_2), except [Bmim][DCA], no other IL gave satisfactory results. [Bmim][DCA] was able to remove complete alkyd paint within three days with the aid of a dry swab. In comparison to other investigated ILs, [Bmim][DCA] due to its low viscosity promoted the swelling most and was the most effective in removal of paint. Through ab initio methods, Hunt et al. [90] had studied the ion pairs in 1-butyl-3- methylimidazolium (Bmim) cation based ILs. As per the study, IL with chloride anion formed strongly connected and structured network to make its viscosity higher then the rest of the anions studied, i.e. BF_4 and NTf_2. [Bmim][NTf_2] which formed a weak network making the viscosity much lower than the chloride anion containing IL. Despite its low viscosity, the bulky anion size of [Bmim][NTf_2] hindered its penetration into the alkyd coating film as well as its removal efficiency. Furthermore, due to higher unsaturation in its structure, IL with [DCA] anion interacted more (additional π-π interaction in addition to electrostatic interaction) strongly with the alkyd paint to promote the removal of the paint.

The intermolecular interactions within ILs govern the viscosity as it describes the internal friction within the ILs. The solvation of paints depends on the viscosity of the ILs and so to be a good candidate for the paint industry, viscosity must be low. The low viscosity ILs are used as the solvent as they increase the mass transfer rates and improve the solvation whereas higher viscosity ILs may be used as a lubricant or in separation processes as a membrane. The viscosity of ILs varies widely based on the type of constituting cations and anions and it is relatively higher as compared to the common organic solvents [41-44].

As far as the cations is concerned, imidazolium-based ILs exhibit the lowest viscosity than the pyridinium and pyrrolidinium based ILs [41-44]. For the NTf2 anion, the viscosity of the [Cnmim][NTf2] ILs is slightly lower than those of [Cnmpy][NTf2]. Within the same cation series, increasing the alkyl chain length, viscosity increases; e.g. from going ethyl to octyl in the [Cnmim][PF_6] ILs, where n = 2,4–8, the viscosity increases from 172.3 to 677.4 cP at controlled the temperature of 298.15 K, the increase is monotonous. Talking of anions, the viscosity of the ILs depends on the anion symmetry i.e. highly symmetric or spherically shaped anions have higher viscosity whereas viscosity decreases with increasing the asymmetry. Viscosity of the ILs increases in the order of [NTf2] < [OTf] < [BF_4] < [C_2SO_4] < [$C1SO_4$] < [PF_6] < [CH_3COO] < Cl. Viscosity of the ILs could be tailored through mixing them with organic solvents. An unlimited number of combinations that arise through the binary and ternary mixtures of ILs with molecular solvents could be selected in such a way that the mixture could have better removal efficiency for the paint than the individual component of the mixture. To be a good solvent for the paint industry, ILs with a short alkyl chain, imidazolium-based cations, and asymmetric anion with a higher tendency to interact with pain or varnishes such as DCA could be used.

The performance of hydrofluorinated ether (HFE) 7100 for the vapor degreasing operations at Air Logistics Centers (ALCs) matched with that of 2-ethylhexyl lactate (2EHL), the lactate based degradable IL [91]. Molybdenum disulfide grease with 50 mg of free soil coated 4130 steel and 2024-T3 aluminum (Al) panels were taken as the test samples to check the cleaning ability of the 2EHL along with 1-ethyl-3-methylimidazolium (EMIM) acetate and EMIM ethyl sulfate. Among the three selected ILs, 2EHL and EMIM acetate demonstrated better cleaning ability than the EMIM ethyl sulfate. Authors didn't comment on the reason for the difference in cleaning ability of the selected ILs. Several other ILs such as 1-ethyl-3-methylimidazolium (EMIM) acetate, EMIM methane sulfonate, EMIM ethylsulfate, and triethylsulfoniumbis(trifluoromethylsulfonyl)imide were also found to be good cleaning agents [91].

ILs were recently used to remove varnishes from paintings. The study was conducted by Pacheco et al. [92] who selected the ILs keeping in view their miscibility with water or solvents with lower toxicity, polarity, and the nature of the paint layers and varnishes used. Among the studied 12 ILs with different polarities, [Bmim][DCA] was the most effective in removing the varnishes (both, the natural and synthetic). Other ILs used in the study had selective removal efficiency for the natural and synthetic varnishes.

Conclusion

Present chapter deals with the role of ionic liquids as the alternative media to be used in the paint industry. Paint composed of the pigments, binders and thinner utilizes various volatile organic compounds as one of the important components. These VOCs have a damaging effect on the eco-system and need to be replaced. Ionic liquids could be the best alternative due to their unique physicochemical properties. The chapter describes the potential use of ILs in the paint industry. Readers could get the insights into designing new low viscous ILs with negligible vapour pressure.

References

[1] S. Solomon, The whole truth what's news (and what's not) about the ozone hole, Nature 427(2004) 289–291.

[2] E.C. Weatherhead, S.B. Andersen, The search for signs of recovery of the ozone layer, Nature 441 (2006) 39–45. https://doi.org/10.1038/nature04746

[3] O. Morgenstern, P. Braesicke, M.M. Hurwitz, F.M. O'Connor, A.C. Bushell, C.E. Johnson, J.A. Pyle, The world avoided by the Montreal Protocol, Geophys. Res. Lett. 35 (2008) 1–5. https://doi.org/10.1029/2008gl034590

[4] A.M. Omer, Energy, environment and sustainable development, Renew. Sustain. Energ. Rev. 12 (2008) 2265–2300.

[5] O.O. Kunle, J. Fortunak, R.D. Rogers, Workshop in green chemistry production of essential medicines in developing countries, Green Chem. 10 (2008) 823–824. https://doi.org/10.1039/b806001k

[6] R.A. Sheldon, E Factors, green chemistry and catalysis: An odyssey, Chem. Comm. (2008) 3352–3365.

[7] W. Leitner, M. Poliakoff, Supercritical fluids in green chemistry, Green Chem. 10 (2008) 730-731.

[8] M. Poliakoff, P. Licence, Sustainable technology: Green Chemistry, Nature 450 (2007) 810–812. https://doi.org/10.1038/450810a

[9] D. Kralisch, D. Reinhardt, G. Kreisel, Implementing objectives of sustainability into ionic liquids research and development, Green Chem. 9 (2007) 1308-1318. https://doi.org/10.1039/b708721g

[10]   R. Lambourne, T.A, Strivens, Paint and Surface Coatings, Theory and Practice, Second edition, Woodhead Publishing Ltd. England, 1999.

[11]   G. Schuerman, R. Bruzan, Chemistry of paint, J. Chem. Educ. 66 (1989) 327–328.

[12]   Kathryn R. Williams, House paint, J. Chem. Edu. 83 (2006) 1448-1449.

[13]   G. Wypych, Handbook of Solvents, ChemTec Publishing, Ontario, Canada, 2014.

[14]   C. Reichardt, Pyridinium N -phenolate betaine dyes as empirical indicators of solvent polarity : Some new findings, Pure Appl. Chem. 76 (2004) 1903–1919. https://doi.org/10.1351/pac200476101903

[15]   C. Reichardt, Solvents and Solvent Effects in Organic Chemistry, Weinheim, Germany, 2003.

[16]   C. Reichardt, Solvatochromic Dyes as Solvent Polarity Indicators, Chem. Rev.94 (1994) 2319-2358. https://doi.org/10.1021/cr00032a005

[17]   E. Buncel, R. A. Stairs, H. Wilson, The Role of the Solvent in Chemical Reactions, Oxford University Press: Oxford, 2003.

[18]   Z.R. Master, Z.S. Vaid, U.U. More, N.I. Malek, Molecular interaction study through experimental and theoretical volumetric, transport and refractive properties of

N-ethylaniline with aryl and alkyl ethers at several temperatures, Phys. Chem. Liq. 54 (2016) 223-244. https://doi.org/10.1080/00319104.2015.1074047

[19] Z.R. Master, N. I. Malek, Molecular interactions study through experimental and theoretical volumetric, acoustic and refractive properties of binary liquid mixtures at several temperatures 1. N, N-dimethylaniline with Aryl, and Alkyl Ethers, J. Mol. Liq.196 (2014) 120-134. https://doi.org/10.1016/j.molliq.2014.03.027

[20] N.I. Malek, S.P. Ijardar, S.B. Oswal, Estimation of speeds of sound in cyclohexane with benzene, benzaldehyde or cyclohexylamine, and cyclohexylamine with benzene https://doi.org/10.1016/j.tca.2012.08.011in the temperature range (293.15–323.15) K employing semi-empirical and theoretical equations. Ind. J. Chem. 52A (2013) 492-497. https://doi.org/10.1016/j.tca.2012.08.011

[21] N.I. Malek, S.P. Ijardar, Z.R. Master, S.B. Oswal, Temperature dependence of densities, speeds of sound, and derived properties of cyclohexylamine+ cyclohexane or benzene in the range (293.15 to 323.15) K, Thermochim. Acta 547 (2012) 106-119. https://doi.org/10.1016/j.tca.2012.08.011

[22] N.I. Malek, S.P. Ijardar, S.B. Oswal, Volumetric and acoustic properties of binary mixtures of Cyclohexane + Benzene and + Benzaldehyde at (293.15 to 323.15) K, Thermochim. Acta 539 (2012) 71– 83.

[23] V. Pandiyan, S.L. Oswal, N.I. Malek, P. Vasantharani, Thermodynamic and acoustic properties of binary mixtures of ethers. V. Diisopropyl ether or oxolane with 2- or 3-chloroanilines at 303.15, 313.15 and 323.15 K, Thermochim. Acta, 524 (2011) 140-150. https://doi.org/10.1016/j.tca.2011.07.005

[24] S.P. Ijardar, N.I. Malek, S.L. Oswal, Studies on volumetric properties of triethylamine in organic solvents with varying polarity, Ind. J. Chem. 50-A (2011)1709-1718.

[25] S.L. Oswal, J.S. Desai, S.P. Ijardar, N. I. Malek, Studies of viscosities of dilute solutions of alkylamine in non-electrolyte solvents. II. Haloalkanes and other polar solvents,Thermochim. Acta, 427 (2005) 51–60. https://doi.org/10.1016/j.tca.2004.08.013

[26] P. Knochel (Ed.) Modern solvents in organic synthesis, Springer-Verlag, Berlin Heidelberg 1999.

[27] D.J. Adams, P.J. Dyson, S.J. Tavener, Chemistry in alternative reaction media, Wiley: Chichester, 2004.

[28] Loupy, Solvent free reactions, in: P. Knochel (Ed) Modern solvents in organic synthesis Springer-Verlag, Berlin Heidelberg 1999.

[29]   M. Steinbacher, M.K. Vollmer, B. Buchmann, S. Reimann, An evaluation of the current radiative forcing benefit of the Montreal Protocol at the high-Alpine site Jungfraujoch, Sci.Total Environ. 391 (2008) 217–223. https://doi.org/10.1016/j.scitotenv.2007.10.003

[30]   O. Morgenstern, P. Braesicke, M.M. Hurwitz, F.M. O'Connor, A.C. Bushell, C.E. Johnson, J.A. Pyle, The world avoided by the Montreal Protocol, Geophys. Res. Lett. 35 (2008) 1–5. https://doi.org/10.1029/2008gl034590

[31]   P.T. Anastas, J.C. Warner, Green chemistry: Theory and practice, Oxford university press: Oxford, 1998.

[32]   A.S. Matlack, Introduction to green chemistry, Marcel Dekker, New York, 2001.

[33]   W.M. Nelson, Green solvents for chemistry- Perspective and practice; Oxford University Press, Oxford 2003.

[34]   P.A. Gricco, Organic synthesis in water; Blackie academic and professional: London, 1998.

[35]   N. Akiya, P.E. Savage, Roles of water for chemical reactions in high-temperature water, Chem. Rev. 102 (2002) 2725-2750. https://doi.org/10.1021/cr000668w

[36]   P.G. Jessop, W. Leitner, Chemical synthesis using supercritical fluids, Wiley-VCH, Weinheim, 1999.

[37]   R. Noyori, Supercritical fluids: Introduction, Chem. Rev. 99 (1999) 353-354.

[38]   N. Akiya, P.E. Savage, Roles of water for chemical reactions in high-temperature water, Chem. Rev. 102 (2002) 2725–2750. https://doi.org/10.1021/cr000668w

[39]   Y. Marcus, Solvatochromic probes in supercritical fluids, J. Phys. Org. Chem. 18 (2005) 373–384. https://doi.org/10.1002/poc.882

[40]   M. Kidwai, Dry media reactions, Pure Appl. Chem. 73 (2001) 147–151.

[41]   G.A. Baker, S.N. Baker, S. Pandey, F. V. Bright, An analytical view of ionic liquids, Analyst 130 (2005) 800–808. https://doi.org/10.1039/b500865b

[42]   P. Wasserscheid, T. Welton, Ionic liquids in synthesis, Org. Proc. Res. Dev. 7 (2003) 223-224.

[43]   J.P. Hallett, T. Welton, Room-temperature ionic liquids. Solvents for synthesis and catalysis.2, Chem. Rev. 99 (2011) 3508-3576. https://doi.org/10.1021/cr1003248

[44]   R.D. Rogers, K.R. Seddon, Ionic Liquids: Industrial Applications for Green Chemistry, J. Am. Chem. Soc. 125 (2003) 7480.

[45]   R.D. Rogers, K.R. Seddon, S. Volkov (Eds.), Green industrial applications of ionic liquids, NATO Science Series II, Springer Netherlands 2003.

[46]   R.D. Rogers, K.R. Seddon, Ionic liquids III: Fundamentals, challenges, and opportunities, American Chemical Society, Washington, 2005.

[47]   H. Ohno, Electrochemical aspects of ionic liquids, 2nd edition, Wiley-Interscience: New York, 2005.

[48]   C.F. Poole, Chromatographic and spectroscopic methods for the determination of solvent properties of room temperature ionic liquids, J. Chromotogr. A. 1037 (2004) 49-82. https://doi.org/10.1016/j.chroma.2003.10.127

[49]   H. Weingärtner, Understanding ionic liquids at the molecular level: Facts, problems, and controversies, Angew. Chem. Int. Ed. Engl. 47 (2008) 654–670. https://doi.org/10.1002/anie.200604951

[50]   T.L. Greaves, C.J. Drummond, Protic ionic liquids: Properties and applications, Chem. Rev. 108 (2008) 206-237. https://doi.org/10.1021/cr068040u

[51]   P. Domínguez De María, "Nonsolvent" applications of ionic liquids in biotransformations and organocatalysis, Angew. Chem. 47 (2008) 6960–6968. https://doi.org/10.1002/anie.200703305

[52]   S. Pandey, Analytical applications of room-temperature ionic liquids: A review of recent efforts, Anal. Chim. Acta. 556 (2006) 38–45. https://doi.org/10.1016/j.aca.2005.06.038

[53]   J.S. Wilkes, A short history of ionic liquids-From molten salts to neoteric solvents, Green Chem. 4 (2002) 73–80. https://doi.org/10.1039/b110838g

[54]   J.G. Huddleston, A.E. Visser, W.M. Reichert, H.D. Willauer, G.A. Broker, R.D. Rogers, Characterization and comparison of hydrophilic and hydrophobic room temperature ionic liquids incorporating the imidazolium cation, Green Chem. 3 (2001) 156–164. https://doi.org/10.1039/b103275p

[55]   H. Ohno (Ed.), Ionic liquids: The front and future of material developments, CMC, Tokyo 2003.

[56]   K.R. Seddon, Ionic liquids: A taste of the feature, Nat. Mater. 2 (2003) 363-365.

[57]   R.D. Rogers, K.R. Seddon, Ionic liquids solvents of the future, Science 32 (2003) 792-793.

[58]   Shah, M. Kuddushi, S. Rajput, O.A. El Seoud, N.I. Malek, Ionic liquids based catanionic coacervates: The novel microreactors for membrane free sequestration of dyes and curcumin, ACS Omega, 3 (2018) 17751-17761. https://doi.org/10.1021/acsomega.8b02455

[59]   M. Kuddushi, N. K. Patel, S.M. Rajput, A. Shah, O.A. El Seoud, N.I. Malek, Thermo-switchable de Novo ionic liquid-based gelators with dye-absorbing and drug-encapsulating characteristics, ACS Omega, 9 (2018) 12068-12078. https://doi.org/10.1021/acsomega.8b01984

[60] S. M. Rajput, K. Gangele, S. Kumar, V.K. Aswal, J.P. Mata, N.I. Malek, S.K. Kailasa, K.M. Poluri, Nano-vehicles for drug delivery using low-cost cationic surfactants: A drug induced structural transitions, Chemistry Select, 3 (2018) 9454-9463. https://doi.org/10.1002/slct.201801111

[61] A.C. Pinheiro, A.B. Gonçalves, W.J. Baader, L.F. Yamaguchi, N.I. Malek, E.L. Bastos, O.A. El Seoud, Biofuels from coconut fat and soybean oil: microwave-assisted synthesis and gas chromatography/mass spectrometry analysis, Quimica Nova, 41(2018) 1200-1204. https://doi.org/10.21577/0100-4042.20170273

[62] T.A. Bioni, N.I. Malek, O.A. El Seoud, Kinetics of cellulose acylation with carboxylic anhydrides and N-acylimidazoles in ionic liquid/molecular solvent mixtures: Relevance to the synthesis of mixed cellulose esters, Lenzinger Berichte, 94 (2018) 57-66.

[63] Z.S. Vaid, S.M. Rajput, A. Shah, Y. Kadam, A. Kumar, O.A. El Seoud, J.P. Mata, N.I. Malek, Salt-induced microstructural transitions in aqueous dispersions of ionic-liquids based surfactants, Chemistry Select, 17 (2018) 4851-4858. https://doi.org/10.1002/slct.201800041

[64] S.M. Rajput, S. Kumar, V.K. Aswal, O.A. EI Seoud, N.I. Malek, S.K. Kailasa, Drug-induced micelle-to-vesicle transition of a cationic gemini surfactant: Potential applications in drug delivery, Chem. Phys. Chem. 19 (2018) 865-872. https://doi.org/10.1002/cphc.201701134

[65] Z.S. Vaid, S.M. Rajput, M. Kuddushi, A. Kumar, O.A. El Seoud, N.I. Malek, Synergistic interaction between cholesterol and functionalized ionic liquid based surfactant leading to the morphological transition, Chemistry Select 3 (2018) 1300-1308. https://doi.org/10.1002/slct.201702561

[66] U.U More, Z.S. Vaid, S.M. Rajput, N.I. Malek, O.A. El Seoud, Effects of 1-alkyl-3-methylimidazolium bromide ionic liquids on the micellar properties of [butanediyl-1,4-bis(dimethyldodecylammonium bromide)] gemini surfactant in aqueous solution, Colloid. Polymer. Sci. 295 (2017) 2351-2361. https://doi.org/10.1007/s00396-017-4210-x

[67] Z.S. Vaid, A. Kumar, O.A. El Seoud, N.I. Malek, Drug induced micelle-to-vesicle transition in aqueous solutions of cationic surfactants, RSC Adv. 7 (2017) 3861-3869. https://doi.org/10.1039/c6ra25577a

[68] S.M. Rajput, U.U. More, Z.S. Vaid, K.D. Prajapati, N.I. Malek, Impact of organic solvents on the micellization and interfacial behavior of ionic liquid based surfactants, Colloid. Surface. Physiochem. Eng. Aspect. 507 (2016) 182-189. https://doi.org/10.1016/j.colsurfa.2016.08.008

[69]   K. Moodley, M. Mabaso, I. Bahadur, G.G. Redhi, Industrial application of ionic liquids for the recoveries of spent paint solvent, J. Mol. Liq. 219 (2016) 206-210. https://doi.org/10.1016/j.molliq.2016.03.035

[70]   K. Kowalczyk, T. Spychaj, Zinc-free varnishes and zinc-rich paints modified with ionic liquids, Corrosion Sci. 78 (2014) 111-120. https://doi.org/10.1016/j.corsci.2013.09.006

[71]   Sarwono, Z. Man, A. Idris, T.H. Nee, N. Muhammad, A.S. Khan, Z. Ullah, Alkyd paint removal: Ionic liquid vs volatile organic compound (VOC), Progr. Org. Coating. 122 (2018) 79-87. https://doi.org/10.1016/j.porgcoat.2018.05.005

[72]   Weyershausen, K. Lehmann, Industrial application of ionic liquids as performance additives, Green Chem. 7 (2005) 15-19. https://doi.org/10.1039/b411357h

[73]   J.F. Liu, N. Li, G.B. Jiang, J.M. Liu, J.A. Jönsson, M.J. Wen, Disposable ionic liquid coating for headspace solid-phase microextraction of benzene, toluene, ethylbenzene, and xylenes in paints followed by gas chromatography-flame ionization detection, J. Chrom. 1066 (2005) 27-32. https://doi.org/10.1016/j.chroma.2005.01.024

[74]   D. Li, O. Sevastyanova, M. Ek, Ionic liquids pretreatment of cellulose fiber materials for improvement of reactivity and value added applications, 16th International symposium on wood, fiber and pulping chemistry-proceedings, ISWFPC, 1 (2011) 503-510.

[75]   Syrotyńska, E. Makarewicz, O. Shyichuk, A study of ionic liquids as biocides in emulsion paints and coatings, Przemysl Chemiczny, 89 (2010) 1528-1532.

[76]   R.D. Roger, K.R. Seddon, Ionic liquids as green solvents: progress and prospects, American Chemical Society: Washington, D.C. 2003.

[77]   K.R. Seddon, A. Stark, M.J. Torres, Influence of chloride, water, and organic solvents on the physical properties of ionic liquids, Pure Appl. Chem. 72 (2000) 2275-2287. https://doi.org/10.1351/pac200072122275

[78]   Y. Chauvin, B. Gilbert, I. Guibard, Catalytic dimerization of alkenes by nickel complexes in organochloroaluminate molten salts, J. Chem. Soc., Chem. Commun. 0 (1990) 1715–1716. https://doi.org/10.1039/c39900001715

[79]   R.T. Carlin, J.S. Wilkes, Complexation of Cp_2MCl_2 in a chloroaluminate molten salt: relevance to homogeneous Ziegler-Natta catalysis, J. Mol. Catal. 63 (1990) 125–129. https://doi.org/10.1016/0304-5102(90)85135-5

[80]   J.S. Wilkes, M.J. Zaworotko, Air and water stable 1-Ethyl-3-methylimidazolium based ionic liquids, J. Chem. Soc., Chem. Commun. (1992) 965-967. https://doi.org/10.1039/c39920000965

[81]   Y. Chauvin, L. Mussmann, H. Olivier, A novel class of versatile solvents for two-phase catalysis: Hydrogenation, isomerization, and hydroformylation of alkenes catalyzed by rhodium complexes in liquid 1,3-Dialkylimidazolium salts, Angew. Chem. 34 (1996) 2698-2700. https://doi.org/10.1002/anie.199526981

[82]   J.S. Wilkes, M.J. Zaworotko, Air and water stable 1-Ethyl-3-methylimidazolium based ionic liquids, J. Chem. Soc., Chem. Commun. (1992) 965-967. https://doi.org/10.1039/c39920000965

[83]   P.B. Hitchcock, T.J. Mohammed, K.R. Seddon, J.A. Zora, C.L. Hussey, E. Haynes Ward, 1-methyl-3-ethylimidazolium hexachlorouranate(IV) and 1-methyl-3-ethylimidazolium tetrachlorodioxo-uranate(VI): Synthesis, structure, and electrochemistry in a room temperature ionic liquid, Inorg. Chim. Acta. 113 (1986) L25–L26. https://doi.org/10.1016/s0020-1693(00)82244-6

[84]   G. Diakun, L. Fairall, A. Klug, Room tempreature ionic liquids as a solvent for electronic absorption spectroscopy of halide complexes, Nature 324 (1986) 698–699.

[85]   Y. Chauvin, B. Gilbert, I. Guibard, Catalytic dimerization of alkenes by nickel complexes in organochloroaluminate molten salts, J. Chem. Soc., Chem. Commun. (1990) 1715–1716. https://doi.org/10.1039/c39900001715

[86]   J.A. Boon, J.A. Levisky, J.L. Pflug, J.S. Wilkes, Friedel-Crafts Reactions in Ambient-Temperature Molten Salts, J. Org. Chem. 51 (1986) 480-483. https://doi.org/10.1021/jo00354a013

[87]   https://ichemeblog.org/tag/quill/(accessed 13 December 2018).

[88]   C. Chiappe, D. Pieraccini, Ionic liquids: Solvent properties and organic reactivity, J. Phys Org. Chem. 18 (2005) 275–297. https://doi.org/10.1002/poc.863

[89]   J.F. Brennecke, E.J. Maginn, Ionic liquids; innovative fluids for chemical processing, AIChE Journal, 11 (2001) 2384-2389. https://doi.org/10.1002/aic.690471102

[90]   P.A. Hunt, I.R. Gould, B. Kirchner, The structure of imidazolium-based ionic liquids: Insights from ion-pair interactions, Aust. J. Chem. 60 (2007) 9-14. https://doi.org/10.1071/ch06301

[91]   National Aeronautics and Space Administration (NASA) Final Report and Deliverables, "Precision Cleaning of Oxygen Systems and Components," NASA/CR-2009-214757.

[92]   M.F. Pacheco, A.I. Pereira, L.C. Branco, A.J. Parola, Varnish removal from paintings using ionic liquids, J. Mater. Chem. A, 1 (2013) 7016-7018. https://doi.org/10.1039/c3ta10679a

Industrial Applications of Green Solvents I
Materials Research Foundations **50** (2019) 125-146

Materials Research Forum LLC
doi: https://doi.org/10.21741/9781644900239-5

Chapter 5

An Overview of Green Solvents in Sustainable Organic Synthesis

D. Devi Priya, Selvaraj Mohana Roopan*

Chemistry of Heterocycles & Natural Product Research Laboratory, Department of Chemistry, School of Advanced Sciences, Vellore Institute of Technology, Vellore 632 014, Tamilnadu, India

*mohanaroopan.s@gmail.com; mohanaroopan.s@vit.ac.in

Abstract

The developing awareness with the demanding requirement for greener, more sustainable innovations has concentrated on the utilization of molecule proficient reactant techniques used for new synthetic compounds and drugs. The use of alternative reaction agents for green, sustainable organic synthesis is reviewed in this chapter. "The best solvent is no solvent'' however, on the off chance that a solvent is required at that point water is a considerable solvent to suggest and catalysis in aqueous biphasic systems. To overcome the acceptable disadvantages of regular natural solvents (harmfulness, non-biodegradability, combustibility, and gathering in the climate), significant emphasis has been given to the substitution of conventional organic solvents by green solvents. In this essence, the safe, non-toxic, bio sustainable and low-cost reaction solvent is a significant objective in natural synthesis. Research concerning green solvents is centered on diminishing natural harms with the utilization of green solvents in natural science. Thus, there have been considerable developments of solvent-free procedures and in addition more effective reuse of conventions in the most recent decades. However, these methodologies have their barriers. Therefore, the emphasis has been on developing reactions utilizing water, ionic liquids, natural carbonates, supercritical carbon dioxide, and in addition bio-solvents rather than regular natural solvents. Along these lines, this special topic on "An overview of green solvents in sustainable organic synthesis" has been expected to exhibit various sub-zones of organic synthesis in green solvents.

Keywords

Bio-Based Solvents, Water, Ionic Liquids, Sustainable Organic Synthesis

Contents

1.   Introduction

Sustainable development is defined as the development that enables the present generation in meeting their needs [1]. Green chemistry or sustainable technology deals with chemical working processes which utilize raw materials, eliminate waste and avoid the use of toxic solvents and reagents. New synthetic developments in organic processes can increase the chemo- and regioselective, functional group tolerance and reaction yields in industries or academics.

Moreover, high product selectivity must be achieved, which in turn limits the amount of solvents, reagents, and promoters [2-4]. From the green chemistry principles, chemical reactions should reduce waste, and reuse materials to increase sustainability. Due to that, the reactions have been, atom-economical [3], catalytic [4], safe for both humans and the environment, and designed with energy efficiency.

In general, there are three approaches for mitigating the discharge of solvent and contaminated water into the environment. First one involves solvent recycling or reduction. Numerous enterprises made exceptional development in executing "closed-loop systems" that decreased the usage of water or solvent thereby enhancing their recycling capability. Secondly, changing to procedures which are free from solvents could be a favorable step towards a "greener environment". Finally, organic solvents and other unstable natural volatile organic compounds were removed.

Solvents have both commercial and domestic uses. Solvents are used in the chemical industries as a media in the production of chemicals and also for chemical purification/separation. Here, we try to show how suitable selection of solvents for chemical handling has been utilized to enhance the supportability of these procedures utilizing patterns. These procedures were collected for helpful purposes and are not extensive collected works of all the presented examples in the literature. Solvents are involved in the day to day life in various processes like industrial emissions (60%) and all other volatile organic compounds (30%) [5]. Most of the reactions cannot be carried out under solventless conditions, so this will lead to our third approach, the green approach, which reduces the release of the solvents into the ecosystem. Even though most of the commonly used solvents can cause serious health and environmental issues, they have also proven to be beneficial for temperature control of the solution regarding boiling point, heat supply for exo- and endothermic reactions, filtration, extraction, recrystallization, azeotropic refining, chromatography, alteration of reaction rates, and selectivity of the reaction [6]. New solvent alternatives have attracted attention in the past few years and they have started to slowly replace the conventional ones. Water, fluorous solvents, ionic fluids, natural carbonates, carbon dioxide and also bio solvents are included in this category. These various types of solvents, with their advantages and disadvantages, complement one another than competing with each other.

2.   Green solvents

There are a number of solvents that have been recognized as green solvents as shown in Fig .1 and listed below.

1. Water [7-14]

2. Supercritical fluids [15-22]

3. Gas-expanded liquids [23]

4. Ionic liquids [24-30]

5. Liquid polymers [31-37]

6. Solvents derived from the biomass [38-47]

Table 1 presents different types of solvents for their cautions use in organic reactions.

Figure.1   Classification of green solvents

Table 1. Various types of the solvents

RECOMMENDED	water ethanol 2-propanol 1-butanol ethyl acetate 2-propyl acetate 1,1-dimethylethyl acetate anisole sulfolane
RECOMMENDED PROBLEMATIC HIGHLY RECOMMENDED OR PROBLEMATIC	methanol *tert*-butyl alcohol benzyl alcohol ethylene glycol acetone butanone 4-methyl-2-pentanone cyclohexanone methyl acetate acetic acid acetic anhydride
PROBLEMATIC	2-methyltetrahydrofuran heptane methylcyclohexane toluene xylenes chlorobenzene acetonitrile 1,3-dimethyltetrahydropyrimidin-2(1*H*)-one dimethyl sulfoxide
PROBLEMATIC HIGHLY RECOMMENDED Or PROBLEMATIC PROBLEMATIC OR HAZARDOUS	2-methoxy-2-methylpropane tetrahydrofuran cyclohexane dichloromethane formic acid pyridine
HAZARDOUS	diisopropylether 1,4-dioxane dimethyl ether pentane hexane dimethylformamide *N,N*-dimethylacetamide 1-methyl-2-pyrrolidone methoxy ethanol triethanolamine
HIGHLY RECOMMENDED OR PROBLEMATIC PROBLEMATIC OR HAZARDOUS HAZARDOUSHAZARDOUS	diethylether benzene chloroform carbon tetrachloride dichloroethane nitromethane

2.1 Water

Water is the universal solvent for all organic reactions. It possessed attracting properties as the reaction medium in a supercritical state, compared to the standard conditions. Water has important physicochemical properties as denoted in Table 2. Supercritical water has the density that continuously changes from high to low (liquid state to gas state) phase transition by varying the temperature and pressure. Supercritical water performs as a non-polar in behavior it's having pressure as 221 bar and the temperature above 374 °C. Water is very cheap, non-toxic in nature and non-flammable also. Generally, it is high polar in nature, so especially utilized for the extraction as a high polar solvent. However, it is not advisable to use water as a solvent in the synthetic organic chemistry reactions in which the removal of solvent and drying out the final products may be very difficult. The drawback of water is that it cannot be used for the non-polar and less polar constituents.

Table 2 Physico-chemical properties of the water

PROPERTY	VALUE
BOLING POINT	100 °C
MELTING POINT	0 °C
CRITICAL TEMPERATURE	374.2 °C
MOLAR HEAT OF VAPORIZATION	40.67 KJ
MOLAR HEAT OF FUSION	6.02 KJ
MOLAR ENTROPY OF VAPORIZATION	109 Jdeg^{-1}
VISCOSITY	1.005 centipoise
SURFACE TENSION	73 dyens cm^{-1}
DIELECTRIC CONSTANT	80.54
DIPOLE MOMENT	1.84 debye
SPECIFIC HEAT	1 Cal g^{-1}C^{-1}
HEAT OF EVAPORATION	540 Cal g^{-1}

Nowadays water-mediated organic reactions are one of the challenges for modern organic chemistry. There are various water-mediated reactions as classified below:

i. Oxidations

ii. Dehydrogenation

iii. Allylations

iv. Coupling reactions

v. Heck reaction

vi. Wittig reaction

vii.   Mannich-type reactions

viii.   Diels-Alder Reaction

2.1.1 Oxidations

Oxidation reactions have been mainly conducted by utilizing various amounts of heavy metals or moisture-sensitive oxidants. Examples include V_2O_5, potassium permanganate, and *N, N*'-dicyclohexylcarbodiimide (DCC) etc. Nowadays water-compatible oxidants (e.g. O_2, H_2O_2) are used for oxidation reactions. These oxidants are cost-effective, clean, safe, and acts efficiently. The oxidation of β- naphthol in the presence of ruthenium produces corresponding biaryl compounds in water (Scheme 1) [48]. Oxidation of alcohols in water has been achieved *via* molecular O_2 using palladium nanoparticles as a catalyst. The 1° and 2° alcohols have produced corresponding ketones in water as a solvent (Scheme 2) [49].

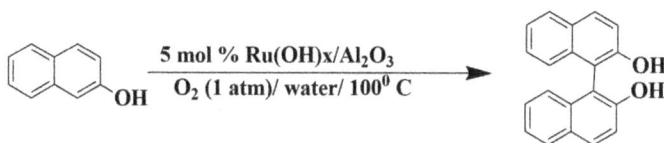

Scheme 1 Oxidation of the β- naphthol

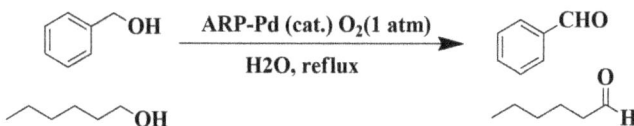

Scheme 2 Oxidation of Alcohols

Selectivity is also carried out in water medium in which the sulphides were converted into sulphoxides*via* β- cyclodexin and N-bromosuccinimide. It is a novel methodology for getting higher yields (Scheme 3) [50].

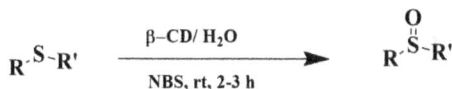

Scheme 3 Oxidation of sulphides

2.1.2 Dehydrogenation

Dehydrogenation is one of the most important processes in synthetic chemistry. It is also known as removal of hydrogen from an organic molecule. Water is one of the sources to remove hydration from the reactant molecules. But it is a very difficult process because the reaction molecules should be detached to shift the equilibrium to the side of the dehydrated product as shown in the scheme 4.

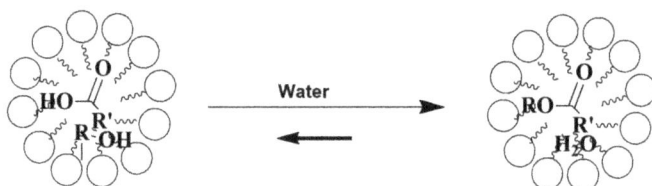

Scheme 4 Dehydrogenation of acid

The primary amines also undergo dehydration to convert as a nitrile group using $K_2S_2O_8$ as an oxidant and $NiSO_4$ as a catalyst. Here $K_2S_2O_8$ act as a cost-effective and stable oxidant in scheme 5 [50].

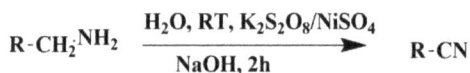

Scheme 5 Dehydration of amines

2.1.3 Allylations

Alkylation process has been done using water as a solvent in a recyclable electrochemical process. Here tin chloride was used as a catalyst for the allylation of aldehydes (Scheme 6) [51].

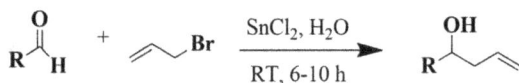

Scheme 6 Allylation of aldehydes

2.1.4 Coupling reactions

For the past ten years, coupling reactions were being performed using transition metal catalysts in water as a medium. Because of this, water molecules acted as a ligand in the metal complexes in their-ordination sites. As shown in scheme 7,the unactivated styrenyl olefins and aryl boronic acids reacted in the presence of water soluble phosphine as a ligand giving the addition hydrolysis products [52] is explained.

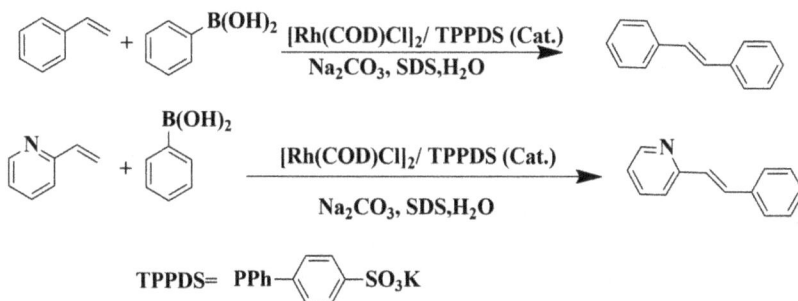

Scheme 7 Coupling reaction of aryl boronic acids and activated olefins

Water has also been used as a solvent for the C-N bond formation using palladium as a catalyst. Here phosphine ligand combined with palladium catalyst. When amines react with aryl chlorides in water, they form amination products [53] (Scheme 8).

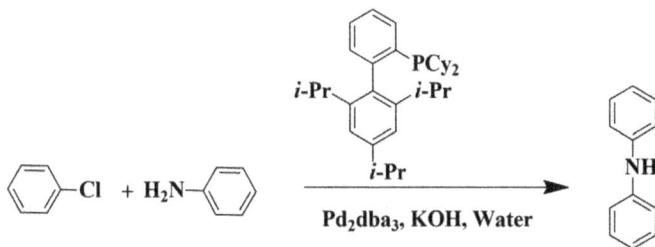

Scheme 8 C-N bond formation of aryl chlorides

Also, a new system in palladium and copper have used as a catalyst for coupling reactions in water medium from acyl halides and alkynes. From this reaction product, an excellent yield of ynones were obtained (Scheme 9) [54].

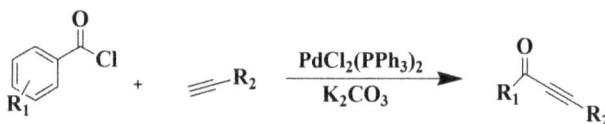

Scheme 9 Coupling reaction of acyl halides and alkynes

2.1.5 Heck reaction

One of the most important palladium-catalyzed reactions is known as Mizoroki-Heck reaction [55] where palladated Kaiser oxime resin was used as a catalyst. 2-Bromo benzaldehyde reacted with 1,2-diphenylacetylene *via* coupling reaction in which K_2CO_3 was used as a base and water as a solvent. The intermolecular Mizoroki-Heck reaction was carried out using the annulation reaction between 2-bromo benzaldehyde and 1,2-diphenylacetylene (Scheme 10).

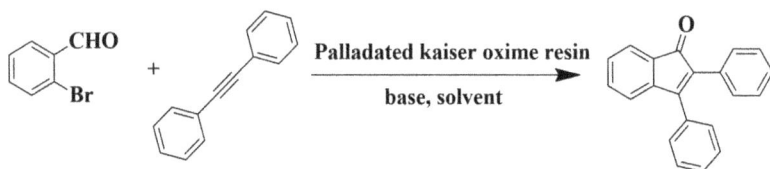

Scheme 10 Annulation reaction of 2-bromo benzaldehyde and 1,2-diphenylacetylene

α,β- Unsaturated carbonyl compounds reacted with aryl iodides in the presence of oxime derived carbapallada cycle catalyst, and water as a solvent to get regioselective diarylation product in fine yields. Here Cy_2NMe was utilized as a base (Scheme 11) [56].

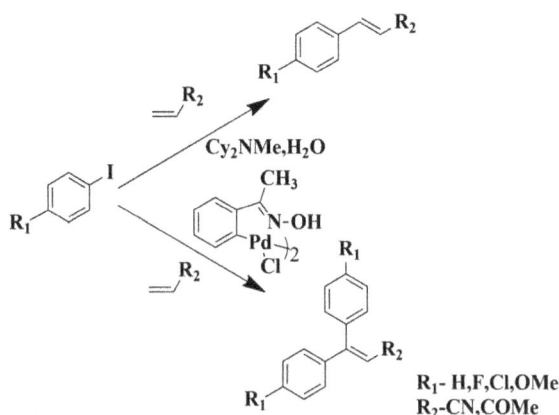

Scheme 11 Heck reactions of aryl iodides

2.1.6 Wittig reaction

The Wittig reaction is one of the important chemical processes for the C-C bond formation in organic chemistry. In non-polar solvents, the reaction rate of this type of reactions is very slow. Water was also utilized as a solvent for the water-soluble phosphonium salts in Wittig reaction. Here Wittig reactions were carried out for aldehydes using water as a medium. The product yield was good when compared to other organic solvents (Scheme 12) [57].

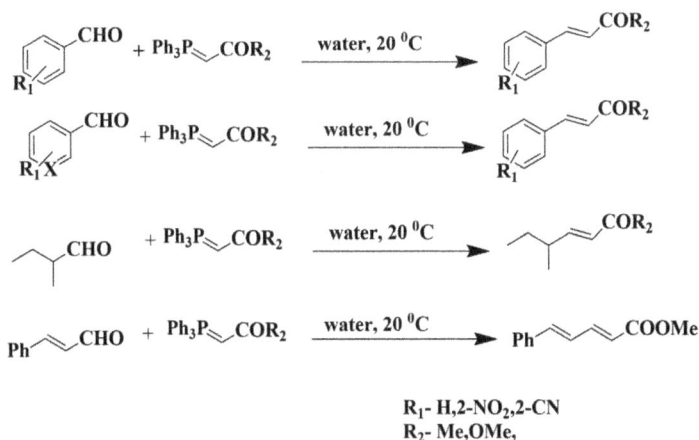

R_1- H,2-NO$_2$,2-CN
R_2- Me,OMe,

Scheme 12 Wittig reaction of aldehydes

2.1.7 Mannich-type reactions

Mannich type reactions give a useful synthetic pathway for nitrogen-containing reactions. Recently most of the organic reactions were carried out in an aqueous medium. But to achieve this type of reactions in water is too difficult. In the following reaction, hydrazono ester reacts with silicon enolates in water as a solvent. Use of water resulted in higher enantioselectivity. The new catalytic combination utilized ZnF_2 and TfOH (Scheme 13) [58].

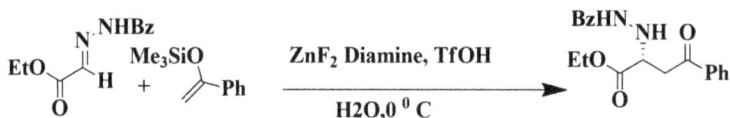

Scheme 13 Reaction of hydrazono ester and silicon enolates

2.1.8 Intramolecular Diels-Alder reaction

The intramolecular Diels-Alder reaction was carried out using indium (III) triflate as a catalyst, water as a solvent and alkenes as reactants in scheme 14 [59]. Diels-Alder reaction is an important reaction in synthetic organic chemistry. Further, this type of reactions was carried out *via* only organic solvents. But water has also been used as eco-friendly solvent, for highly stereoselective reactions.

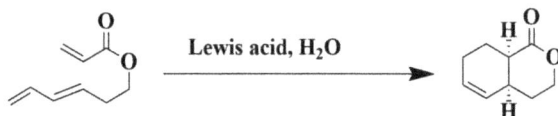

Scheme 14 Diels-Alder reaction of indium (III) triflate

2.2 Supercritical fluids

If the fluid temperature and pressure are above its critical point, it is called a supercritical fluid. The properties of the supercritical fluids in the liquid and gaseous state are highlighted in Table 3

Most of the solvents including pentane, ethylene, dimethyl ether butane, and nitrous-oxide, etc. have been investigated for the supercritical properties. Most of the solvents including pentane, ethylene, dimethyl ether butane, and nitrous-oxide, etc., have been investigated for the supercritical properties [59-61]. Super liquid fluids commonly used are:

a. Carbon dioxide

b. Water

Table 3 The properties of the supercritical fluids in liquid and gaseous state

State of solvent	Density (gcm^{-3})	Diffusivity (cm^{2}s^{-1})	Viscosity (gcm s^{-1})
Gas	10^{-3}	10^{-1}	10^{-4}
Liquid	1	5.10^{-6}	10^{-2}
Supercritical fluid	0.3	10^{-3}	10^{-4}

2.2.1 Carbon dioxide

CO_2 has been used as one of the common solvents in the supercritical state. CO_2 mediated reactions have been faster compared to the normal conventional organic solvent-mediated reactions. It is very simple and linear molecule and utilized for the alteration of the toxic freons. Supercritical CO_2 showing fine solvent properties has been also utilized for the extraction of some hydrocarbons, also CO_2 has been used to dissolve some of the polar compounds like ketones, esters, and aldehydes.

Its good solvent power were characterized in following postulates.

2.2.1.1 Applications of supercritical CO_2

Significant uses of the CO_2 have been documented in various review articles [62-75] and here categorized as follows:

- Utilized for the food industry in the various process like coffee, herbs, and spices, flavors, antioxidants, and seed oils etc.

- Very useful in the nutraceuticals and pharmaceuticals industries for the synthesis of drugs. Examples are carotenoids, lycopene, astaxanthin, and sterols etc.

- Used for numerous chemical reactions, such as polymerization, hydrogenation, destruction of toxic organics, enzymatic reactions etc.

- It also has a lot of applications in material processing methods. Examples include microencapsulation, coating, dyeing, crystallization, aerogels, particle formation, and impregnation.

- One of its most important applications has been in the cleaning process which includes (dry cleaning, soil reclamation, removal of undesired substances)

- It finds application in cosmetics preparation as it has active ingredients for cosmeceutical applications and in fragrances

- Most specific applications in various fields include membrane-based separation in biochemistry, microwave-induced supercritical fluid extraction, sterilization, preparative supercritical fluid extraction, thin film extraction etc.,

2.2.1.2 Chemical reactions of CO_2

The hydrogenation process of carvone in the presence of ScCO2 under mild condition shows good and higher yield as shown in scheme 15 [76].

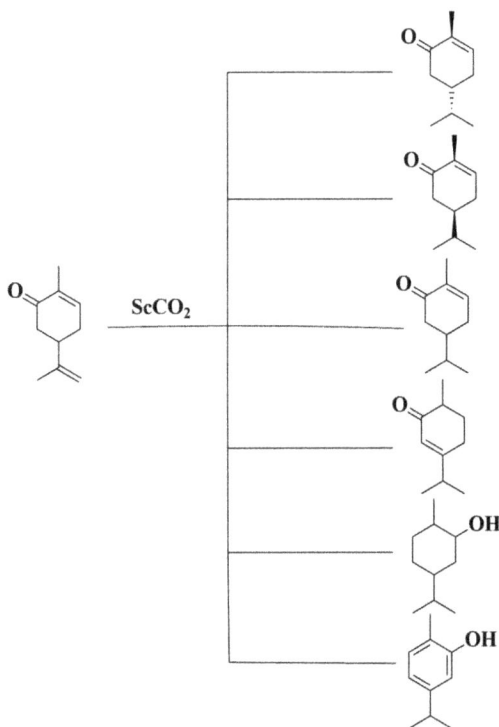

Scheme 15 Hydrogenation process of carvone

Further, the significance of ScCO2 is in epoxidation of the propylene converted into propylene oxide. Here propylene dissolves in supercritical carbon dioxide, which also

Materials Research Forum LLC
doi: https://doi.org/10.21741/9781644900239-5

increases the catalytic activity and increases the amount of the product as in scheme 16 [77].

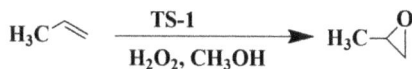

$$H_3C\diagdown \quad \xrightarrow[\text{H}_2\text{O}_2,\ \text{CH}_3\text{OH}]{\text{TS-1}} \quad H_3C\diagdown^{\text{O}}$$

Scheme 16 Epoxidation of propylene

2.3 Ionic liquids

ILqs (Ionic liquids) are also called as molten salts, in lower vapor pressure. Also, many of them have low instability, increased thermal stability, solvating characteristics and are electrically conductive [78]. Subsequently, there is an expanding enthusiasm for interchanging volatile natural solvents with ionic liquids. They fill in as solvents or reaction media for some isolation or synergist forms, as there exist an extensive variety of natural, inorganic and polymeric atoms which are well dissolvable in ionic liquids [79]. Ionic liquids have solvation property depending upon the nature of the anions and cations present in the organic compounds. Some of the example for anions and cations are formate, benzoate, phosphate, methanesulfonate, thiocyanate, and imidazolium, ammonium, phosphonium, pyridinium pyrrolidinium respectively.

2.3.1 General properties and nature of ILqs

ILqs have low electrical conductivity, non-ionizing ability, and low vapor pressure etc. Its other properties include,

a. Low combustibility

b. Temperature stability

c. Catalytic property

d. Viscosity

e. Polarity

f. Molar conductivity

g. Vapor pressure

2.3.2 Application of ILqs

ILqs are used in electrochemistry as they are, electrochemically inert over a wide potential range. They are also utilized in various fields such as chemistry, biotechnology, analytics, process technology and pharmaceutical fields, etc.

2.3.3 Ionic liquids in organic synthesis

Despite the fact that the ionic liquids don't comply fully with green science principles, they are extremely capable as options in contrast to organic solvents. In the scientific literature, there are countless research papers for the utilization of ionic liquids in synthetic routes and various applications. They are also useful for the condensation reactions, for an example for the condensation of indoles with benzaldehyde under microwave conditions with 1-benzyl-3-methylimidazolium hydrogen sulfate ([bnmim][HSO$_4$]) as a catalyst [80] Scheme 17.

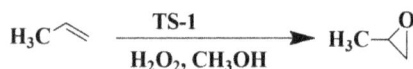

Scheme 17 Condensation of indoles

ILqs have been also utilized for the Friedel Craft acylations reaction, halomethylated of β-enaminones, aldol reaction, Knoevenagel reaction, Michael reaction, Biginelli and Hantzsch reaction or the Doebner modificationof the Knoevenagel reaction, Mannich reaction is shown in Scheme 18.

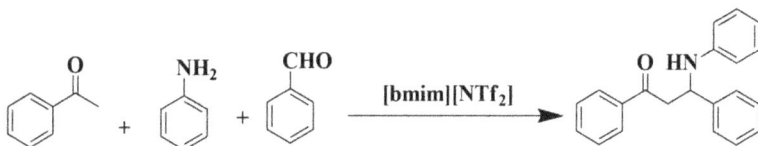

Scheme 18 Mannich reaction of Ionic liquid as a catalyst

Conclusion

During recent years, solvents from green and renewable sources have been found to be the most promising as current solvent innovation. From the above discussion of the various green solvents like water, ionic liquids, and supercritical solvents, it is found that they all gave the same good results compared to conventional solvents. Main contents of this chapter deal with green solvents and their various applications in synthetic chemistry.

Acknowledgment

The authors would like to thank Vellore Institute of Technology, Vellore to providing help for this book chapter. Also, we would like thank to CSIR-SRF for funding.

References

[1]   C.G. Brundtland, Our common future, Oxford, UK: World Commission on Environment and Development, Oxford University Press, 1987.

[2]   I.T.Horvàth, Fluorousbiphase chemistry. Acc. Chem. Res.31(1998) 641- 650.

[3]   S.J. Taverner, J.H. Clark, Can fluorine chemistry be green chemistry? Journal of Fluorine Chemistry, 123 (2003) 31-36. https://doi.org/10.1016/s0022-1139(03)00140-4

[4]   R.A. Sheldon, Green solvents for sustainable organic synthesis: state of the art, Green Chem. 7 (2005) 267-278. https://doi.org/10.1039/b418069k

[5]   P. Anastas, N. Eghbali, Green chemistry: Principles and practice, Chem. Soc. Rev. 39 (2010)301-312. https://doi.org/10.1039/b918763b

[6]   L. Orha, G. R. Akien, I.T.Horvath, Synthesis in green solvents,in: C.J Li, Green synthesis, Handbook of green chemistry, Wiley-VCH Verlag, 7 (2010) pp. 93-120. https://doi.org/10.1002/9783527628698.hgc075

[7]   C.J. Li, T.K. Chan, Organic reactions in aqueous media, Wiley, New York, 1997.

[8]   U.M. Lindstrom (Ed.), Organic reactions in water: principles, strategies, and applications, Wiley-Blackwell, 2007.

[9]   C.J. Li, Reactions in water, in: P.T. Anastas (Ed.) Handbook of green chemistry, Wiley-VCH Verlag, 5 (2010).

[10]   N. Akiya, P.E. Savage, Roles of water for chemical reactions in high-temperature water, Chem. Rev. 102 (2002) 2725–2750. https://doi.org/10.1021/cr000668w

[11]   M.O. Simon, C.J. Lee, Green chemistry oriented organic synthesis in water, Chem. Soc. Rev. 41 (2012) 1415–1427. https://doi.org/10.1039/c1cs15222j

[12]   U.M. Lindstrom, Stereoselective organic reactions in water, Chem. Rev. 102 (2002) 2751-2772.

[13]   C.J. Li, L. Chen, Organic chemistry in water, Chem. Soc. Rev. 35 (2006) 68–82.

[14]   H.C. Hailes, Reaction solvent selection: the potential of water as a solvent for organic transformations, Org. Process Res. Dev. 11(2007) 114-120. https://doi.org/10.1021/op060157x

[15]   D.Dallinger, C.O. Kappe, Microwave-assisted synthesis in water as solvent, Chem. Rev. 107(2007) 2563-2591. https://doi.org/10.1021/cr0509410

[16]   A.A. Clifford, Fundamentals of supercritical fluids, Oxford University Press, U.K. 1998.

[17]   P.G. Jessop, W. Leitner (Eds.), Chemical synthesis using supercritical fluids, Wiley, New York, 1999.

[18]   W. Leitner, P.G. Jessop, (Eds.), Supercritical solvents, in: P.T. Anastas,Handbook of green chemistry,Wiley-VCH Verlag, 5 (2010).

[19]   J.A. Hyatt, Liquid and supercritical carbon dioxide as organic solvents, J. Org. Chem. 49(1984) 5097–5101. https://doi.org/10.1021/jo00200a016

[20]   E.J. Beckman, Supercritical and near-critical CO_2 in green chemical synthesis and processing, J. Supercrit. Fluids, 28 (2004) 121–191.

[21]   C.M. Rayner, The potential of carbon dioxide in synthetic organic chemistry, Org. Process Res. Dev. 11 (2007) 121–132. https://doi.org/10.1021/op060165d

[22]   X. Han, M. Poliakoff, Continuous reactions in supercritical carbon dioxide: problems, solutions and possible ways forward, Chem. Soc. Rev. 41 (2012) 1428–1436. https://doi.org/10.1039/c2cs15314a

[23]   C. Boyere, C. Jerome, A. Debuigne, Input of supercritical carbon dioxide to polymer synthesis: an overview, Eur. Polymer J. 61 (2014) 45–63. https://doi.org/10.1016/j.eurpolymj.2014.07.019

[24]   P.G. Jessop, B. Subramaniam, Gas-expanded liquids, Chem. Rev. 107 (2007) 2666–2694. https://doi.org/10.1021/cr040199o

[25]   P. Wasserscheid, T. Welton, (Eds.)Ionic liquids in synthesis, 2nd (Ed.) Wiley-VCH, Germany, 2008.

[26]   P. Wasserscheid, A. Stark (Eds.), Ionic liquids, in: P.T. Anastas, Handbook of green chemistry, Wiley-VCH, Germany, 2010.

[27]   T. Welton, Room-temperature ionic liquids. Solvents for synthesis and catalysis, Chem. Rev. 99 (1999) 2071-2084. https://doi.org/10.1021/cr980032t

[28]   V.I. Parvulescu, C. Hardacre, Catalysis in ionic liquids, Chem. Rev. 107(2007) 2615-2665. https://doi.org/10.1021/cr050948h

[29]   F. van Rantwijk, R.A. Sheldon, Biocatalysis in ionic liquids, Chem. Rev. 107(2007)2757–2785. https://doi.org/10.1021/cr050946x

[30]   N.V. Plechkova, K.R. Seddon, Applications of ionic liquids in the chemical industry, Chem.Soc. Rev. 37 (2008) 123–150. https://doi.org/10.1039/b006677j

[31]   H. Olivier-Bourbigou, L. Magna, D. Morvan, Ionic liquids and catalysis: recent progress from knowledge to applications. Appl. Catal. A. 373 (2010) 1–56. https://doi.org/10.1016/j.apcata.2009.10.008

[32]   J.P. Hallett,T.Welton, Room-temperature ionic liquids: solvents for synthesis and catalysis, Chem. Rev. 111 (2011) 3508–3576. https://doi.org/10.1021/cr1003248

[33]   M.J. Naughton, R.S. Drago, Supported homogeneous film catalysts, J. Catal. 155(1995) 383–389. https://doi.org/10.1006/jcat.1995.1220

[34]   S. Chandrasekhar, C. Narsihmulu, S.S. Sultana, N.R. Reddy, Poly(ethylene glycol) (PEG) as a reusable solvent medium for organic synthesis. Application in the Heck reaction, Org. Lett. 4(2002) 4399–4401. https://doi.org/10.1021/ol0266976

[35] N.F. Leininger, R. Clontz, J.L. Gainer, D.J. Kirwan, Polyethylene glycol-water andPolypropylene glycol-water solutions as benign reaction solvents, Chem. Eng. Comm. 190 (2003) 431–444. https://doi.org/10.1080/00986440302082

[36] P.C. Andrews, A.C. Peatt, C.L. Raston, Indium metal mediated synthesis of homoallylic amines in poly (propylene) glycol (PPG). Green Chem. 6 (2004) 119–122. https://doi.org/10.1039/b311944k

[37] J. Chen, S.K. Spear, J.G. Huddleston, R.D. Rogers, Polyethylene glycol and solutions of polyethylene glycol as green reaction media. Green Chem. 7 (2005) 64–82. https://doi.org/10.1039/b413546f

[38] D.J. Heldebrant, H.N. Witt, S.M. Walsh, T. Ellis, J. Rauscher, P.G. Jessop, Liquid polymers as solvents for catalytic reductions, Green Chem. 8 (2006) 807–815. https://doi.org/10.1039/b605405f

[39] K.S. Feu, A.F. de la Torre, S. Silva, M.A.F. de Moraes, A.G. Correa, M.W. Paixao, Polyethylene glycol (PEG) as a reusable solvent medium for an asymmetric organocatalytic Michael addition. Application to the synthesis of bioactive compounds, Green Chem. 16 (2014)3169–3174. https://doi.org/10.1039/c4gc00098f

[40] R.T. Mathers, K.C. McMahon, K. Damodaran, C.J. Retarides, D.J. Kelley, Ring-opening metathesis polymerizations in D-limonene: a renewable polymerization solvent and chain transfer agent for the synthesis of alkene macromonomers, Macromolecules 39 (2006) 8982–8986. https://doi.org/10.1021/ma061699h

[41] S.K. Spear, S.T. Griffin, K.S. Granger, J.G. Huddleston, R.D. Rogers, Renewable plant-based soybean oil methyl esters as alternatives to organic solvents, Green Chem. 9 (2007) 1008–1015. https://doi.org/10.1039/b702329d

[42] I.T. Horváth, Solvents from nature, Green Chem. 10 (2008) 1024–1028.

[43] B. Schaffner, F. Schaffner, S.P. Verevkin, A. Borner, Organic carbonates as solvents in synthesis and catalysis, Chem. Rev. 110 (2010) 4554–4581. https://doi.org/10.1002/chin.201046256

[44] Y.Gu, F.Jérôme, Glycerol as a sustainable solvent for green chemistry, Green Chem. 12 (2010) 1127–1138. https://doi.org/10.1039/c001628d

[45] L. Lomba, B. Giner, I. Bandrés, C. Lafuente, M. Rosa Pino, Physicochemical properties of green solvents derived from biomass. Green Chem. 13 (2011) 2062–2070. https://doi.org/10.1039/c0gc00853b

[46] C.S.M. Pereira, V.M.T.M. Silva, A.E. Rodrigues, Ethyl lactate as a solvent: properties, applications, and production processes-a review, Green Chem. 13 (2011) 2658–2671. https://doi.org/10.1039/c1gc15523g

[47] V. Pace, P. Hoyos, L. Castoldi, P.D. de Maria, A.R. Alcantara, 2-Methyltetrahydrofuran(2-MeTHF): a biomass-derived solvent with broad application

in organic chemistry.ChemSusChem,5 (2012) 1369–1379.
https://doi.org/10.1002/cssc.201100780

[48] M. Matsushita, K. Kamata, K. Yamaguchi and N. Mizuno,Heterogeneously catalyzed aerobic oxidative biaryl coupling of 2-naphthols and substituted phenols in water, J. Am. Chem. Soc. 127 (2005) 6632-6637.
https://doi.org/10.1002/chin.200537098

[49] Y. Uozumi and R. Nakao, Templated synthesis of inorganic hollow spheres with a tunable cavity size onto core–shell gel particles,Angew. Chem. Int. Ed.42 (2003) 194-199. https://doi.org/10.1002/anie.200250443

[50] D. Biondini Lucia, B. Raimondo, G. Laura Goracci, G. Savelli, Dehydrogenation of amines to nitriles in aqueous micelles, Eur. J. Org. Chem. (2005) 3060-3063.
https://doi.org/10.1002/ejoc.200500047

[51] K.Surendra, N.S. Krishnaveni, V.P. Kumar, R. Sridhar, K. RamaRao, Selective and efficient oxidation of sulfides to sulfoxides with N-bromosuccinimide in the presence of β-cyclodextrin in water, TetrahedronLett. 46 (2005) 4581-4583.
https://doi.org/10.1016/j.tetlet.2005.05.011

[52] K. Yamaguchi, M. Matsushita, N. Mizuno, Efficient hydration of nitriles to amides in water, catalyzed by ruthenium hydroxide supported on alumina, Angew. Chem. Int. Ed. 43(2004) 1576-1580. https://doi.org/10.1002/anie.200353461

[53] X. H. Huang, K. W. Anderson, D. Zim, L. Jiang, A. Klapars, S. L. Buchwald, Expanding Pd-catalyzed C-N bond-forming processes: the first amidation of aryl sulfonates, aqueous amination, and complementarity with Cu-catalyzed reactions, J. Am. Chem. Soc.125 (2003) 6653-6655. https://doi.org/10.1021/ja033450a

[54] L.Chen, C.J. Li, A remarkably efficient coupling of acid chlorides with alkynes in water, Org. Lett.6 (2004) 3151–3153. https://doi.org/10.1021/ol048789w

[55] E. Alacid, Carmen Nájera, The Mizoroki-Heck reaction in organic and aqueous solvents promoted by a polymer-supported Kaiser oxime-derived palladacycle, Archive for organic chemistry, 8(2008) 50-67.
https://doi.org/10.3998/ark.5550190.0009.806

[56] L. Botella, C.Nájera, Mono-and β,β-double-Heck reactions of α,β-unsaturated carbonyl compounds in aqueous media, J. Org. Chem. 70 (2005), 4360–4369.
https://doi.org/10.1021/jo0502551

[57] J. Dambacher, W. Zhao, A. El-Batta, R. Anness, C. Jiang, M. Bergdahl, Water is an efficient medium for Wittig reactions employing stabilized ylides and aldehydes, Tetrahedron Lett. 46 (2005) 4473–4477. https://doi.org/10.1016/j.tetlet.2005.04.105

[58] T. Hamada, K. Manabe, S.Kobayashi, Enantio-and diastereoselective, stereospecific Mannich-type reactions in water, J. Am. Chem. Soc, 126 (2004) 7768–7769. https://doi.org/10.1021/ja048607t

[59] H. Yanai, A.Saito, T. Taguchi, Intramolecular Diels–Alder reaction of 1,7,9decatrienoates catalyzed by indium(III) trifluoromethane sulfonate in aqueous media. Tetrahedron, 61 (2005) 7087–7093. https://doi.org/10.1016/j.tet.2005.05.062

[60] F. Favati, J.W. King, M. Mazzanti, Supercritical carbon dioxide extraction of evening primrose oil, J. Am. Oil. Chem. Soc.68 (1991) 422-427. https://doi.org/10.1007/bf02663760

[61] V. Illes, O. Szalai, M. Then, H.G. Daood, S. Perneczki, Extraction of hip rose fruit by supercritical CO_2 and propane, J.Supercrit. Fluids,10(1997) 209-218. https://doi.org/10.1016/s0896-8446(97)00018-1

[62] F. Temelli, Perspectives on supercritical fluid processing of fats and oils, J Supercrit. Fluids,47 (2009) 583-590. https://doi.org/10.1016/j.supflu.2008.10.014

[63] M.J.H. Akanda, M.Z.I.Sarker, S.Ferdosh, M.Y.A. Manap, N.N.N. Rahman, M.O.A. Kadir, Applications of supercritical fluid extraction (SFE) of palm oil and oil from natural sources, Molecules,17(2012)1764-1794. https://doi.org/10.3390/molecules17021764

[64] G. Brunner, Supercritical fluids: technology and application to food processing, J. Food.Eng. 67 (2005) 21–33.

[65] F. Temelli, Perspectives on supercritical fluid processing of fats and oils, J.Supercrit.Fluids, 47 (2009)583-590. https://doi.org/10.1016/j.supflu.2008.10.014

[66] Q. Lang, C.M. Wai, Supercritical fluid extraction in herbal and natural product studies-A practical review, Talanta, 53 (2001)771–782. https://doi.org/10.1016/s0039-9140(00)00557-9

[67] S. Jokic, S. Vidovic, K. Aladic, Supercritical fluid extraction of edible oils, in supercritical fluids: fundamentals, properties and applications, J. Osborne, (Eds.) NovaScience Publishers, Inc. New York, USA (2014) pp. 205-228.

[68] C.G. Pereira, M. Angela, A. Meirles, Supercritical fluid extraction of bioactive compounds: fundamentals, applications and economic perspectives, Food Bioprocess. Tech. 3 (2010)340- 372. https://doi.org/10.1007/s11947-009-0263-2

[69] E. Reverchon, I. De Marco, Supercritical fluid extraction and fractionation of natural matter, J.Supercrit. Fluids. 38 (2006)146-166. https://doi.org/10.1016/j.supflu.2006.03.020

[70] M.M.R. De Melo, A.J.D. Silvestre, C.M. Silva, Supercritical fluid extraction of vegetable matrices: applications, trends and future perspectives of a convincing green

technology, J.Supercrit. Fluids. 92 (2014)115–176.
https://doi.org/10.1016/j.supflu.2014.04.007

[71]   C.A. Eckert, B.L. Knutson, P.G. Debenedetti, Supercritical fluids as solvents for chemical and materials processing, Nature 383 (1996)313-318.
https://doi.org/10.1038/383313a0

[72]   N.L. Rozzi, R.K. Singh, Supercritical fluids and the food industry, Comprehensive review in food science food safety, 1 (2002)33-44. https://doi.org/10.1111/j.1541-4337.2002.tb00005.x

[73]   F. Sahena, I.S.M.Zaidul, S. Jinap, A.A. Karim, K.A. Abbas, N.A.N. Norulaini, A.K.M. Omar, Application of supercritical CO_2 in lipid extraction-A review, J. Food Eng. 95,(2009)240–253. https://doi.org/10.1016/j.jfoodeng.2009.06.026

[74]   M. Sovilj, B. Nikolovski, M. Spasojevic, Critical review of supercritical fluid extraction of selected spice plant materials, Macedonian Journal Chemistry and Chemical Engineering, 30 (2011) 197–220.

[75]   K.K. Darani, M.R. Mozafari, Supercritical fluids technology in bioprocess industries: A review, Biochem. Tech. 2 (2009)144-152.

[76]   C.I. Melo, R. Bogel-Aukasik, M. Gomes da Silva, E. Bogel-Aukasik, Advantageous heterogeneously catalysed hydrogenation of carvone with supercritical carbon dioxide, Green Chem. 13 (2011) 2825-2830.
https://doi.org/10.1039/c1gc15495h

[77]   Y. Chen, Y. Wu, Y. Zhang, L. Long, L. Tao, M.Yang, M. Tang, Epoxidation of propylene to propylene oxide catalyzed by large-grain TS-1 in supercritical CO_2, J. Mol.Catal. Chem. 352 (2012)102-109. https://doi.org/10.1016/j.molcata.2011.10.020

[78]   N.V. Plechova, K.R. Seddon, Applications of ionic liquids in the chemical industry, Chem. Soc. Rev. 37 (2008) 123–150.

[79]   J. DeSimone, Practical approaches to green solvents, Science, 297 (2002) 799-803.
https://doi.org/10.1126/science.1069622

[80]   H. Zang, M. Wang. B.W. Cheng, J. Song, Ultrasound-promoted synthesis of oximes catalyzed by a basic ionic liquid [bmIm] OH, Ultrason.Sonochem.16 (2009) 301-303. https://doi.org/10.1016/j.ultsonch.2008.09.003

Industrial Applications of Green Solvents I
Materials Research Foundations **50** (2019) 147-164

Materials Research Forum LLC
doi: https://doi.org/10.21741/9781644900239-6

Chapter 6

Application of Supercritical Carbon Dioxide in the Leather Industry

Surya Pratap Goutam[1], Asheesh Kumar[2], Devesh Kumar[2], Rajkamal Shastri[1], Anil Kumar Yadav[1]*

[1] Advanced Materials Research Laboratory, Department of Physics, Babasaheb Bhimrao Ambedkar (Central) University, Lucknow 226 025, Uttar Pradesh, India

[2] Department of Physics, Babasaheb Bhimrao Ambedkar (Central) University, Lucknow 226 025, Uttar Pradesh, India

akyadavbbau@gmail.com

Abstract

In numerous processes and synthesis, supercritical carbon oxide can be used as a nontoxic and green solvent. This chapter described the supercritical carbon dioxide and its application in the leather industry to develop an eco-friendly technology. Supercritical carbon dioxide has several unique properties and has great potential for advanced processing of materials. Supercritical carbon dioxide as a potential solvent for leather processing hopes to reduce environmental burden by avoiding the parallel pollutions by leather processing.

Keywords

Supercritical Carbon Dioxide, Solvent, Extraction, Leather Industries, Toxicity

Contents

Materials Research Forum LLC
doi: https://doi.org/10.21741/9781644900239-6

1.   Introduction

Leather production is common worldwide as leather industries are the key players in the economy of developing nations. These industries are also considered to be a major source of environmental pollution caused by their potentially toxic and hazardous liquid waste, which creates a negative image of leather technologies in society.

Hides and skins of buffaloes, bovines, pigs, and goats are generally used as primal matter in leather industries for leather production [1]. Nowadays, the global challenge for scientists and engineers is to use their skills and creativity to design, develop, and implement procedures that allow environmental and economic sustainability [2-3]. Green processes are environment-friendly processes in which the chemical activities-including chemical design, manufacture, use, and disposal are such that hazardous substances will not be used and generated [4]. With increasing emphasis placed on environmental responsibility, there is a clear need for reactions and processes that can eliminate or significantly reduce harmful residues such as used organic solvents, salt streams, and dangerous organic or inorganic by-products.

Supercritical CO_2 with its moderate critical constants, nonflammable nature, and low cost provides an attractive alternative for replacing organic solvents that were traditionally used in chemical manufacturing processes which were the main cause of pollution in leather processing. Minimizing liquid waste generation, easy separation of solutes and fast reaction rates are some of the advantages of the supercritical CO_2 technology over conventional solvent extraction methods.

License et al. [5] in 2003 published a research paper on the chemical reaction in supercritical CO_2 from the laboratory to the commercial plant. In this research paper, the development of supercritical carbon dioxide application from the laboratory to plant scale was highlighted and presented a progress report about an ongoing green chemistry initiative.

In 2014, Deng et al. [1] reviewed the developments of CO_2 deliming for leather manufacture. In this chapter, the authors analyzed carbon dioxide properties including solubility in water, deliming activity, and mass transfer. Conclusively, this chapter provided an overview of fundamentals, process optimization, occupational safety, deliming in leather. Hu and Deng [6] in 2015 published a review on the application of supercritical CO_2 for leather processing. This review described the properties of supercritical CO_2 and applications of it for industries associated with the technology of leather production. In this review, the economic position of supercritical CO_2 technology for leather production on the industrial level was evaluated on the basis of published research work.

1.1   What is supercritical carbon dioxide?

In a CO_2 molecule, two oxygen atoms are covalently bonded to a single carbon atom with double bonds. The structure of the molecule is linear and has zero electrical dipole. CO_2 molecule observed with asymmetric stretching at 1243 cm^{-1} in the Raman spectrum. Optimized structure of CO_2 using Gauss View and Raman spectrum of optimized geometry of CO_2 obtained from Gaussian 09 [7] have been shown in Fig.1 and Fig.2 successively. The critical temperature and critical pressure of CO_2 are 30.95 °C and 72.8 atm successively, to obtain supercritical state [8]. The pressure-temperature phase of supercritical fluid of a pure component is shown in Fig.3.

Supercritical CO_2 works extremely well as processing media for an extensive variety of chemical and biological extraction processes. Supercritical carbon dioxide has a significant ability to precise control over components of a complex matrix to be extracted [9]. This fruitful ability of supercritical carbon dioxide can be achieved by controlling various parameters like pressure, temperature, processing time and flow rates [10].

Industrial Applications of Green Solvents I   Materials Research Forum LLC
Materials Research Foundations **50** (2019) 147-164   doi: https://doi.org/10.21741/9781644900239-6

Diffusivity of supercritical CO_2 is higher than liquid so the extraction rates are generally much higher than the corresponding extraction by other systems of solid and liquid [11].

So, supercritical CO_2 has several advantages as a solvent, including, lower pressure drops, better mass-transfer, and ease of solvent recovery. Supercritical CO_2 has unique physical properties which can be exploited to control chemical reactivity.

Fig. 1 *Optimized Structure of CO_2*

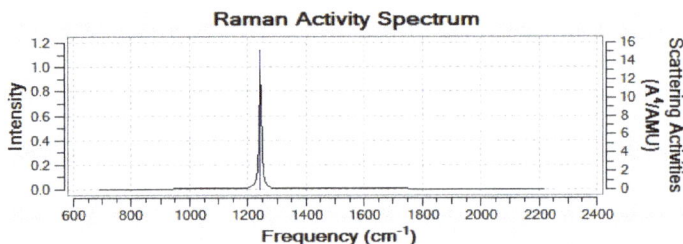

Fig. 2 *Raman spectrum of the optimized geometry of CO_2*

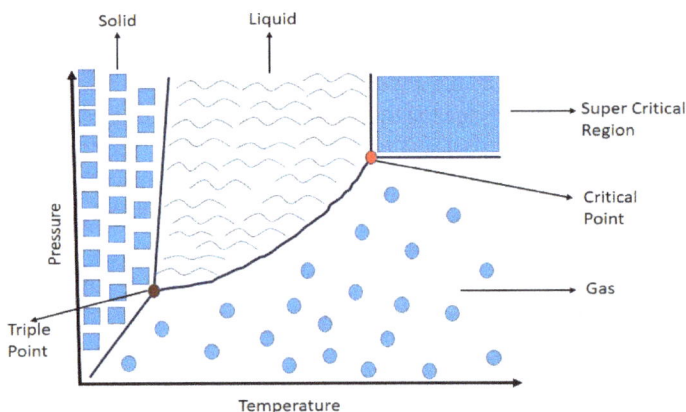

Fig. 3 *Pressure-Temperature phase diagram of supercritical carbon dioxide*

1.2 Leather processing

Leather processing can be understood under three main processes: preparative, tanning and crusting stages. An additional sub-process, the surface coating could also be added into the sequence. The list of operations that animal skins endure varying with the sort of leather. Fig.4 presents the main leather processes.

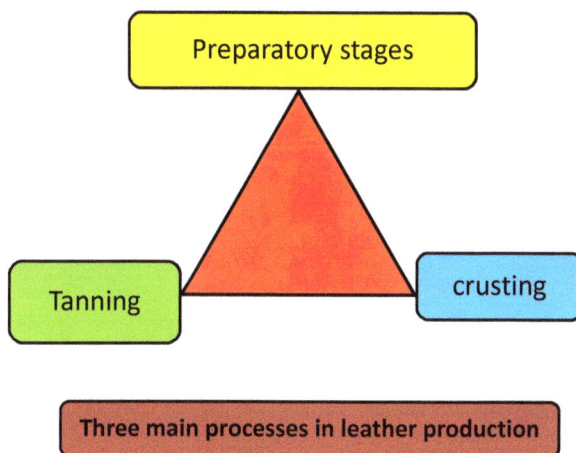

Fig. 4 *Diagram for the leather processes*

1.2.1 Stages for preparation

The stages of preparation start when the skin is ready for tanning [12]. During the stages of preparation, several of the unwanted raw skin elements are removed. Several choices for pretreatment of the skin exist. Not all of the choices could also be performed. Stages of preparation might embody preservation, soaking, liming, unhairing, fleshing, splitting, reliming, deliming, bating, degreasing, freezing, bleaching, pickling, unpickling [13]. Stages for preparation of leather involved in preparatory stages are shown in Fig.5.

1.2.2 Tanning

Tanning is the method that converts the macromolecule of the primal skin into a constant material which cannot decompose and is appropriate for a good sort of finish applications. The principal distinction between raw skins and tanned skins is that raw skins dry resolute type a tough inflexible material which will decompose once re-wetted, whereas tanned material dries resolute a versatile type that doesn't become rotten once wetted back. An oversized variety of various tanning ways and materials are often used; the selection is ultimately obsessed with the top application of the animal skin. The generally used tanning material is Cr, that leaves the animal skin when the product is tanned once it has a pale blue color and the product is generally called "wet blue" product.

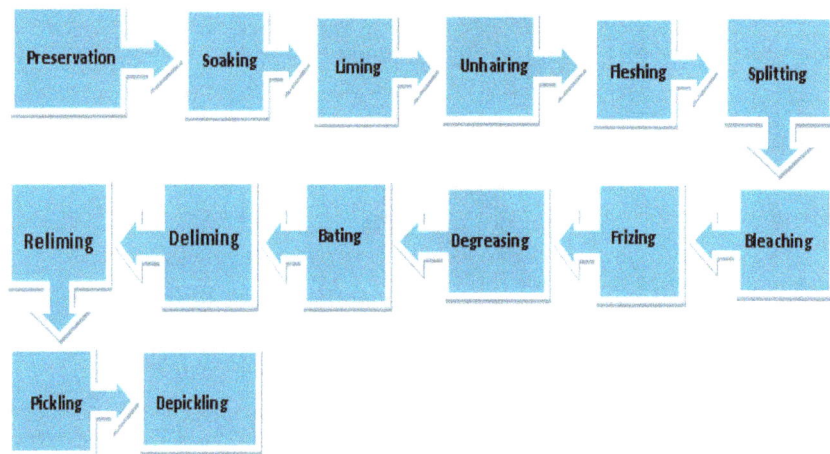

Fig. 5 *Flow diagram of preparatory stages of leather production*

1.2.3 Crusting

Crusting is a very important process of leather production. Crusting is the process to make the skin thinned, retanned and lubricated. The fate of the crusting sub-process is the drying and softening operations of skin.

2.   The journey of CO_2 to supercritical carbon dioxide

In 2002, initial, multi-reaction, supercritical flow reactor was made as a fruitful collaboration between the Clean Technology Group at the University of Nottingham and the fine chemicals manufacturer, Thomas Swan & Co. Ltd. In November 1995, a project was started in that project first reaction involved the hydrogenation of cyclohexene in supercritical CO_2 (Plan 1).

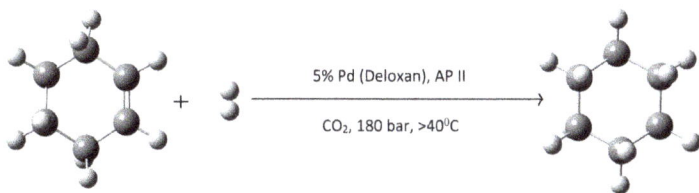

Plan 1 Hydrogenation of cyclohexene under supercritical conditions.

In Nottingham, a detailed investigation was performed on chemical process of acetophenone which showed that supercritical CO_2 allowed the reaction conditions to be optimized terribly effectively to maximize the yield of specific chemical process merchandise (Plan 2) [14-16].

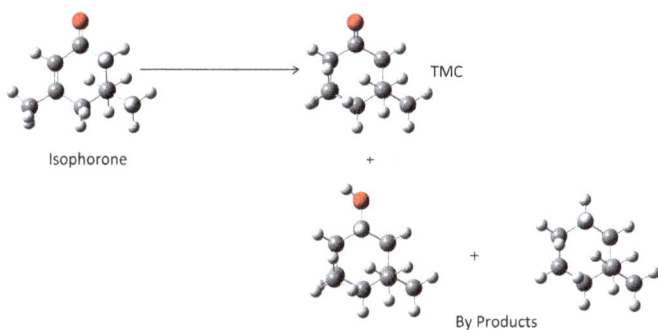

Plan 2 Different products obtained during the hydrogenation of acetophenone.

Supercritical CO_2, as a solvent is found different from conventional solvents because of its most effective dependency of density on pressure and it is easily compoundable with H_2 [17]. Licencse et al. proposed that, despite the weird options of supercritical CO_2 as a solvent, the general property of a given catalyst wasn't essentially modified compared to its behavior in typical solvents however rather that conditions may optimize a lot of effects in supercritical CO_2 [5,18-19] (Plan 3).

It was important to identify a model compound, which could be used by the two laboratories, Nottingham and Thomas Swan & Co., as the basis for developing a viable supercritical fluid process. Further, the hydrogenation reaction of isophorone to trimethyl cyclohexanone (TMCH) was chosen for potential commercial interest where the ease of optimization in supercritical CO_2 could be exploited [14] (Plan 4).

Above reaction transferred to the Thomas Swan & Co. laboratory where the industrial environment was better suited to investigating the feasibility of scale-up to a production scale. The general procedure of supercritical CO_2 extraction is shown in Fig.6.

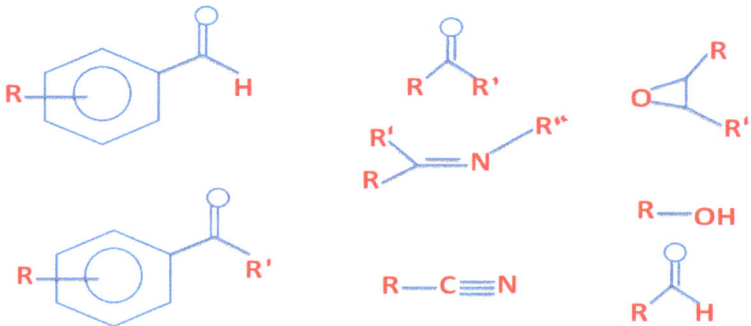

Plan 3 Some important functionalities acquired from the successfully hydrogenated under supercritical conditions [Licence et al]

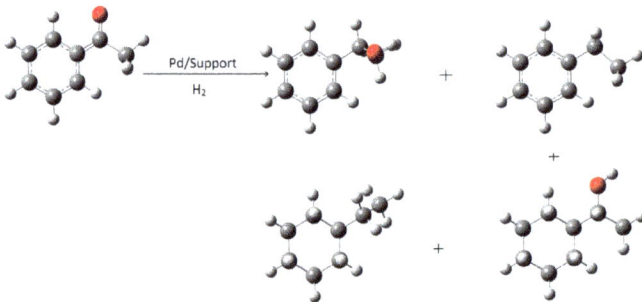

Plan 4 Different products obtained in the hydrogenation of isophorone.

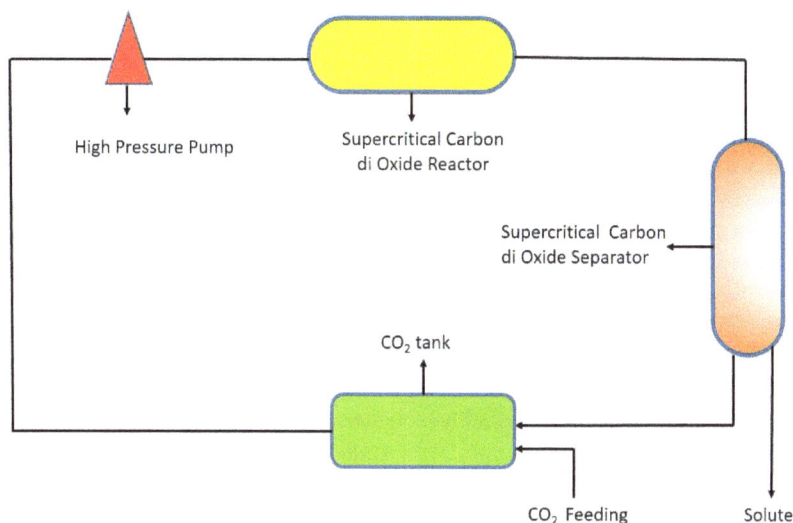

Fig. 6 *Pictorial diagram for supercritical carbon dioxide extraction.*

2.1 Environmental challenges of leather production

The leather industry is one of the most polluting industries. Conventional leather tanning technology produces a large amount of organic and chemical pollutants. The leather processing is responsible for unfavorable impact on the environment. The pollutants, which are mostly contained in the effluents discharged by tanneries, are a serious threat to the environment. The tannery effluent, if not treated properly, can cause a serious threat to soil and water bodies. The global production of leather is about 24bn m^2 that presents a substantial challenge to the leather industry.

In the leather industry, solid waste includes animal hairs, keratin wastes, buffing dust, animal skin trims, and flesh wastes. These wastes contain protein and if this protein is not utilized well than it can cause serious pollution problems to the environment. Tanneries produce harmful gases, dust and a large amount of solid waste like chrome throughout chromate reduction and from the buffing procedure, ammonia during deliming and unhairing, sulfides during liming.

2.1.1 Ammonium Nitrogen (NH₃-N)

Ammonium chemical element (NH3-N) is the major waste material of animal skin [19]. In 2011 Wang et al. [20] conjointly reported that seventy-eight percent NH_3-N of the entire emission comes from a typical ammonia salt deliming procedure. Professionally, long contact with ammonia (NH_3) may cause hepatic encephalopathy [21] and methemoglobinemia diseases [22]. Free ammonia is an additional active substrate providing nutrient for plant bacterium in surface water and buffering anaerobic digestion system can cause serious diseases [23]. Various researchers suggested that prevention from NH_3-N can be an improved key for greener manufacture of leather.

2.1.2 Chromium (Cr)

Compounds of chromium are the increasing environmental burden and serious concern of environmental pollution. Most typical ways in which to get rid of ionic Cr from the waste waters are the chemical precipitation and surface adsorption on different materials. Various researchers and scientists have reported the successful methodologies for Cr compound removal from tannery wastewater [24-26].

3. Applications of supercritical carbon dioxide

Supercritical carbon dioxide is turning into a very important business because of its role in chemical extraction additionally to its low toxicity and environmental impact. The comparatively coldness of the method conjointly the stability of carbon dioxide also permits most compounds to be extracted with very little injury or denaturing. Due to the unique properties of supercritical CO_2, it has various applications in different fields. Applications of supercritical carbon dioxide in the field of pharmaceuticals and in the leather industry are discussed below.

3.1 Supercritical carbon dioxide in the pharmaceutical industry

In the pharmaceutical industry, supercritical CO_2 has been used extensively due to its extraordinary properties like low and easily accessible critical temperature (31.2 °C) and pressure (7.4 MPa), non-flammability, non-toxicity [27]. The planning of active pharmaceutical ingredients (APIs) to create a solid quantity with appropriate chemical properties is extremely relevant. Controlling the properties of particles, size, crystal structure and morphology needed to optimize the formulation. The bioaccessibility of orally applied medication depends upon the speed of absorption and dissolution. The polymorphic management vital is crucial to attaining important levels of APIs within the blood [27–29]. For the micronization of drugs, numerous supercritical fluid techniques

Industrial Applications of Green Solvents I   Materials Research Forum LLC
Materials Research Foundations **50** (2019) 147-164   doi: https://doi.org/10.21741/9781644900239-6

have been acquired. A large number of APIs can be dissolved in supercritical CO_2 to produce fine drug particles.

3.2 Supercritical carbon dioxide in the leather industry

The conventional leather making processes involve some important steps like soaking, unhairing, deliming, tanning, fatliquoring and finishing [3]. The process of soaking is performed to renovate the natural swollen condition of hides to give away the soluble dirt and other unwanted ingredients. The epidermis, wool, and hair are eradicated in the process termed as unhairing. Liming involves loosening up of collagen fiber texture and to convert the natural grease partially. Since liming and unhairing are usually performed using sodium sulfide and lime in the same bath, therefore, it completely destroys the hair and epidermis. However, unhairing and liming are collectively termed as hair burning liming, that is quite similar to liming [30]. After deliming, the process of tanning is carried out. Pickling is to acidify the pelts to such a pH that it reduces astringency in tanning. Therefore, tanning is one of the main processes of leather production. In tanning, the material is transformed such that it prevents microbial attack [2,31]

All the processes have their own aim to be achieved using various chemicals. These chemicals get absorbed by the leather and hence, pollution is reduced. Supercritical CO_2 that is used in the process of degreasing, mitigates the pollution level while other degreasing agents like alkylphenol ethoxylates(APEO), lime, amine, boric acid, ammonium salt and naphthalene which were conventionally used as degreasing agent increase the pollution level. The supercritical CO_2 technology is quite advanced in comparison to the conventional one.

In 1955, the first supercritical CO_2 application in leather production was reported [32, 33]. Liao et al. [34] reported the first supercritical CO_2 equipment for leather and introduced this technology [34-37]. Liao et al. [34] used supercritical CO_2 as a medium for various preliminary stages of leather production.

3.2.1 Degreasing

Degreasing with supercritical CO_2 was first executed on animal skins in the leather process [32]. This process involved the extraction of fat from sheepskins. Liao et al. confirmed the dehydration effect of supercritical CO_2 [34]. The traditional supercritical CO_2 equipment comprised of a reactor in which CO_2 was kept in contact with the material to recover desirable products [38]. The degreasing method is used to precisely degrease the hides irrespective of toxic solvent.

3.2.2 Fiber separation

Leather fibers separation and loosening are obtained by the liming [2]. In supercritical CO_2 processes, the fiber separation is carried out to explore the feasibility of supercritical CO_2 leather fiber separation [37-40]. In conclusion, supercritical CO_2 can be used to separate out leather fibers [38].

3.2.3 Deliming

In the leather production, deliming is carried out after unhairing and liming. Not only CO_2 gas [41] but also supercritical CO_2 may be utilized for ammonium free deliming, thus reducing pollution. Liao et al. [34] found that deliming process was efficient using supercritical CO_2. Thus, it can be concluded that deliming by utilizing the supercritical CO_2 has an upper hand over the previously used deliming methods.

3.2.4 Chrome tanning

Chromium pollution is one of the major pollutions from tanneries. The reduction of chromium discharge is an emerging area of environmental research [2]. Chrome tanning using supercritical CO_2 is found to be an innovative technique that should be adopted so as to minimize the chromium pollution in the leather field [42].

3.2.5 Dyeing

The time in the dyeing process can be reduced by using the supercritical CO_2 [43]. It can be observed that while using the supercritical CO_2, the dye uptake had reached almost 100%. One of the main benefits of using supercritical CO_2 in dyeing is that it provides better penetration and evenness.

3.2.6 Fatliquoring

In this process, the leather fibers are lubricated to attain the required flexibility [2]. As a raw material, mineral oil with saturated hydrocarbon, vegetable oil, and animal fats are used as fatliquoring agents. Liao group [34] proved that supercritical CO_2 can be used for fatliquoring. In the future, some novel fatliquoring agents based on supercritical CO_2 application could be developed. The fatliquor penetration and dispersion can be tuned by using the supercritical CO_2.

3.2.7 Finishing

For the finishing process, leather is dried and prepared after the completion of wet processing. By using the appropriate finishing agents, the dried crust is coated to make

Industrial Applications of Green Solvents I   Materials Research Forum LLC
Materials Research Foundations **50** (2019) 147-164   doi: https://doi.org/10.21741/9781644900239-6

the leather ready for the use. Impregnation is carried out for coating of layers which remunerates the surface leather properties that provide it the necessary capacity to absorb.

Conclusion

The main motive of the supercritical carbon dioxide application in leather industry is to replace the non-biodegradable surfactants and organic solvents. In this way, the supercritical CO_2 reduces pollution due to solvent and surfactants. Although supercritical CO_2 has a number of advantageous factors for its applications in the different process yet very few reports are available on supercritical carbon dioxide also it has an upper hand over the conventional processing methods. Thus, leather technology can be more safe, clean and green by using supercritical carbon dioxide. It is hoped that, this technology will encourage researchers to explore new opportunities.

Table 1 Some important properties like molecular weight, the temperature of liquid, critical pressure, the critical temperature of some important supercritical fluid.

Fluid	Molecular weight, M_w (gm/mol)	Density of Liquid(gm/mol)	Temp. of Liquid (^0K)	Critical Volume, V_c (cm³/mol)	Critical Pressure (P_c) (bar)	Critical Temp. (^0K)
Acetonitrile (CH₃CN)	41.05	0.782	293	173.0	48.3	548.0
Ammonia (NH₃)	17.03	0.639	273	72.5	113.5	405.4
Benzene (C₆H₆)	78.11	0.885	289	260.0	48.9	562.2
Chloroform (CHCl₃)	119.38	1.489	293	238.9	53.7	536.4
Carbon dioxide (CO₂)	44.01	-	-	94.0	73.8	304.2
Carbon disulfide (CS₂)	76.13	1.293	273	160.0	79.0	552.0
Dichloromethane (CH₂Cl₂)	84.93	-	-	-	63.0	510.0
Ethane (C₂H₆)	30.07	0.548	183	148.3	48.8	305.4
Methane(CH₄)	16.04	0.425	112	98.7	46.0	190.6
Methanol (CH₃-OH)	32.04	0.791	293	118.0	80.9	513.1
Nitrous Oxide (N₂O)	44.01	1.226	184	97.4	72.4	309.6
n-Butane (C₄H₁₀)	58.12	0.579	293	255.0	38.0	425.2
n-Heptane (C₇H₁₆)	100.21	0.648	293	432.0	27.4	540.3
Propane (C₃H₈)	44.09	0.582	231	203.0	42.5	369.8
Sulfur Hexafluoride (SF₆)	146.05	1.83	223	8198.8	37.6	318.7
Water (H₂O)	18.02	0.998	293	55.3	221.2	647.4

Table 2 Density, viscosity and diffusion coefficients of supercritical CO_2 different states

State	Density, viscosity and diffusion coefficient of liquid, gaseous and supercritical CO_2		
	Density(gcm^{-3})	Viscosity($gcm^{-1}s^{-1}$)	Diffusion(cm^2s^{-1})
Liquid	1	10^{-2}	$<10^{-5}$
Gas	10^{-3}	10^{-4}	10^{-1}
Supercritical	10^{-1} to 1 (Liquid Like)	10^{-4} to 10^{-3} (Gas Like)	10^{-4} to 10^{-3} (Liquid Like)

References

[1]   W. Deng, D. Chen, M. Huang, J. Hu, L. Chen, Carbon dioxide deliming in leather production: a literature review, J. Clean. Prod.87 (2015) 26-38. https://doi.org/10.1016/j.jclepro.2014.09.066

[2]   E. Heidemann. E. R. KG-Darmstadt, Fundamentals of leather manufacture, Polymer Int. 37 (1993) 149-150.

[3]   A.D. Covington, Tanning chemistry: The science of leather, First Edition, Royal Society of Chemistry, 2009.

[4]   W.H. Hauthal, Advances with supercritical fluids, Chemosphere, 43 (2001) 123-135.

[5]   P. Licence, J. Ke, M. Sokolova, S.K. Ross, M. Poliakoff, Chemical reactions in supercritical carbon dioxide: from laboratory to commercial plant, Green Chem. 5 (2003) 99-104. https://doi.org/10.1039/b212220k

[6]   J. Hu, W. Deng, Application of supercritical-carbon dioxide for leather processing, J. Clean. Prod.113 (2016) 931-946. https://doi.org/10.1016/j.jclepro.2015.10.104

[7] J. A. Montgomery, M. J. Frisch, J. W. Ochterski, and G. A. Petersson, A complete basis set model chemistry. VI. Use of density functional geometries and frequencies, 110 (1999) 2822-2827. https://doi.org/10.1063/1.477924

[8]   P. Munshi, S. Bhaduri, Supercritical CO_2: a twenty-first century solvent for the chemical industry, Curr. Sci. 97 (2009) 63-72.

[9]   S.Bowadt, S.B. Hawthorne, Supercritical fluid extraction in environmental analysis, J.Chrom. 703 (1995) 549-571. https://doi.org/10.1016/0021-9673(95)00051-n

[10] S.B. Hawthorne, Analytical-scale supercritical fluid extraction, Anal. Chem. 62 (1990) 633A-642A. https://doi.org/10.1021/ac00210a001

[11] L.H. McDaniel, M.A. Khorassani, L.T. Taylor,Supercritical fluid extraction of wood pulp with analysis by capillary gas chromatography-mass spectrometry, The Journal of Supercritical Fluids, 19 (2001) 275-286. https://doi.org/10.1016/s0896-8446(00)00097-8

[12] V. Kumar, C. Majumdar, P. Roy, Effects of endocrine disrupting chemicals from leather industry effluents on male reproductive system, J. Steroid Biochem. Mol.Biol. 111 (2008) 208-216. https://doi.org/10.1016/j.jsbmb.2008.06.005

[13] R. Laurenti, M.Redwood, R.Puig, B. Frostell, Measuring the environmental footprint of leather processing technologies, J. Ind. Ecol. 21 (2017) 1180-1187. https://doi.org/10.1111/jiec.12504

[14] K. Joseph, N. Nithya, Material flows in the life cycle of leather, Journal of Cleaner Production 17(2009) 676-682. https://doi.org/10.1016/j.jclepro.2008.11.018

[15] M. Harrod, P. Moller, Hydrogenation of fats and oils at supercritical conditions, Process Technology Proceedings. 12 (1996) 43-48.

[16] M.G. Hitzler, F.R. Smail, S.K. Ross, M. Poliakoff, Selective catalytic hydrogenation of organic compounds in supercritical fluids as a continuous process, Org. Proc. Res. Dev. 2 (1998) 137. https://doi.org/10.1021/op970056m

[17] M.A. McHugh, V J. Krukonis, Supercritical fluid extraction: principles and practice, Second ed., Elsevier, Boston, 1994.

[18] E. Bach, E. Cleveland, E. Schollmeyer, Past, present and future of supercritical fluid dyeing technology-an overview, Rev. Prog. Color, 32(2002) 88-102. https://doi.org/10.1111/j.1478-4408.2002.tb00253.x

[19] J. Ludvik, United Nations industrial development organization (UNIDO), The scope for decreasing pollution load in leather processing, US/RAS/92/120/11-51.

[20] Y. Wang, Y. Zeng, X. Liao, Q. He, B. Shi, Ammonia nitrogen in tannery wastewater: distribution, origin and prevention, J. Am. Leather Chem. Assoc. 107 (2011) 40-50.

[21] A.Lemberg, M.A. Fernandez, Hepatic encephalopathy, ammonia, glutamine and oxidative stress, Ann. Hepatol. 8 (2009) 95-102.

[22]   A. Julio, A. Camargo, A. Alvaro, Ecological and toxicological effects of inorganic nitrogen pollution in aquatic ecosystems: a global assessment, Environ. Int. 32 (2006) 831-849. https://doi.org/10.1016/j.envint.2006.05.002

[23]   D. Paredes, P. Kuschk, T.S.A. Mbwette, F. Stange, R.A. Muller, H. Koser, New aspects of microbial nitrogen transformations in the context of wastewater treatment-A review, Eng. Life Sci. 7 (2007) 13-25. https://doi.org/10.1002/elsc.200620170

[24]   N.F. Fahim, B.N. Barsoum, A.E. Eid, M.S. Khalil, Removal of chromium(III) from tannery wastewater using activated carbon from sugar industrial waste, J. Hazard. Mater. 136 (2006) 303-309. https://doi.org/10.1016/j.jhazmat.2005.12.014

[25]   G. Arslan, E. Pehlivan, Batch removal of chromium(VI) from aqueous solution by Turkish brown coals, Bioresour. Technol. 98 (2007) 2836-2845. https://doi.org/10.1016/j.biortech.2006.09.041

[26]   S.P.Goutam, G. Saxena, V. Singh, A. K.Yadav, R.N.Bharagava, K.B. Thapa, Green synthesis of TiO2 nanoparticles using leaf extract of Jatropha curcas L. for photocatalytic degradation of tannery wastewater, Chem. Eng. J.336 (2018) 386-396. https://doi.org/10.1016/j.cej.2017.12.029

[27]   I. Pasquali, R. Bettini, F. Giordano, Solid-state chemistry and particle engineering with supercritical fluids in pharmaceutics, Eur. J. Pharm. Sci. 127 (2006) 299-310. https://doi.org/10.1016/j.ejps.2005.11.007

[28]   M. Perrut, J. Jung, F. Leboeuf, Enhancement of dissolution rate of poorly soluble active ingredients by supercritical fluid processes. Part I. Micronisation of neat particles, Int. J. Pharm. 288 (2005) 3-10. https://doi.org/10.1016/j.ijpharm.2004.09.007

[29]   R.H. Muller, C. Jacobs, O. Kayser, Nanosuspensions as particulate drug formulations in therapy: rationale for development and what we can expect for the future, Adv. Drug Deliv. Rev. 47 (2001) 3-19. https://doi.org/10.1016/s0169-409x(00)00118-6

[30]   J. Fages, H. Lochard, J.J. Letourneau, M. Sauceau, E. Rodier, Particle generation for pharmaceutical applications using supercritical fluid technology, Powder Technol. 141 (2004) 219-226. https://doi.org/10.1016/j.powtec.2004.02.007

[31]   A.D. Covington, Modern tanning chemistry, Chem. Soc. Rev. 26 (1997) 111-126.

[32] A.Marsal, P.J.Celma, J.Cot, M.Cequier, Application of the supercritical CO_2 extraction technology on the recovery of natural fat from the sheepskin degreasing process, The Journal of Supercritical Fluids, 18 (2000) 65-72. https://doi.org/10.1016/s0896-8446(00)00062-0

[33] Process for the treatment of skins, hides or sheet materials containing collagen by a dense, pressurized fluid, US Patent.

[34] L. Liao, Y. Feng, M. Chen, Z. Li, Study on non-polluting leather making technology, using CO_2 supercritical fluids. Feasibility study on CO_2 supercritical fluids clean production technology of leather, People Republic China. 28 (1999) 14-16. https://doi.org/10.1016/s0896-8446(99)00031-5

[35] B. Diaz-Reinoso, A. Moure, H. Dominguez, J.C. Parajo, Supercritical CO2 extraction and purification of compounds with antioxidant activity, J. Agric. Food Chem. 54 (2006) 2441-2469. https://doi.org/10.1021/jf052858j

[36] Y. Ahn, S.J. Bae, M. Kim, S.K. Cho, S. Baik, J. IkLee, J.E. Cha, Review of supercritical CO_2 power cycle technology and current status of research and development, Nuclear Engineering and Technology, 47 (2015) 647-661. https://doi.org/10.1016/j.net.2015.06.009

[37] Z.Li, L. Liao, Y. Feng, Method for leather-making with CO_2 supercritical fluid as medium, China Patent. https://patentimages.storage.googleapis.com/ea/9d/6a/52adeda12df4a4/CN1673394A. pdf/, 2005.

[38] A. Marsal, P.J. Celma, J. Cot, M. Cequier, Supercritical CO_2 extraction as a clean degreasing process in the leather industry. The Journal of Supercritical Fluids, 16 (2000) 217-223. https://doi.org/10.1016/s0896-8446(99)00031-5

[39] M. Renner, E. Weidner, B. Jochems, H. Geihsler, Free of water tanning using CO_2 as process additive-An overview on the process development, The Journal of Supercritical Fluids, 66 (2012) 291-296. https://doi.org/10.1016/j.supflu.2012.01.007

[40] Q. Yang, S. Qin, J. Chen, W. Ni, Q. Xu, Supercritical carbon dioxide-assisted loosening preparation of dry leather, J. Appl. Polym. Sci. 113 (2009) 4015-4022. https://doi.org/10.1002/app.30476

[41] D. Weijun, L. Chen, H. Jing, Development of carbon dioxide gas in leather deliming process, China Leather, 42 (2013) 51-55.

[42]   R. Manfred, W. Eckhard, J. Bjorn, G. Helmut, Free of water tanning using CO_2 as process additive-An overview on the process development, The Journal of Supercritical Fluids 66 (2012) 291-296. https://doi.org/10.1016/j.supflu.2012.01.007

[43]   P. Swidersky, D. Tuma, G. M. Schneider, High-pressure investigations on the solubility of anthraquinone dyestuffs in supercritical gases by VIS-spectroscopy. Part II-1,4-Bis-(n alkylamino)-9,10-anthraquinones and disperse Red 11 in CO_2, N_2O, and CHF3 up to 180 MPa, The Journal of Supercritical Fluids, 9 (1996) 12-18. https://doi.org/10.1016/s0896-8446(96)90039-x

Industrial Applications of Green Solvents I   Materials Research Forum LLC
Materials Research Foundations 50 (2019) 165-241   doi: https://doi.org/10.21741/9781644900239-7

Chapter 7

Green Solvents in Chemical Reactions

Manviri Rani[1] and Uma Shanker[2]

[1]Department of Chemistry, Malaviya National Institute of Technology Jaipur, Rajasthan-302017-INDIA

[2]Department of Chemistry, Dr. B R Ambedkar National Institute of Technology Jalandhar-144011, Punjab, INDIA

manviri.chy@mnit.ac.in; shankeru@nitj.ac.in

Abstract

Due to environmental toxicity, harmful impact and prolonged exposure of traditional solvents on all living organisms, the trend of using green technology in scientific work has been in progress. Reducing the use of solvents or replacing them with less toxic/green ones, are two of the most important ambitions of green chemistry. Water, ionic liquids (imidazolium-based), supercritical fluids, deep eutectic solvents or bio-based solvents, non-toxic liquid polymers and their varied combinations have been extensively used as part of the class of green solvents in organic synthesis. They are characterized by low toxicity, convenient accessibility, and the possibility of reuse as well as great efficiency. Moreover, green organic solvents have been used in analytical extraction and chromatographic separation processes. The use of natural ingredients to synthesize nanomaterials and design environmentally benign synthetic processes has been extensively explored.

Keywords

Green Solvents, Chemical Reaction, Water, Ionic Liquids, Organic Synthesis, Nanomaterials, Analytical Studies

Contents

1.   Introduction

A huge amount of volatile, flammable and poisonous organic solvents have been used in reaction systems, separation steps, and commercial applications. Nowadays, worldwide approximately 15 billion kilograms of the annual production of organic and halogenated

Materials Research Forum LLC
doi: https://doi.org/10.21741/9781644900239-7

solvents by the majority of manufacturing industries is reported. Pharmaceutical companies use more than 80% of those in manufactures of the active ingredient. These horrendous numbers finally raised international concern in the 1960s and lead to the U.S. Pollution Prevention Act in 1990. The environmental concern of those solvents arises mainly depends on three factors: synthesis and precursor of the solvent itself; accidental discharge of solvent; and finally its disposal. Regardless of restrained, restriction being manageable in laboratory research, the environmental unfeasibility of organic solvents would limit their commercial applications in view of their toxicity and recovering. The use of volatile organic compounds and their effect on the environment recognized to re-examine various processes by the Montreal Protocol [1].

Annually, more than 20 million tons of waste residues from organic solvents are discharged to the environment, producing undesired waste of solvents and contaminating the environment [2]. In spite of the established fact that solvents, like benzene, toluene and chlorinated solvents dimethylsulfoxide, dimethylformamide, acetone, contributes to environmental pollution, but still these solvents continue to be used in bulk quantities [3]. Solvents have a detrimental effect on prolonged exposure to all systems present in living organisms, mainly damaging respiratory and nervous systems [3-4]. Furthermore, the use of harmful solvents is toxic to organs, e.g. CCl_4 and $CHCl_3$ are hepatotoxic [5-6]. Use of glycol ethers and chlorinated solvents may cause kidney failure [5]. Halogenated hydrocarbons, petroleum distillates, and diethylene glycol may cause renal tubular necrosis, even after a short period of time [7]. Long term exposure to environmental pollutants causes about one-fourth of current diseases as per data available by WHO. The toxic levels of various pollutants are increasing day by day due to the discharge of synthetic chemicals or an accumulation of natural chemicals. The number of wildlife is decreasing due to increasing levels of pollutants, damage the ecosystem and possess a threat to human health [5].

All the foregoing discussions create prodigious interest among the researchers from academia and industries to implement green chemistry in order to make a more environmentally friendly alternative to hazardous petrochemical solvents."Green Chemistry" was developed as regulations on pollution prevention to diminish their discharges and waste. The concept of twelve principles of green chemistry seems regulation for environmentally benign solvent chemistry. The fifth principle "The use of auxiliary substances (i.e. solvents) should be made unnecessary whenever possible and, when used, innocuous" is directly refers to solvents selection [5]. The use of green solvent or replacing them with less hazardous one should be preferred to protect the environment is the main aim of green chemistry (Fig.1) [8].

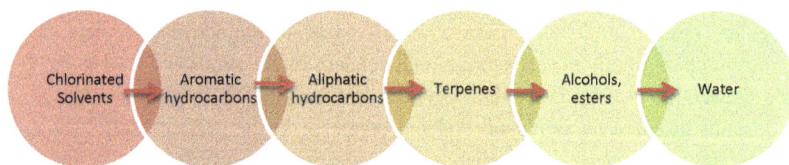

Recommended	Usable	Not recommended
Water	Toulene	Dimethylformamide
Isopropyl alcohol	IsoparG	Dioxane
Acetone	Heptane	Chloroform
Ionic liquids	Dimethylsulfoxide	Hexane
Supercritical CO_2	Tetrahydrofuran	Dimethylacetamide
Perfluorinated liquids	Acetonitrile	N-methylpyrrolidinone
Organic Carbonates		
Biosolvents		
Polyethylene glycol solvents		

Fig 1. Depiction of green chemistry involving role of green solvents

A greener environment can be obtained by solvent reduction or recycling (by using "closed-loop systems") and switching to solvent-free processes (e.g., dry powder coatings and solid ultraviolet curable coatings, polycarbonate production using a melt phase polymerization technology). However, many reactions need solvent or liquid of some kind that can be sorted out by the use of alternatives.

Green technologies are designing new, environmentally-friendly or biosolvents (derived from agricultural crops) and tunable solvents for industrial and economic needs. Since last two decades, a extensive literature of solvents used suggests that solvents and solvent classes that have been regarded as 'green' solvents include water [9-17], supercritical

Industrial Applications of Green Solvents I   Materials Research Forum LLC
Materials Research Foundations **50** (2019) 165-241   doi: https://doi.org/10.21741/9781644900239-7

fluids [18-20], ionic liquids [21-37], non-toxic liquid polymers [38-42], solvents derived from biomass [43-45] and their diverse combinations. Such solvents are categorized by lesser toxicity, easy approachability and the opportunity of reusability as well as high efficiency or generally called 'environmental, health and safety (EHS) properties [46]. A model green solvent also intercedes reactions, separations or recycling of catalyst [47]. Solvent selection guides (SSGs) developed by pharmaceutical industry companies [48]. Reichardt 1988; assessed each solvent with a labeled of either "recommended", "problematic" and "hazardous" to "highly hazardous. The outcome and process efficiency of the reaction greatly depends on the choice of solvents and green solvents can improve the chemical processes by decreasing the number of steps of the process [5, 8, 49,50].

Several authors have mentioned the importance of green solvents in several applications [51-53]. Hackl and Kunz [52] reviewed green solvents in organic synthesis. Several green solvent alternatives to traditional solvents used are 2-methyltetrahydrofuran, ethyl lactate, cyclopentyl methyl ether, limonene, and p-cymene, as well as solvent-free organic synthesis. Gu [54] summarized multicomponent reactions (Knoevenagel condensation, activation of carbonyl group with water, imine formation, synthesis of dithiocarbamate, Isocyanide, transition metal catalysis etc) in traditional solvents like polyethylene glycol, ionic liquids, water, and bio-based solvents. Green Solvents (supercritical carbon dioxide, fluorous solvents, water, ionic liquids, organic carbonates, and biosolvents) in organic synthesis for combating environmental damages due to the use of toxic solvents in organic chemistry. Tobiszewski and Namiesnik [55] reported the analytical applications of ionic liquids and supercritical fluids which are greener organic solvents. They suggested the use of ethanol or acetone (greener mobile phases) instead of acetonitrile in liquid chromatography and bio-organic solvents like alcohols, esters, or terpenes in solvent-based extractions. Importance of greener solvents like ionic liquids in the synthesis of various inorganic nanomaterials synthesis has been reported [56]. Lu and Ozcan [57] suggested ways to improve commercial readiness and sustainable future with current advances and challenges in green nanotechnology. In this chapter, we have summarized updated information on the use of environmentally benign solvents in the synthesis of organic compounds and inorganic nanomaterials.

2.   Types of green solvents used

2.1   Water

Water being a protic and polar solvent, safe, noninflammable, and abundance is the most favorable solvent ever found [58,59]. Water as greener solvent exhibits fascinating aspects concerning to reactivity: rare selectivity's, effect of hydrogen-bonding network on reaction behavior, application of biphase reaction systems tunable pH values, and the use of salts for in/out salting-out effect. Due to reactivity and selectivity, it is possible to recover and recycle the catalyst via phase extraction in organometallic catalysis [53]. It includes not only liquid (ionic liquid), but also supercritical (fluids e.g. water, CO_2) and on-water [60,61].

Pioneering works include the Ruhrchemie/Rhône-Poulenc process, hydroformylation of propylene using a rhodium complex catalyst [62] as well as other hydrogenations and hydroformylations with water-soluble organometallic complexes in aqueous biphasic systems [63,64].

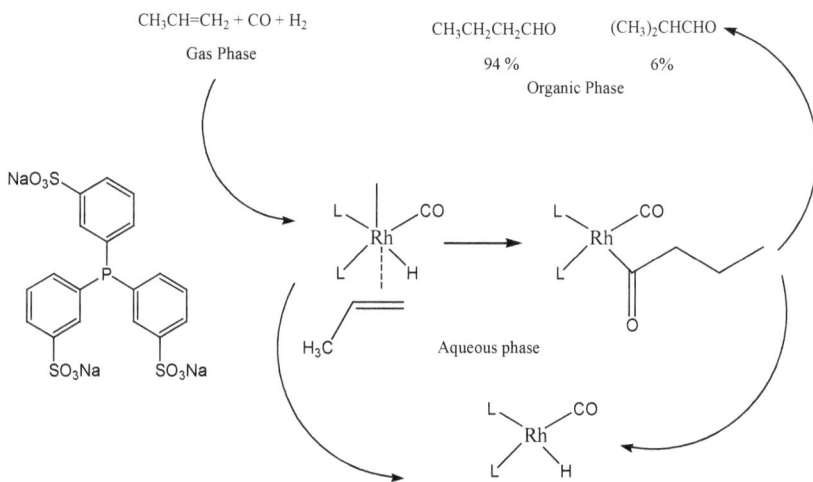

Fig 2. Rhone-Poulene process for aqueous biphasic hydroformylation

Next, water-based palladium catalyzed carbonylations were studied, where the carbonylation catalyst was formed in situ from a rhodium-tppts complex or even faster, with a Pd(II) salt [65]. Of medical interest was the synthesis of ibuprofen via carbonylation in an aqueous biphasic reaction system [66] (Fig. 3). The recycling and recovering of the catalyst can be achieved by phase extraction and additionally, the aqueous bottom layer can be used for several reaction processes. As an oxidizing agent, oxygen from the surrounding air is sufficient which makes the process even greener [67].

Industrial Applications of Green Solvents I   Materials Research Forum LLC
Materials Research Foundations **50** (2019) 165-241   doi: https://doi.org/10.21741/9781644900239-7

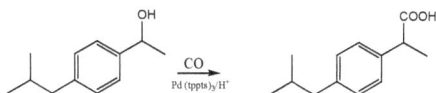

Fig 3. Synthesis of ibuprofen via carbonylation in an aqueous biphasic reaction system

Examples for the use of biocatalysis are the production of 6-amino penicillanic acid (APA, Fig 4) and cephalosporines based on penicillin G **1** as a starting material [68], as well as the synthesis of aspartame which is also based on highly selective enzymatic biocatalysts [69].

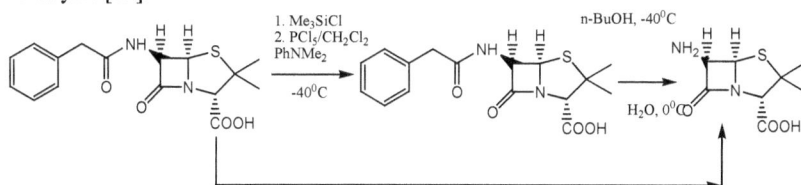

Fig 4. Enzymatic versus chemical deacylation of pencillin G

Glycolic acid, which is needed in large amounts in different chemical processes, can also be biocatalyzed via oxidase and dioxygen, generating hydrogen peroxide in situ as an oxidant Fig 5 [70].

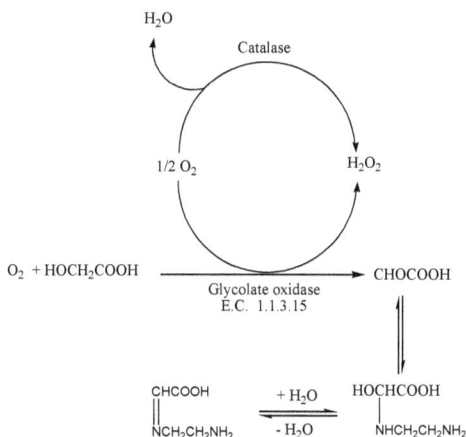

Fig 5. Biocatalysis of Glycolic acid

2.1.1 MCRs in water based on Knoevenagel condensation

The base-catalyzed condensation reaction of 1,3-dicarbonyl compounds with aldehydes is compatible using water as solvent [71,72]. Heterogeneous catalysts developed in multicomponent reactions were found water-compatible and also able to keep a high activity during recycling. Gu et al. [73] have reported the water-mediated good yields synthesis of 2,5,6-trisubstituted dihydropyrans from 1,3-dicarbonyl compounds, formaldehyde, and α-methyl styrene [74-77] (Fig 6). 1,3-dicarbonyl compound and formaldehyde produced methylene intermediate which can then be trapped by α-methyl styrene by oxo Diels–Alder reaction. In this reaction, water helps in the generation of methylene intermediate and enhances the speed of reaction [78].

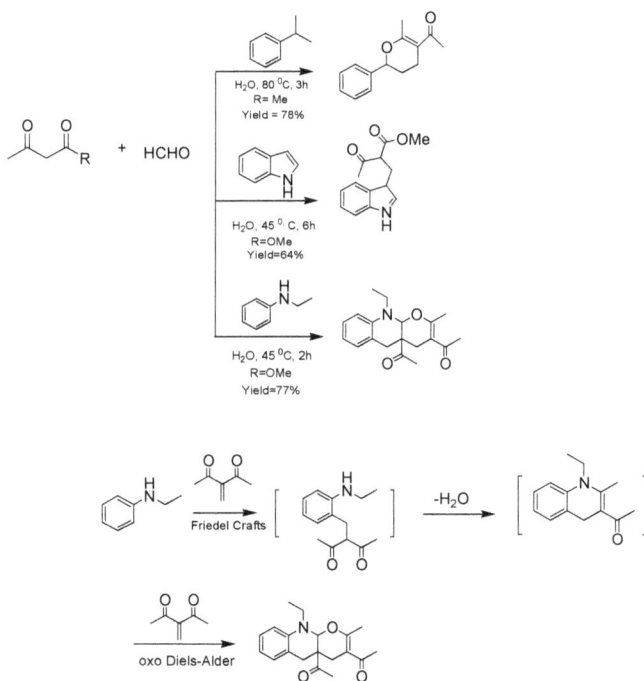

Scheme 1. MCRs of beta-dicarbonyl compound and formaldehyde in water

Fig 6. Reactions of β-carbonyl compound and formaldehyde in water

However, reaction faced many limitations such as involvement of many side reactions like formaldehyde with trapping reagent; methylene intermediate and 1,3-dicarbonyl compounds or C–H acids. In order to solve this, methylene intermediate was generated by oxidation of Baylis-Hillman alcohol with iodoxybenzoic acid [73]. This will be conducted in water to allow easy trapping of the methylene intermediate with several nucleophiles.

Kumar and Maurya [79] reported boric acid-catalyzed reaction of formaldehyde, 1,3-dicarbonyl compound, and N, N-dialkyl aniline, in aqueous micelles made by SDS and water (Fig 7).

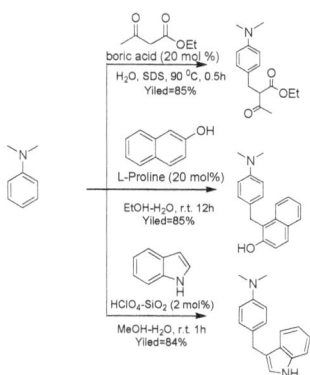

Fig 7. Reaction of N,N-dimethylaniline and formaldehyde in water

The condensation reaction of an aldehyde with two different nucleophiles is usually difficult to control. An efficient three-component reaction of malononitrile, indole, and aldehyde, using a copper(II) sulfonato Salen complex as a catalyst in water solvent yielded 3-indole derivative in high yield (Fig 8). Reaction selectivity and controlling the pH of the aqueous solution was improved by KH_2PO_4.

Importance of water was shown by developing a three-component reaction of 3-nitrobenzaldehyde N, N-diethyl barbituric acid, and NaCN in water. The important outcome of this reaction was that poor yields were found with the other solvents, like toluene and ethanol under similar reaction conditions. A similar observation has also been reported by Li et al. [80] on the good yield of β-mercapto diketone derivative 11 in water and poor with organic solvents. The products in the form of liquid crude were separated from water by extraction method. Sometimes catalysts are recyclable in aqueous conditions.

Industrial Applications of Green Solvents I   Materials Research Forum LLC
Materials Research Foundations **50** (2019) 165-241   doi: https://doi.org/10.21741/9781644900239-7

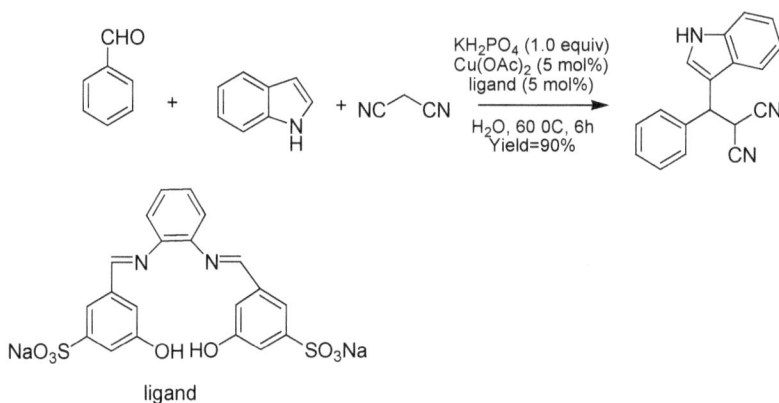

Fig 8. Three component reaction of indole, aldehyde and mononitrile in water

Solhy et al. [81] described an environmentally benign synthesis of 2-amino-chromene through 1-naphthol, malononitrile, and aldehyde in water (**Fig 9**). Addition of varying volumes of the yield of the reaction was increased considerably.

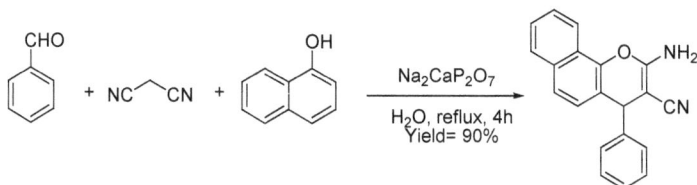

Fig 9.Three component reaction of 1-naphthol, benzaldehyde and mononitrile in water

Et₃N mediated 3-component reaction of 5-aminopyrazoles, aromatic aldehydes, and dimedone in water under microwave irradiation, yielding pyrazolo[3,4-b]quinolin-5-one derivatives in good to high yields. Liu et al. [82] carried out the synthesis of 3-methyl-4-arylmethylene- isoxazol-5(4H)-one 18 in H$_2$O in the presence of sodium benzoate as a low-cost and green catalyst. High yield of the product observed with H$_2$O while with other solvents or under solvent-free conditions sharp decline in yields seen. Hydrophobic

Industrial Applications of Green Solvents I   Materials Research Forum LLC
Materials Research Foundations **50** (2019) 165-241   doi: https://doi.org/10.21741/9781644900239-7

interactions and basicity of sodium benzoates enhance the rate of the reaction and purity of the product in water. The similar reaction could also be accomplished in H_2O in the presence of sodium silicate or pyridine with or without ultrasonic irradiation [83-84] (Fig 10).

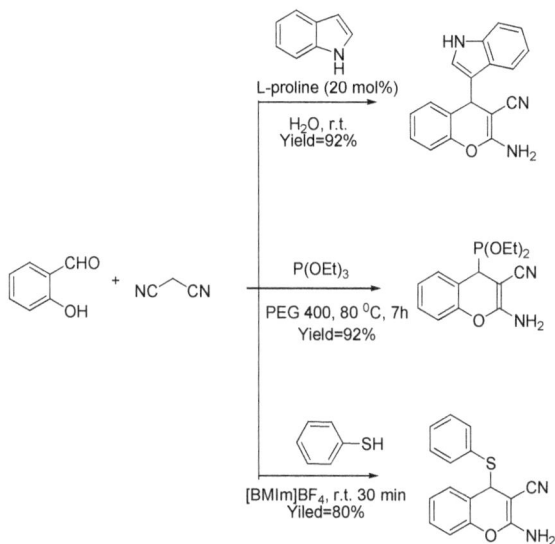

Fig 10. Three component reaction of salicyaldehyde and malononitrile

Kumaravel and Vasuki [85] reported the great power of water as a solvent for the creation of molecular complexity and diversity by catalyst-free synthesis of highly functionalized 4-pyrazolyl- 4H-chromene in water at ambient temperature (Fig 11). A series of 2-aryl-5-methyl-2,3- dihydro-1H-3-pyrazolones have been synthesized in the presence of p-toluene sulphonic acid in water in good yields [86].

Industrial Applications of Green Solvents I   Materials Research Forum LLC
Materials Research Foundations **50** (2019) 165-241   doi: https://doi.org/10.21741/9781644900239-7

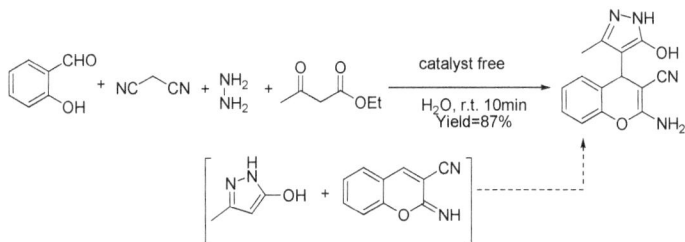

Fig 11. Three component reaction of hydrazine hydrate, ethyl acetoacetate, 2-hydroxybenzaldehyde and molononitrile

Several 2-amino-3,5- dicarbonitrile-6-thio-pyridines were synthesized in considerable yields by using boric acid [87] or basic alumina [88] as a catalyst in water. On same method, 2-amino-4-aryl(alkyl)-6-sulfanyl pyridine-3,5-dicarbonitrile was synthesized [89]. Here, the clean product was easily achieved by simple filtration followed by recrystallization with CH$_3$CN or C$_2$H$_5$OH. Similarly, Zhou et al. [90] synthesized a 3-amino-1-aryl-8-bromo-2,4-dicyano-9H fluorene derivative in aqueous media under microwave irradiation (MWI) by using sodium hydroxide as a base (Fig 12).

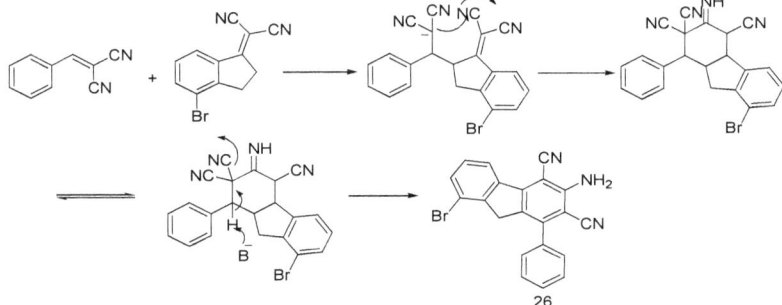

Fig 12. Possible reaction mechanism for synthesizing 3-amino-1-aryl-8-bromo-2,4-dicyano-9Hfluorene derivative in aqueous media

Mukhopadhyay et al. [91] showed the synthesis of 1,2-dihydro[1,6] naphthyridine via reaction of methyl ketone, amine, and malononitrile under catalyst-free conditions. Lu et al. [92] reported good yields of dihydrothiophene ureido formamides in water and avoided the use of organic amine catalyst used in conventional systems. The four-component reaction of an aromatic aldehyde, dimethyl acetylenedicarboxylate, benzylamine, and malononitrile undergoes easily at ambient conditions in H_2O in the presence of catalytic amount (20%) of $(NH_4)_2HPO_4$ (Fig 13) [93].

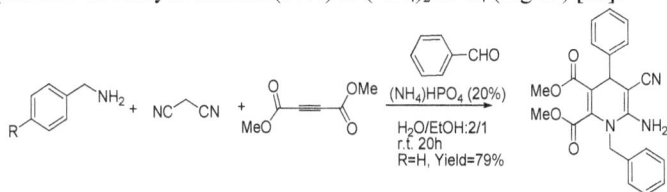

Fig 13. Four component reaction of dimethyl acetylenedicarboxylate, benzylamine and malonitrile in water

L-proline was used as a catalyst in the four-component sequential reaction of phenylhydrazine, 3-aminocrotononitrile, aromatic aldehydes, and 2-hydroxynaphthalene-1,4-dione in aqueous media [94]. In such reactions, water was established as better solvent as compared several other organic solvents due to fact that all precursor materials were insoluble in aqueous media. Low yield, 45% after 10 h of the reaction was reported in water without any catalyst. Zou et al. [95] reported the synthesis of dihydropyrano[2,3-c]-pyrazoles under sonicator in aqueous media in the presence of a catalyst (Fig 14). The output showed decent to brilliant antibacterial activity.

Chai et al. [96] observed that water (10–20%) was essential in biocatalysis reaction for high yield. Example, excellent catalytic activity of lipase from porcine pancreas was found in aqueous media. In general, reaction depends on catalysis with enzyme while expensive enzyme catalyst limits their use.

Spirooxindoles derivatives in decent yields under water micellar media were generated using sodium stearate as a catalyst [97]. Sodium stearate also exhibits the supportive role as a surfactant in such reactions, and hence it may behave as a base to activate the substrate molecules and a surfactant to form stable colloidal dispersion with water-insoluble matrices.

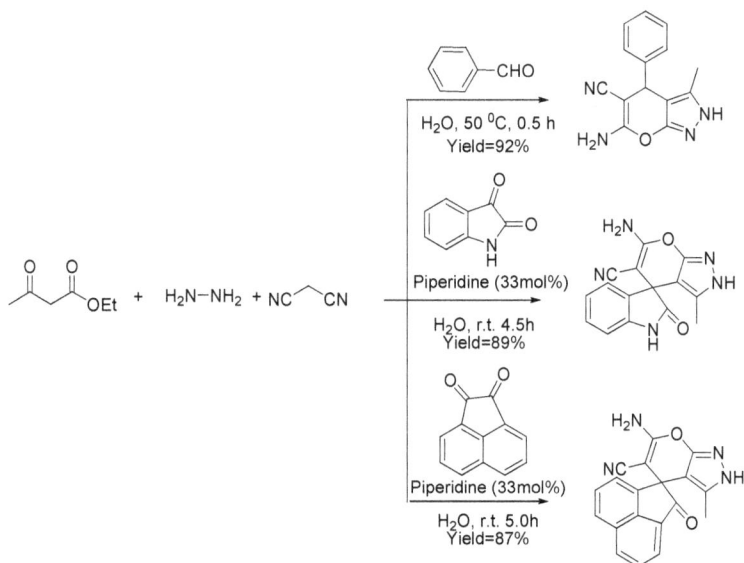

Fig 14. Four component reaction of hydrazine, ethylacetate and malononitrile in water

Li et al. [98] reported L-proline catalyzed the synthesis of spirooxindole derivatives in an aqueous medium. Since L-proline is soluble in the reaction medium and the desired products are less soluble in water, the products can be directly separated by cooling to room temperature and filtering after the reaction was completed.

A three-component reaction of 3-hydroxy-1H-phenalen-1-one, isatin, malononitrile and also proceeded well-using p-TSA as a catalyst in aqueous media [99]. Spirooxindole derivatives were synthesized in H_2O using ZnS nanoparticles under ultrasonic irradiation [100]. The reusability of ZnS NPs endowing an important characteristic of green chemistry to this system.

2.2.2 MCRs based on activation of carbonyl group with water

Water as a solvent can activate carbonyl group electrophilically of aldehyde through H_2 bond interaction. Numerous researchers have reported good yield and feasibility of reaction using H_2O as solvent e.g. one-pot Hantzsch condensation (Fig 15) of an aromatic aldehyde, ammonium acetate, and dimedonein without the use of a catalyst. Rimaz and

Industrial Applications of Green Solvents I   Materials Research Forum LLC
Materials Research Foundations **50** (2019) 165-241   doi: https://doi.org/10.21741/9781644900239-7

Khalafay [101] synthesized alkyl 6-aryl-3-methylpyridazine-4-carboxylate in water avoiding the use of an acidic condition or metallic catalysts.

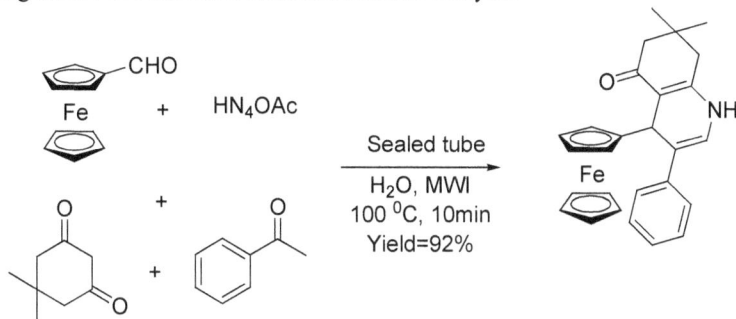

Fig 15.Four component reaction of ferrocenecarboxyladehyde, ketone, dimedone and ammonium acetate in water.

2, 3-dihydroquinazolin-4(1H)-ones were synthesized in aqueous ethanol using $SrCl_2 \cdot 6H_2O$, Fe_3O_4 nanoparticles, and ethylenediamine diacetate as a catalyst [102-104]. It was observed that when 2-aminobenzothiazole was used as substrate instead of the 1^0 amines, the three component reaction progressed very well in ethanol in the absence of a catalyst [105]. Recently, several spiro [indole-thiazolidinone] compounds were synthesized under ultrasonication using CTAB as catalyst (Fig 16).

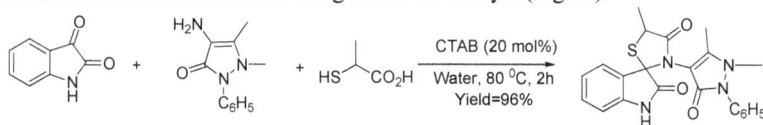

Fig 16 Three component reaction of isatin, 4 -amino antipyrine and 2-mercaptopropionic acid in water.

Spirooxindolepyrazoline derivatives synthesized in good yields in a pseudo-five-component reaction of isatin, 1,1-bis-(methylthio)-2-nitroethylene and hydrazine hydrate under catalyst-free conditions in water [106]. A three-component reaction of N-methylimidazole, methyl propiolate, and butyraldehyde was demonstrated to proceed well in aqueous water yielding products with good yields [107].

Industrial Applications of Green Solvents I   Materials Research Forum LLC
Materials Research Foundations **50** (2019) 165-241   doi: https://doi.org/10.21741/9781644900239-7

2.2.3 MCRs in water based on imine formation

Based on observations of Tanaka and Shiraishi [108], for facile synthesis of imine from aldehyde and aniline in water, Mannich-type three-component reaction has been comprehensively examined in H_2O [109]. Some homogeneous acids have also been utilized for the same reactions [110-113] though; difficulty in reusability of catalyst limits its application in useful synthesis. Hong et al. [114] joined $Yb(OTf)_3$ and a PEG-supported quaternary ammonium salt, for utilization of Mannich reaction in aqueous media. Stereoselective synthesis of 100% anti-β-amino ketones in water at room temperature in excellent yield has recently synthesized using polyaniline doped (PANI–$AgNO_3$–p-TSA) reusable solid acid catalyst. Furthermore, it was concluded that water as solvent provides high yields as compared to the other solvents systems [115-116]. Various α-amino phosphonates synthesized in high yields using potential and cheap catalyst like magnesium dodecyl sulfate (10 mol%), Kabacknik–Fields reaction of aldehydes, amines, and triethyl or diphenyl phosphate in H_2O [117]. The similar reaction may also be performed in aqueous media using K_2PdCl_4 as a catalyst [118] and with SO_3H-functionalized ionic liquid that generates α-aminonitriles in good yields with reusability of catalyst [119-121].

One-pot, facile and direct route for diastereoselective synthesis piperidine derivatives performed in aqueous media in the presence of catalytic amount of $Bi(NO_3)_3$ at room temperature [122]. 1,2,3,6-tetrahydropyrimidine, 1,3,4,5-tetrasubstituted and 1,3,3-trisubstituted 4,5-dioxopyrrolidine were formed in good to high yields using water as solvent at room temperature in dilute HCl with the addition of indium (Fig 17) [123].

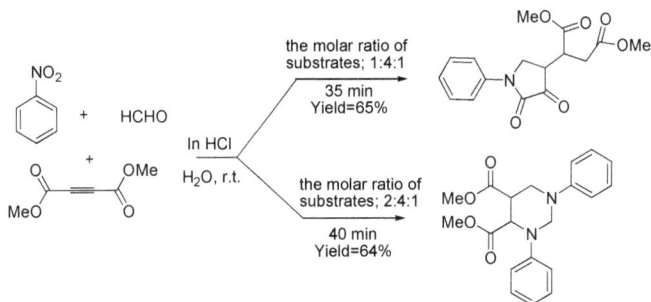

Fig17. Three component reactions of nitroarenes, formaldehyde and acetylenedicarboxylate in water

García-Tellado et al. [124] established a Strecker reaction of acetyl cyanide, aniline and ketones in water showed an added advantage in evolving environmentally benevolent systems (Fig 18).

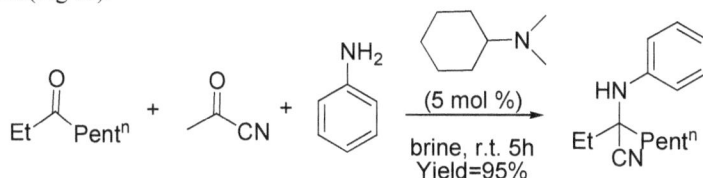

Fig 18. Strecker reaction of ketones, aniline and acetyl cyanide in water

Galletti et al. [125] developed a simple and facile scheme for the synthesis of α-amino nitriles via a one-pot, three-component reaction of an amine, the carbonyl compound and acetone cyanohydrin in aqueous media has been established owing to the immiscibility of the substrates and products with H_2O.

Mathew and M. Nath [126] reported a three-component reaction of phenols, formaldehyde, and amines in water in absence of any catalyst, yielding product in a 30 min to 1 h at 25 °C in water with high yields. It is important to mention here that the same reaction takes several hours to days to complete with other organic solvents. Zanardi et al. [127] reported Mukaiyama–Mannich type reaction of a pyrrole ketene silyl-N, anilines, O -acetal and various aldehydes in water and solvent-free conditions. Excellent yields, with γ-site selectivity and chemo-selectivity, and moderate to high diastereoselectivity in favor of anti-configured adducts have been observed in this reaction. It is noteworthy that one-pot synthesis in ultrasonic condition was found more effective than the traditional two-step method [128]. Water was found to be more effective than other organic solvents, like dichloromethane, ethanol, 2-propanol, acetonitrile under similar conditions.

Overall, it was concluded that most of the three component reaction performed under aqueous media and in the absence of a catalyst and resulted in a high yield of products.

2.2.4 Synthesis of dithiocarbamate through MCRs in water

Alizadeh and Zohreh [106] reported catalyst-free three-component reaction of fumarylchloride, carbon disulfide and benzylamine in water at room temperature. The main function of water was in the concluding step of hydrolysis of intermediate leads to the excellent yields of product in 12 h. Excellent yields of alkyl dithiocarbamates was obtained with one-pot reaction of primary and secondary amines and carbon disulfide with alkyl vinyl ethers in pure water under catalyst-free conditions at room temperature (Fig 19) [129-130]. Although the good performance of water as a solvent in this reaction was established, but cannot be ascribed to an accelerating effect of water on the reaction since the other solvents, like THF and ethanol, may also produce the desired products with similar yield.

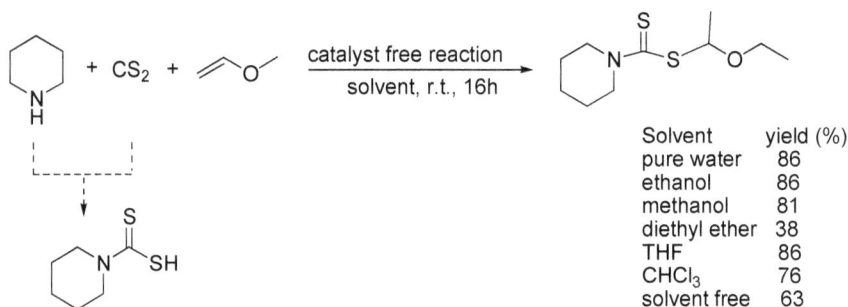

Solvent	yield (%)
pure water	86
ethanol	86
methanol	81
diethyl ether	38
THF	86
CHCl$_3$	76
solvent free	63

Fig 19. Three component reaction of amine, CS$_2$ and alkyl vinyl ether in water

Ranu et al. [131] synthesized S-aryl dithiocarbamates, by a one-pot condensation of aryl diazonium fluoroborate, carbon disulfide, and amine in the catalyst-free conditions in aqueous media at room temperature (Fig 20). This method has benefits of using non-toxic phenyl diazonium tetrafluoroborates and water in the propagation of the green chemistry requirements of current organic synthesis.

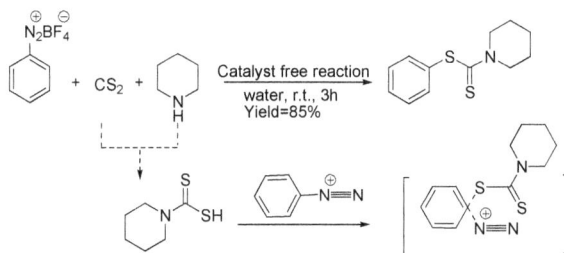

Fig 20. One pot condensation of aryl diazonium fluoroborate, CS2 and amine

2.2.5 Isocyanide-based MCRs in water

Isocyanide is an extraordinary functional group in Passerini and Ugi reactions. Santra and Andreana [132] realized a novel Ugi-type four-component reaction of p-hydroxybenzaldehyde, benzylamine, fumaric acid monoethyl ester, and tert-butyl isocyanide in water under microwave irradiation, yields to products similar to natural-product such as 5,5,6-fused azaspiro tricycle. It was concluded that water as a solvent has the excessive capacity for generating molecular complexity. An effective methodology for the development of ketenimine sulfonamides derivatives in the presence of alkyl or aryl sulphonamides in aqueous media under catalyst-free conditions [133] (Fig 21). Essentially, the products may finally get hydrolyzed to the sulfonamide-butanamide derivatives at 80 °C in water with good yields under catalyst-free conditions. In the same pattern, the zwitterion resulted from the reaction of an alkyl isocyanide and a dialkyl acetylenedicarboxylate, reacts freely with phenacyl halides in water to yield γ-iminolactone derivatives with high yields [134]. 3,4-dihydrocoumarin derivatives in decent to high yields under catalyst-free conditions or activation by simple one-pot protocol at room temperature[133].

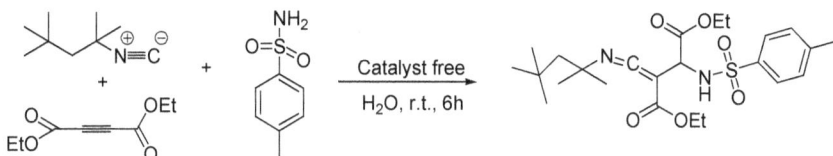

Fig 21. Three componenet reaction of isocyanides, diethyl acetylenedicarboxulate and alkyl or aryle sulfonamide in H_2O

A three-component reaction of an isocyanide, dialkyl acetylenedicarboxylate, and 4-hydroxycoumarin in water at 80 °C, in the presence of a phase-transfer catalyst, tetrabutylammonium bromide, yielded a pyrano[3,2-c]-coumarin with good yields [135]. An effective three-component reaction of isocyanides, heterocyclic thiols and gem-dicyano olefins having electron-withdrawing groups in an aqueous solution of acetonitrile, produced imino-pyrrolidine-thione scaffoldin high yields have been reported [136] (**Fig 22**).

Passerini reaction organic solvents reported the formation of the product with a 50% yield after 18 h, while the use of water yielded the product 25 quantitatively within 3.5 h

Industrial Applications of Green Solvents I Materials Research Forum LLC
Materials Research Foundations **50** (2019) 165-241 doi: https://doi.org/10.21741/9781644900239-7

on. The products of Ugi- Passerini reaction in water could be collected by filtration as the only purification [137].

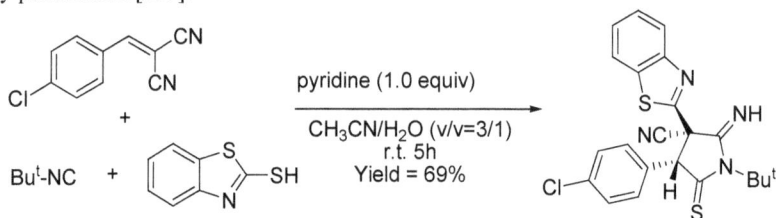

Fig 22. Three-component reaction of isocyanides, heterocyclic thiols and gem-dicyano olefins in water.

3.   MCRs in water based on transitional metal catalysis

Carrying out a transitional-metal-catalyzed reaction in aqueous offers an easy approach to expand the efficacy and user-friendliness of this process [138] to utilize a biphasic system, in such a way that the homogeneous catalyst is soluble in one phase (e.g. water) and the other in the organic phase [139]. Simple phase separation of the product can be obtained. With such an approach, manufacturing applications have been rapidly recognized, like the Ruhrchemie AG/Rhone-Poulenc process for the hydroformylation of propylene. Reusable design of homogeneous metal complex catalysts in aqueous has received great attention. An ammonium salt-tagged [(SIPr)CuCl] complex have recently developed, showed a high catalytic potential for the extensive arrange of benzyl bromides and alkynes in water at room temperature [140]. Since the catalyst is water soluble, therefore, it could be easily reusable up to three times without a substantial loss of its activity with catalyst loading as low as 2.0 mol% in the new run. High yield of 1,4-disubstituted 1,2,3-triazole in water using structurally well-defined efficient catalyst copper (I) isonitrile complex was reported [141]. Liu and Xia [142] synthesized three solid Cu catalysts to be used as effective catalysts for the one-pot synthesis of 1,4-disubstituted-1,2,3-triazoles by the reaction of alkyl halides with sodium azide and terminal alkynes in water at room temperature. Due to the high proton-providing ability of water in catalyst systems, water was demonstrated to be a more appropriate solvent than others.

The nano-ferrite-supported copper catalyst was synthesized under ultrasonication of ferrites nanoparticles with glutathione in H_2O for 2 h, after that the $CuCl_2$ in alkaline aqueous conditions was added [143]. Under microwave irradiation, the catalyst was

Industrial Applications of Green Solvents I   Materials Research Forum LLC
Materials Research Foundations **50** (2019) 165-241   doi: https://doi.org/10.21741/9781644900239-7

reusable (n=3) and efficiently catalyze Huisgen 1,3-dipolar cycloaddition reactions in aqueous media.

A novel mixed catalyst of oxidized copper nanoparticles on activated carbon (CuNPs/C) for the multicomponent Huisgen 1,3-dipolar cycloaddition in water was developed and found that the catalyst was recyclable and apparently activates under heterogeneous conditions (Fig 23) [144].

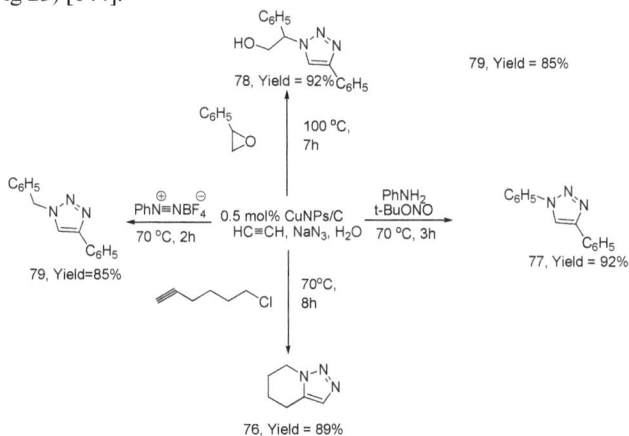

Fig 23.Scheme 44 Multicomponent Huisgen 1,3-dipolar cycloaddition in water catalyzed by CuNPs/C

Since azetidinone may react freely with alkyl azide in water [145]. MCR of azetidinone, sodium azide, and terminal alkyne goes well in water using various Cu based catalysts. With a yield of 95% copper iodide was found the best choice of catalyst. It was concluded that water was the perfect solvent capable of supporting a Cu(I) acetylide species formed during the reaction. Using one-pot pseudo-five-component reaction between numerous azides, dimedone, aromatic propargylated aldehydes, and anilines a number of triazolyl methoxyphenyl 1,8-dioxo-decahydroacridine derivatives were synthesized in of ethanol [146]. A mixture of Cu(OAc)$_2$/sodium ascorbate and 1-methylimidazolium trifluoroacetate was utilized as a catalyst in order to get high yield.

Palladium supported catalyst (Pd-CPSIL) have been found as an efficient heterogeneous catalyst for the generation of several α,β-alkynyl ketones in good to high yields by aryl

Materials Research Forum LLC
doi: https://doi.org/10.21741/9781644900239-7

iodides with terminal alkynes in aqueous media (Fig 24). Use of such catalytic system evades the use of noxious phosphine ligands and also answered the recovery and reuse of pd based catalysts which were a common issue with such types of reactions.

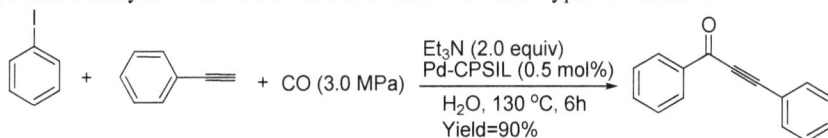

Fig 24.Carbonylative Sonogashira coupling reaction of aryl iodides with terminal alkynes in water

Synthesis of N-substituted pyrrolo[2,3-b]quinoxalines in water using one-pot Pd/Cu-catalyzed, the four-component reaction of 1,2- dichloroquinoxaline, hydrazine hydrate, phenylacetylene, and aromatic aldehydes has been reported (Fig 25).

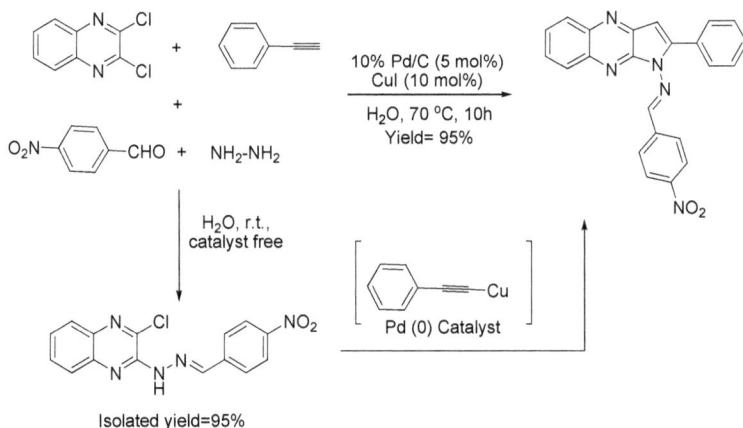

Fig 25.Pd/Cu catalysed four component reaction of 1,2 dichloroquinoxaline, hydrazine hydrate, phenylacetylene and aromatic aldehyde in water

Industrial Applications of Green Solvents I   Materials Research Forum LLC
Materials Research Foundations **50** (2019) 165-241   doi: https://doi.org/10.21741/9781644900239-7

One-pot gold mediated t-coupling reaction in water obtained high aldehyde conversion by single-site incorporation of amines and alkynes into aldehyde-containing oligosaccharides [142].

Various nitrogen-containing biologically active compounds and natural products were synthesized using propargylamines as intermediates. Adapa and co-worker [147] established a convenient and efficient route for good to excellent yields production of propargylamines through one pot three-component coupling reaction of an aldehyde (aromatic, hetero or aliphatic), amines, and alkyne using Zn(OAc)$_2$/2H$_2$O zinc salts in a catalytic amount without any use of additives or base.

Kumar et al. [148] synthesized dihydrothiophenes and tacrine derivatives hexahydrothieno[2,3-b]quinoline-2-carboxamide with good to high yields in aqueous media.

4.   Fluorous solvents (perfluorinated liquids)

They are "aqueous" to perfluorinated alkanes, dialkyl ethers, and trialkyl amines (Fig 26) [149-150]. Lack of intermolecular interactions in ethers and amines is credited to their no residual basicity [151]. These have useful and attractive properties (chemical inertness, high thermal stability, nonflammability, an extreme nonpolar character, and small intermolecular attraction) for organic synthesis [152]. Their immiscibility with common organic solvents at room temperature allows the formation of biphasic systems which have been extensively studied in stoichiometric and catalytic transformations [153-155]. It dissolves catalyst by attaching fluorocarbon moieties such as linear or branched perfluoroalkyl chains.

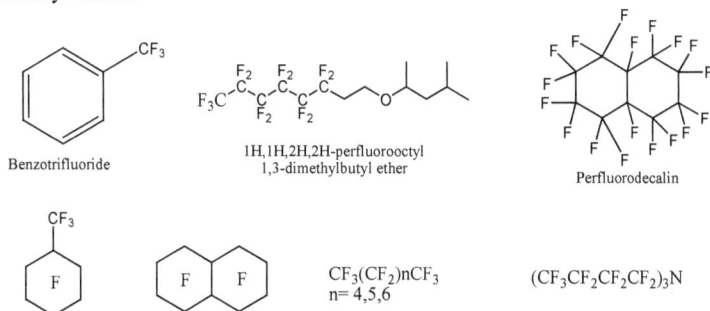

Fig 26. Various fluorous solvents

Industrial Applications of Green Solvents I   Materials Research Forum LLC
Materials Research Foundations **50** (2019) 165-241   doi: https://doi.org/10.21741/9781644900239-7

At higher temperature, fluorous biphasic systems can become miscible forming a homogenous liquid phase reaction media. Thus, the advantages of both, a single-phase media for the reaction and a biphasic system for separation of the products can be exploited [58, 150]. Fluorous solvents are not yet widely used commercially, probably due to drawbacks like rather high cost, environmental persistence, and biological half-lives, especially of C7- and C8-perfluoroalkyl group-containing compounds [58, 156].

Aldehydes or ketones, amines, and trimethylsilyl cyanide or trimethyl phosphate reacted efficiently in recyclable 1,1,1-trifluoroethanol in a one-pot coupling reaction to give the corresponding amino nitriles or α-amino phosphonates with high yields, without the requirement of any acid or base catalyst [157].

Hydroformylation of 1-octene was achieved with Rh-fluorous phosphine $P(C_6H_4-OCH_2C_7F_{15})_3$ as an active catalyst in a fluorous biphasic system. The authors reported high selectivity for aldehydes (up to 99%) and high regioselectivity for the linear product (Fig 27) [158].

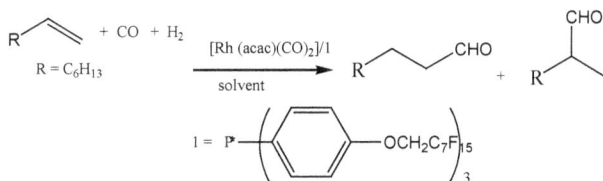

Fig 27. High selectivity for aldehydes (up to 99%) and a high regioselectivity for the linear product

Caballero et al. [159] have developed the efficient functionalization of various linear and branched alkanes by inserting a carbene group from ethyl diazoacetate and a complex catalyst (TpBr3M [Ag; TpBr3=hydrotris (3,4,5-tribromopyrazolyl)borate]) in a fluorous medium. This method provides the advantage that the catalyst can easily be reusable and employed for several runs without significant loss of activity.

5. Supercritical fluids

A number of physical properties, like thermal conductivity, density, and diffusion coefficients are different and variable between ''gas-like'' and ''liquid-like'' states through an easy change of pressure and/or temperature at supercritical state [160]. The synthesis of nanoparticles could be wisely influenced in terms of size, morphology, structure, composition and architecture with the change in operating parameters, like reaction solvent type, temperature, time, and concentration of reagent.

5.1   Supercritical water (scWater)

Supercritical water behaves like a nonpolar solvent and exists at pressures above 221 bar and temperatures above 374 °C [161]. The nonpolarity is caused by the loss of hydrogen bonding, hence salts are no more soluble in scWater, while O_2 and nonpolar solvents are soluble. Therefore, oxidation reactions in scWater have been developed and studied since the 1970s, mainly for the disposal of organic waste. Unfortunately, there has not been a solution for the occurring corrosion problems due to halogen traces yet, leading to the failure of commercialization of this method [60]. Sue et al. [163] established that the variation of dielectric constant is the main factor on which solubility of materials depends. By tuning the dielectric constant, the size of the nanoparticles can be controllable (narrow) in supercritical hydrothermal synthesis [162-163]. Complexation between the organics and the surface of nanoparticles facilitated due to dispersity of nanoparticles in various solvents.

5.2   Supercritical carbon dioxide (scCO$_2$)

Supercritical carbon dioxide exists at a pressure of 74 bar and temperature of 304 K, Fig. 11a) is extra fever than those of water (221 bar and 646 K) owing to excellent properties like nonflammable, renewable, safe, readily evaporating and chemically inert towards many substances [58]. It is featuring gas-like viscosities and liquid-like densities but with outstanding solvent wetting stuff. It was concluded that ''coffee rings'' formed by evaporation in many solvents could be reduced using scCO2 to generate ordered lattices of nanoparticles [160]. Changing either the temperature or the pressure leads to dramatic changes in its viscosity, density, and dielectric properties. This makes supercritical carbon dioxide an unusually tunable, environmentally friendly, versatile and selective solvent [60]. These characteristics, in addition, to easy recovery and reusability via depressurization, leads to significant reduction of energy responses related to industrial production [161].

Reactions in scCO2

The synthesis of nanoparticles could be controllable by a change in the temperature and pressure because the values of dielectric constant, density and viscosity are significantly sensitive to the variation of temperature and pressure [164]. Pai and co-workers [165] successfully altered the solubility of reactants as well as the catalysts by modifying $scCO_2$ pressure in the synthesis of mesoporous metal oxides using micelles as the template of the block copolymer. The simplification of this methodology lies in the variability of $scCO_2$ and the flexibility of morphology of the templates. The potential of supercritical CO_2 as an environmentally preferable solvent is now being realized in several areas. One of the main use of $scCO_2$ is in the food and nutrition industry, where it aids as an extraction medium for decaffeination processes for coffee beans and tea, thus effectively substituting earlier used chlorinated organic solvents. The other main applications of $scCO_2$ implementation are pharmaceutical processing, dry cleaning, metal degreasing, and polymer modification. A feature of supercritical carbon dioxide is its high miscibility with gases, which is especially useful in reactions such as hydrogenation with H_2, oxidation with O_2 and hydroformylation [166-168] with syngas (CO/H_2), leading to great efficiency and often excellent selectivity [169]. Furthermore and due to its plasticizing and swelling effects, $scCO_2$ can easily be separated from catalysts and products by simple depressurization and recapture. It is therefore repeatedly applied as a green reaction medium for catalysis [170].

Recently, Melo et al. [171] reported some catalyzed hydrogenation methods of carvone with scCO2 thereby varying the pressure of $scCO_2$. By using mild reaction conditions, great conversion, and diverse and high product selectivity could be achieved. Furthermore, short reaction time depends on the catalyst was comprehended in supercritical CO_2 compared to hydrogenation occurring in traditional organic solvents. Also Endalkachew et al. [172] reported the catalytic hydrogenation of anthracene in scCO2 over Ni supported on Hβ-zeolite catalyst thereby yielding 100% yield due to reduced mass transfer limitations and increased solubility of H_2 and substrate in dense CO_2 medium (Fig 29).

The epoxidation of propylene to propylene oxide via hydrogen peroxide, catalyzed by Pd/titanium silicalite in supercritical CO_2 was reported [173]. The use of $scCO_2$ as the reaction medium considerably enhanced the catalytic activity of the large-grain in particular but also the small-grain TS-1 and increased the propylene oxide yield, as propylene has high miscibility in $scCO_2$. Transport limitations could be eliminated successfully thereby increasing catalytic performance due to mass-transfer and diffusion performance of $scCO_2$. Jiang et al. [174] have reported an effective procedure for generation of high yields tetrasubstituted alkenes in $scCO_2$. Addition of methanol as co-solvent with $scCO_2$ was proven to be an efficient medium for such a Pd-catalyzed system. Products were easily separated from catalyst due to the low viscosity of $scCO_2$ while catalyst was partly dissolved in the suitable quantities of methanol in $scCO_2$. Substituting $scCO_2$ with other solvents, like dioxane, tetrahydrofuran, water, ethanol, dimethylsulfoxide, acetonitrile, toluene, and dimethylformamide resulted in significant decline the yields of reaction, thus showing the vital effect of $scCO_2$ in stimulating the reaction.

6.   Ionic liquids (ILs)

Ionic liquids are electrically conductive liquids comprised of charged organic and inorganic ion pairs, highlighting liquid state at room temperature. Ionic liquids have low combustibility, exceeding thermal stability as well as solvating qualities [152]. Whenever, ILs are used as solvents there is no need of extra capping agents that will further lessen the material usage and made reactions easy. Since a number of organic, inorganic as well as polymeric molecules which are well soluble in ILs hence they used as solvents or reaction medium for many separations or catalytic processes [60]. The solvating properties of ILs are depending on the smaller anions and larger organic cations, the ILs consist of: anions like acetate, benzoate, formate, halides, hexafluorophosphate, hydrogensulfate, methanesulfonate, nitrate, phosphate, tetrafluoroborate, thiocyanate, tosylate and trifluoromethanesulfonate anions and cations: ammonium, imidazolium, phosphonium, pyridinium and pyrrolidinium cations [153].

ILs as such is not green but as solvents, it makes the whole reaction process greener in some cases [61]. Earlier it was considered green because of its negligible vapour pressure and not releasing of volatile organic compounds. Many ILs are highly biodegradable and exhibited substantial toxicity [22-24]. Moreover, they are expensive, with high to very high viscosity and their synthesis required special effort. ILs are stable over a good range of temperature and can be used in biphasic systems due to the immiscibility with some organic solvents. After extraction with an organic solvent, the catalyst leftovers in the IL and can easily be recycled. ILs might be used as catalysts themselves, as well as ligands or as solvents simultaneously.

Widely used ILs are N, N0-dialkylsubstituted imidazolium cations, choline hexanoate (potentially green due to the capacity to dissolve suberin, the main component of cork and a natural polyester) [29] and sodium 2,5,8,11-tetraoxatridecan-13-oate ([Na][TOTO]) containing ethylene oxide (EO) groups [30]. These TOTO complexes are low-cost, obtainable in bulk amounts, of less toxicity and, as compared to choline ILs, even added in cosmetic products. Additionally, they persuade with great thermal and electrochemical stability.

Many ionic liquids have been used for the synthesis of thiazoles which are imperative in organic and medicinal chemistry [175-177]. Condensation of aldehydes and ketones with hydroxylamine hydrochloride using 1-butyl-3-methylimidazolium hydroxide ([bmim]OH) or the condensation of indoles with benzaldehydes under microwave conditions with 1-benzyl-3-methylimidazolium hydrogensulfate ([bnmim][HSO$_4$]) as a catalyst has been reported [178]. The reaction products are used in tumor chemotherapy [179]. Also, Friedal-Crafts acylations can be carried out with inorganic Lewis acidic ionic

Industrial Applications of Green Solvents I   Materials Research Forum LLC
Materials Research Foundations **50** (2019) 165-241   doi: https://doi.org/10.21741/9781644900239-7

liquid catalysts like [bmim]Cl-AlCl$_3$[180]. In regard to agricultural and medicinal chemistry [181], halomethylated ß-enaminones were produced with [bmim]BF$_4$.

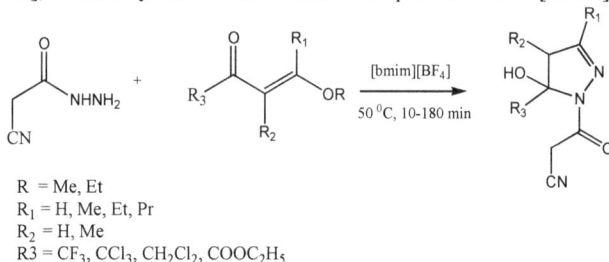

R = Me, Et
R$_1$ = H, Me, Et, Pr
R$_2$ = H, Me
R3 = CF$_3$, CCl$_3$, CH$_2$Cl$_2$, COOC$_2$H$_5$

Fig 28. Use of the ionic liquid [bmim]BF4 for the production of halomethylated ß-enaminones

The synthesis of xanthenes for laser technology, biological and pharmaceutical use could be realized with the Brønsted acidic 1- methyl-3-propanesulfonic imidazolium hydrogensulfate ([mimps][HSO$_4$]) with five-fold recycling of the catalyst [182]. Concerning the treatment of several severe illnesses such as cancer, AIDS and other 1,4-benzodiazepine-2,5-diones could be synthesized in a green reaction using 1-butyl-3-methylimidazolium bromide in the presence of isatoic anhydrides and amino acids [183]. Pyrazolines [184] and olefins [185] are other examples for the successful use of ILs as catalysts and/or solvents in green reaction conditions.

Suresh and Sandhu [186-188] refer to other standard reactions in organic chemistry where ionic liquids can replace conventional solvents. For example, considerable efforts have been made for the Knoevenagel reaction using mild Lewis acid catalysts like LiBr, KI or water, partly in solvent-free reactions or, for instance, in THF, TEAA or MeCN [186-193]. The Doebner modification could be carried out with BiCl$_3$ as a mild catalyst or in ionic liquids such as [bmim]BF4 or [bpyr]BF4 for synthesizing unsaturated carboxylic acids [194]. The Michael addition could be recently performed in isopropyl alcohol with (S)-pyrrolidine sulfonamide, as a catalyst [195] and with copper nanoparticles in 1-Htetrazole-5-acetic acid [196], both at room temperature. The Biginelli and the Hantzsch reaction, both multi-component tandem or cascade reactions, could also be recently modified in view of green chemistry [197]. Additionally, efforts have been made to carry out the Biginelli reaction in solvent-free conditions with mild Lewis acid catalysts [198]. As a recent example, tri-(2-hydroxyethyl) ammonium acetate was used as a catalyst under microwave conditions (200W, 4-8min) as well as under conventional conditions at 90 °C for 5-8h [199]. As products of the Hantzsch synthesis are of interest in clinical use, some green chemistry process contributions have been made [200].

Other important standard procedures in organic chemistry which can be modified to comply with the requirements of green chemistry are the Mannich reaction (Fig 29), Reformatsky reaction, Strecker reaction, Barbier reaction, Pechmann reaction, Henry reaction and the Cannizzaro reaction [201].

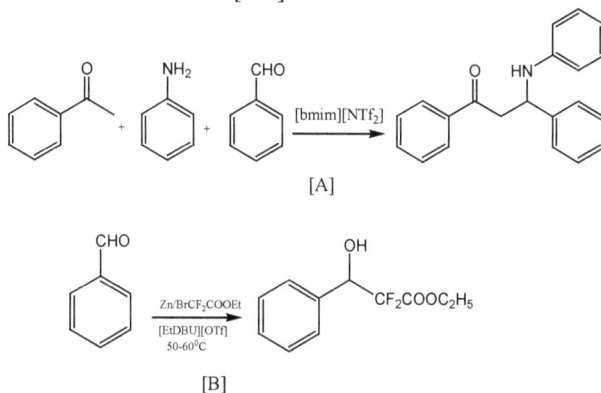

Fig 29 Modified Mannich reaction with [bmim][NTf2]. [B] Modified Reformatsky reaction with [EtDBU][OTf].

Beginning with imidazolium cations, pyridinium, ammonium, phosphonium, thiazolium, and triazolium species are the main cationic component. Generally, all above-mentioned cations may be joined with weakly coordinating anions, though not all weakly coordinating anions form RTILs. General examples are triflimide, hexafluorophosphate, tetrafluoroborate, triflate, and dicyanimide. Among them hexafluorophosphate, tetrafluoroborate have been mostly explored. It is important to mention that they should be treated with great care because they can be easily hydrolyzed to phosphate and boric acid.

Various types of nanoparticles have been synthesized in ILs using chemical reduction in which numerous used reducing agents such as organic (ascorbic acid), gases (H_2), and inorganic ($NaBH_4$) are available. The other important process is decomposition of metal carbonyls with metal atoms already in the zero-valent oxidation states. Moreover, to the use of ILs in conventional thermal heating, there is an attractive trend to combine ILs with more environmentally beneficial heating technologies, i.e. microwave and ultrasound. Since their essential characteristics are according to the principles of green chemistry and also for their unexpected synergetic effects resulting in the enhancement in reaction abilities, in terms of the product quality and reaction rate and time. Due to their high ionic charges, polarities, and dielectric constant, ILs are capable to efficiently absorb the microwave energy resulting into a quick and homogeneous temperature raise in the reaction systems without the need for heat transfer. Consequently, reaction time and energy consumptions could be considerably diminished. Furthermore, the harvesting of nuclei could be attainable at the early phase of the reaction, favoring the generation of nanoparticles with small and narrow size distribution. Synthesis of uniform sized metal nanoparticles (5 nm) from their metal carbonyl precursors dissolved in ILs using microwave irradiation. The reaction efficiency (10 W, 3 min) was found superior to that of UV-photolysis (1000W, 15min) or conventional thermal heating (250 1C, 6–12 h). In order to short out dispersion of the nanoparticles during synthesis another environmental friendly combination, ILs/low-frequency ultrasonic irradiations could be used. Implementation of such systems greatly not only reduces the reaction time but also the crystallization degree of the resulting nanoparticles was often improved as compared with traditional methods.

6.1 MCRs in ionic liquids

Ionic liquids are capable to produce an internal pressure and encourage the union of reactants in the solvent void during the activation process. Consequently the ILs are fairly appropriate for MCRs where the entropy of the reaction is reduced in the transition state. ILs creates a nice prospect for governing the selectivity of MCRs owing to their excellent physical properties. A three-component reaction of malononitrile aromatic aldehyde and 2-(2,3-dihydrothiochromen-4-ylidene) malononitrile, where the product might be transformed by changing the anion of ILs. High yields of products at room temperature with BMIm]BF$_4$ have been reported by numerous researchers [202-204]. No reaction happened in the absence of ZrCl$_2$, NaNH$_2$, [BMIm]BF$_4$ or elements of these under sonication. Using ethyl acetate the reactions were completed within 20 min and the products can be separated by extraction. It was observed that the residual [BMIm]BF$_4$ was retrieved and reused in subsequent reactions with a measured reduction in the activity.

Good to excellent yields of products have been reported for a four-component, one-pot reaction of aromatic aldehyde, cyclohexanone, malononitrile, and amines performed in a basic ionic liquid [BMIm]OH. An effective one-pot multicomponent synthesis of 2,4-diamino-5-pyrimidine carbonitrile derivatives using [BMIm]OH ionic liquid as a base under microwave irradiation (100 W) at 60 °C has been reported. The ionic liquid could be recycled up to five cycles in the reaction medium without substantial loss in the yields. Some three-component reactions of substituted hydrazines, 1,3-diketones and benzaldehydes have also been executed in [BMIm]BF$_4$ by using Yb(OTa f)$_3$ as a catalyst [205]. Such scheme provided an effective and green route to generate the 2-substituted tetrahydroindazolone derivatives. Yb(OTf)$_3$/[BMIm]BF$_4$ is an efficient and green source for the synthesis of β-acetamido ketones and 2-substituted tetrahydroindazolone derivatives [206].

A one-pot synthesis of 6-aminouracils via in situ produced ureas and cyanoacetylureas (Fig 30) 1,1,3,3-tetramethylguanidine acetate ([TMG]Ac), ILs as a recyclable solvent and catalyst [207]. The IL was recovered under reduced pressure and found reusable up to five cycles. The ([TMG]Ac), catalyze different sequential steps in one pot reaction, that not only cuts the reaction time but also improves the financial and environmental features of reaction process.

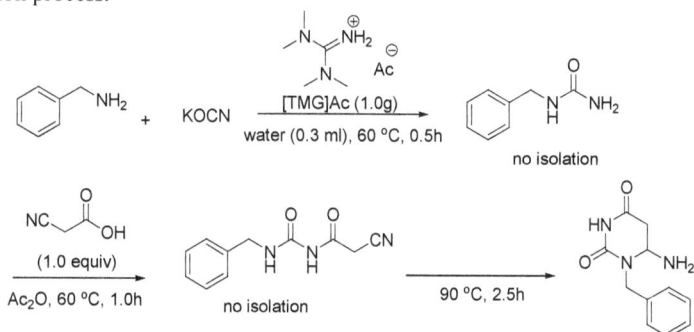

Fig 30. One pot synthesis of 6-aminouracil in [TMG]Ac

Hasaninejad et al. [208] reported a one-pot effective method for the production of 1,2,4,5-substituted imidazoles under simple heating and microwave irradiation using [BMIm]Br. Similar reactions in different solvents such as CH$_3$CN, DMSO, DMF and toluene, yield little amount of product, revealed the presence of a great promoting effect of [BMIm]Br in this reaction. It was concluded that [BMIm]Br performs two tasks: (i) activates the aldehyde carbonyl electrophilically via hydrogen bonding to the carbonyl oxygen; and (ii) improving the nucleophilicity of the amine over deprotonation of the N–

Industrial Applications of Green Solvents I   Materials Research Forum LLC
Materials Research Foundations **50** (2019) 165-241   doi: https://doi.org/10.21741/9781644900239-7

H bond. A straightforward and an efficient, three-component reaction of alkynes, aldehydes, and amines using ionic liquid under microwave irradiation was reported for the synthesis of quinolines by use of the Yb(OTf)$_3$-catalyzed by [186]. The catalyst and the [BMIm]BF$_4$ ionic liquid was found reusable up to four cycles deprived of significant loss of its catalytic activity. [BMIm]- BF$_4$ and [BMIm]Br as a recyclable medium were also proved to be a suitable medium for high yield [209-212]. The similar [BMIm]BF$_4$ has also credentials for efficient preparation of a spiro[4H-pyran-oxindole] heterocycle [BMIm]BF$_4$ was a suitable medium for good to excellent yields of product in three-component reaction of aldehydes, 1-naphthylamine and barbituric acid and two-step excellent yields synthesis of 4(1H)-quinolones in ionic liquids [213-215]. Shekouhy and Hasaninejad [216] synthesized good yields of 2H-indazolo[2,1-b]phthalazinetriones by reaction of an aldehyde, phthalic anhydride, hydrazinium hydroxide, and dimedone under ultrasonic irradiation in [BMIm]Br (Fig 31). Though same product could be obtained with other solvents under the same circumstances but yield was relatively poor indicating the key role of both cation and anion of [BMIm]Br in supporting the reaction.

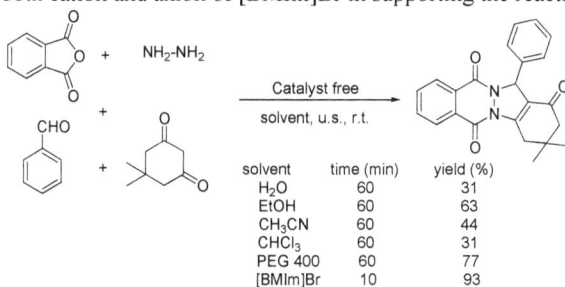

solvent	time (min)	yield (%)
H$_2$O	60	31
EtOH	60	63
CH$_3$CN	60	44
CHCl$_3$	60	31
PEG 400	60	77
[BMIm]Br	10	93

Fig 31 Four component reaction of aldehyde, pthalic anhydride, hydrazinium hydroxide and dimedone in diggerent solvents

Constantieux et al. [217] synthesized efficiently the novel heterocyclic complex with the reaction of β-ketoamide, acrolein and 2-aminophenol in different ionic liquids (Fig 32).

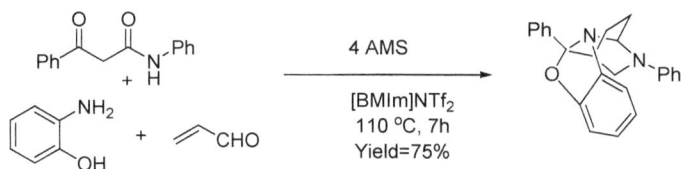

Fig 32 Three component reaction of beta-ketoamide, acrolein and 2-aminophenol in ionic liquid

Hitherto, the equivalent reaction has been achieved in toluene but replacing toluene with recycled and reusable ionic liquid [BMIm]NTf$_2$ not only provided competitive yields as well correspondingly improved the reaction rate. Dabiri et al. [218] obtained 4-Substituted-spiro-1,2-dihydroquinazolines derivatives in the presence of 1-methylimidazolium trifluoroacetate ([HMIm]TFA) that played dual role of catalyst and solvent. [HMIm]TFA was recyclable (n=3) with significant activity and unglued easily by modest extraction. Soleimani et al. [219] synthesized high reaction yields of benzo[b][1,4]oxazines with recyclable [BMIm]Br under mild conditions and at ambient temperature. The one-pot MCR of anthranilic acid, carboxylic acid and aniline to give high yields of 4(3H)-quinazolinones run well in presence of [BMIm]BF$_4$ under ultrasound irradiation [220]. Xu et al. [221] observed [BMIm]PF$_6$ as an proficient and recyclable medium for extremely chemoselective preparation of 2,2-dimethyl-6-substituted 4-piperidones (Fig 33). Here, [BMIm]PF$_6$ also considerably boosts the chemoselectivity via formation of imine that resulted in the privileged manifestation of Mannich reaction over aldol reaction. Moreover, the ionic phase comprising L-proline catalyst may have considerable reactivity and chemoselectivity over recyclability.

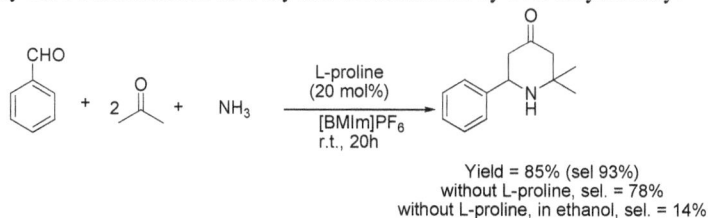

Fig 33. Chemoselective synthesis of 2,2-dimethyl-6-substituted 4-piperidones in ionic liquid.

Industrial Applications of Green Solvents I   Materials Research Forum LLC
Materials Research Foundations **50** (2019) 165-241   doi: https://doi.org/10.21741/9781644900239-7

Furthermore, Brønsted-acidic functional groups such as SO_3H into the cations or anions enhanced the acidities and water miscibility of ionic liquids. Consequently, Brønsted-acidic ionic liquids e.g., PEG1000-based dicationic acidic (PEG1000-DAIL) have been used as highly effectual and green catalysts for extensive research as substitutes for H2SO4, HF, and AlCl₃ in chemical processes [222] PEG1000-DAIL were efficiently recoverable (simple decantation) and recyclable (n=10) with achievement of significant catalytic activity. 2,4,5-trisubstituted imidazoles was synthesized by using a basic ionic liquid, [BMIm]OH [223-225]. One-pot reactions of 2-naphthol, aldehydes and 1,3-cyclohexanediones continued very well to give good yields of 8,9,10,12-tetrahydrobenzo [a]xanthen-11-one derivatives in catalytic presence of Brønsted acidic dialkylimidazolium with SO3H-functionalization [226-227]. The alike reaction can also be accomplished with neutral ionic liquid and IBX or p-TSA[228-229]. Several acidic ionic liquids based on Polyethylene glycol dication or SiO_2-supported SO_3H-functionalised benzimidazolium, brønsted-acidic $[(CH_2)_4SO_3HMIm]HSO_4$ and $[(CH_2)_4SO_3HPy]HSO_4$ were used as catalyst under solvent-free conditions [230-233]. Hajipour et al., (2011) [234] used N-(4-sulfonic acid) butyl triethyl ammonium hydrogen sulfate ([TEBSA]HSO₄) as an effective and recyclable catalyst for generating various pyrimidinone derivatives in good to excellent yields. [BMIm]Br with TMSCl (additive) can also be employed for the reaction [235]. However, large amounts of catalysts were used in those reactions. Mirzai and Valizadeh [236] stated a novel nitrite-functionalized dialkylimidazolium ionic liquid (IL-ONO) as a weak Lewis base catalyst for the productive yields in Biginelli reaction. The same reaction can also be resourcefully catalyzed with an equivalent phosphinite ionic liquid (IL-OPPh₂) [237] Recently, Brønsted acidic ionic liquid,1-methyl-2-pyrrolidinone hydrosulfate ([HNMP]HSO4), as catalyst were employed for solvent-free synthesis of several derivatives of 1H-pyrano[2,3-d]pyrimidin-2(8aH)-one (Fig 34).

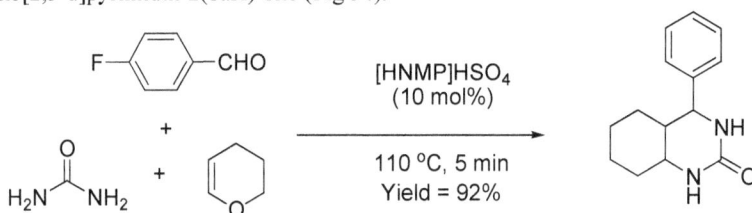

Fig 34. Three component reaction of aromatic aldehydes, urea or thiourea and 3,4-dihydro-2H-pyran in ionic liquid

Industrial Applications of Green Solvents I   Materials Research Forum LLC
Materials Research Foundations **50** (2019) 165-241   doi: https://doi.org/10.21741/9781644900239-7

Li et al. [238] used [BMIm]BF4 in high yield Biginelli-like reaction of aldehydes and 5-aminotetrazole with ethyl 4,4,4-trifluoro-3-oxobutanoate. To find an efficient ionic liquid for Biginelli-like reaction, Chakraborti et al. [239] have investigated the catalytic efficiency of [BMIm]-based compounds. They concluded that both counter anion and 1,3-disubstituted imidazolium moiety are important for catalytic activity and out of that [BMIm]MeSO$_4$, was selected with a very small amount of the ionic liquid (1 mol%) at 100 °C for 30 min to give good yields of dihydropyrimidinones. However, the postulated mechanism for [BMIm]MeSO$_4$ might be diverse from the reactions occurred in aqueous media.

Fang et al. [240] synthesized α-amino phosphonates with a geminal dicationic ionic liquid such as N,N,N',N'-tetramethyl-N,N'-dipropanesulfonic acid ethylenediammonium hydrogen sulfate ([TMEDAPS]HSO$_4$). Advantage of process was efficient as well as recyclable catalyst and room temperature. Recently, synthesis of α-aminophosphonates was performed with efficient and eco-friendly catalyst of choline chloride· ZnCl$_2$. High yield of 2,4,6-triarylpyridines (Kröhnke pyridines) was obtained with reusable catalyst of 3-methyl-1-(4-sulfonylbutyl)imidazolium hydrogen sulfate [HO$_3$S-(CH$_2$)$_4$MIm]HSO$_4$ under solvent-free surroundings [241]. Concluded facts included the reusable catalyst (after simple work-up) with insignificant decline of their activity. [(CH$_2$)$_4$SO$_3$HPy]HSO$_4$ as catalyst was also used in high productive reactions performed [242]. Fang et al. [243] used original functionalized 3-(N,N-dimethyldodecylammonium) propanesulfonic acid hydrogen sulfate ([DDPA]HSO$_4$) for Mannich-type reaction to give good yields of various β-amino carbonyl compounds at room temperature. Mannich reaction (asymmetric and direct) also go on well with [BMIm]PF$_6$. Equivalent enantioselectivities was even achieved by siloxy serine organocatalyst in [BMIm]PF6 on three times recycle [244]. [BHP-OMe]Br can promote diastereoselective reaction to give good to excellent yields of products in water and revealed good reusability at least in four following reactions.

7.   Ephemeral solvents or deep eutectic solvents

Ephemeral solvents are created only during the reaction by addition of organic salt with an H-bond donor followed by mingling them at raised temperature to get consistent liquid forms e.g., choline chloride and urea [33, 245]. They have properties parallel to ILs except synthesis because the method for preparation of DESs is very unpretentious and cheap. DESs are considered eco-friendly solvent though toxicity and biocompatibility are individual components dependent [34-37]

Industrial Applications of Green Solvents I   Materials Research Forum LLC
Materials Research Foundations **50** (2019) 165-241   doi: https://doi.org/10.21741/9781644900239-7

7.1   Natural DESs

They are low melting mixtures or "natural hydrotropic mixtures" with "natural" or of "biological origin" and used as solubiliers in biological systems, especially in plants [246]. DESs are liquid at low temperature and frequently mixed with a substantial amount of water [247-248] e.g., typical mixture of malic acid-choline chloride-water (molar ratio of 1:1:2). The mixture of malic acid-choline chloride displays significant depression in melting point while water (cosolvent) reduced viscosity. Association of these three indicated the solubilization of malic acid and choline chloride with water. For plant extraction practices, these ternary mixtures of NADESs with water are most valued and useful, and hence, considered as 'designer solvents'. Adeyemi et al. (2017) [249] investigated the viability of amine based DESs (Monoethanolamine, diethanolamine and methyldiethanolamine) for carbon capture. Results revealed that amine-based DESs have much higher absorption capacity than that of both aqueous amine (30 wt%) and conventional DESs. Additionally, broadening of the IR peak corresponding to O–H and N–H stretching specified the formation of H-bonds between choline chloride and – monoethanolamine (1:6) before CO_2 absorption.

7.2   Hydrotropes

Neuberg 2009, [250] coined the term for amphiphilic compounds that showed increased solubility for frugally soluble organic compounds in water [32] e.g., ternary macroscopically uniform mixture of water, hydrotrope and hydrophobic compound at any compositions.

8   Switchable solvents

In 2010, Jessop et al. [251] used the term of "switchable solvents" for water soluble or hydrophobic solvent depending on the presence or absence of CO_2 e.g., water enamidine systems. Because of reversible and tuned process (by controlling air and CO_2 pressure) a hydrophobic water-immiscible phase can be temporarily produced and removed on completion of reaction. Moreover, non-polar organic product can also be collected by separation from hydrophilic solvent i.e., simple isolation of the product. In addition, uncomplicated recovery of solvent from water by eliminating carbonate from the reaction mixture was also advantage of those solvents. Only few solvents are considered green but they are clever system for development of green and solvent systems. Temperature can also be used instead of CO_2 [252] in water/IL mixtures. Tetrapentylammonium bromide (Pe4NBr) is showed remarkable phase behavior when mixed with water at close to $100\,^{\circ}C$ and splited into water-rich and organic phase in equilibrium. Hence, organic compound can be dissolved in latter phase and at the end of the reaction, the hydrophobic phase can be easily taken out simply by cooling down. Salt precipitated out without polluting the water phase. However, salts may have bromide that is undesirable for industrial processes.

Bergbreiter et al. [253] used monophasic solvent polymer mixture near immiscible and separated the phase by simply addition of a small amount of solvent. Ludmer et al. [254] developed temperature induced "sediments remediation phase transition extraction" coupled with a phase transition cycle. The mixture has characteristics of the green solvents like water, ethanol and ethyl acetate Hence, it was successfully used to depollute heavily contaminated sludge.

9. Organic Carbonates

Organic carbonates (open chain and cyclic esters of carbonic acids) are easily available in large amounts, inexpensive, possess low (eco) toxicity and are complete biodegradable. They are widely used for extraction purposes, pharmaceutical and medical applications, and also in batteries. Cyclic carbonates fulfill the requirements of green solvents which have low flammability, volatility, wider temperature range in the liquid state and low toxicity. Especially, carbonate of propylene (PC) is an aprotic, highly dipolar solvent with low viscosity and a very large liquid state range (mp. -49 °C, bp. 243 °C) [255]. Since PC has a high molecular dipole moment (4.9 D) it is susceptible to microwave irradiation and can be considered a very interesting solvent for microwave assisted organic synthesis, which unfortunately has been hardly investigated yet [256].

Rhodium-catalyzed asymmetric hydrogenations of some usually employed functionalized olefins (20) were carried out in propylene carbonate, butylene carbonate, and conventional solvents for comparison. Both carbonates showed similar or better results than "standard" reagents, such as MeOH, THF, and CH_2Cl_2. Butylene carbonate even enhanced enantioselectivities at longer reaction times compared to propylene carbonate [256].

10. Biosolvents

Biosolvents (esters of naturally occurring acids and fatty acids, terpenic compounds, bioethanol, isosorbide, glycerol and derivatives) offer the advantage of being produced from renewable sources such as vegetable, animal or mineral raw materials [257]. Many biobased solvents are approved by governmental legislations for use. They are already widely used in cosmetics, cleaning agents, paint, inks, and agricultural chemicals [257-258].

Industrial Applications of Green Solvents I   Materials Research Forum LLC
Materials Research Foundations **50** (2019) 165-241   doi: https://doi.org/10.21741/9781644900239-7

Bio-based solvents have successfully been employed in multicomponent reactions with seemingly synergistic effects. Condensation of dimedone, formaldehyde, and styrene in glycerol as solvent with superior yields of product compared to standard solvents (H_2O, CH_3NO_2, toluene, aceticacid). The authors concluded that glycerol exercises a promoting effect on the oxo-Diels-Alder reaction of the intermediate methylendimedone with styrene due to its polar and protic properties. Glycerol was proven to be an indispensible reaction medium, providing excellent results in multicomponent reactions [259]. Quan et al. [260] reported the promoting effect of glycerol combined with a polystyrene-poly(ethylene glycol) (PS-PEG) resin supported sulfonic acid catalyst in Biginelli reaction, and the reaction of amide, 2-naphthol, and aldehydes.

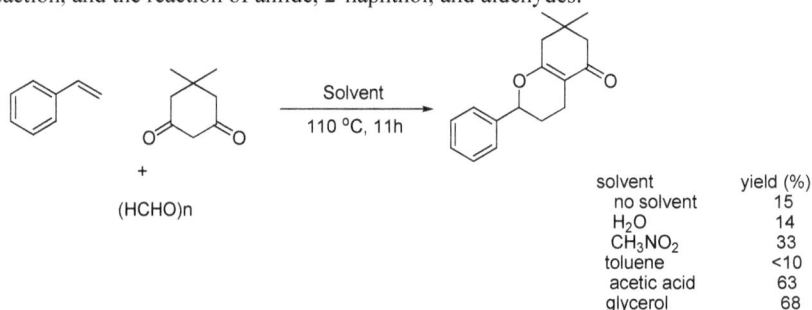

solvent	yield (%)
no solvent	15
H_2O	14
CH_3NO_2	33
toluene	<10
acetic acid	63
glycerol	68

Fig 35 Three component of dimedone, formaldehyde and styrene in glycerol

Due to synergistic properties of glycerol, MCRs can go on well with greater selectivity than that in other systems. Of late, glycerol was found to be a vital solvent for one-pot sequential reaction of styrenes β-ketone esters, formaldehyde and arylhydrazines. Glycerol has unique ability for encouraging an electrophilic alkylation of carbonyl functional group [258]. Good yields of products were obtained via hydroxymethylation of formaldehyde with β-ketosulfone in mixture of biosolvents (meglumine and gluconic acid 50 wt%) while low yield in other solvent systems showed the capability of biosolvents used [258].

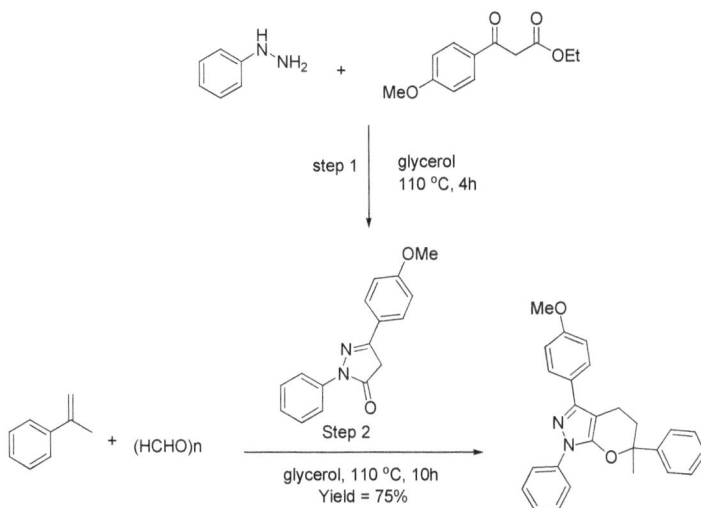

Fig 36. One pot step sequential reaction of arylhydrazines, beta-ketone esters, formaldehyde and styrenes.

The formed hydroxymethylation product is assumed to further react with a nucleophile. Based on this, one-pot stepwise reactions of β-ketosulfone with formaldehyde have been developed in this binary mixture. These developed bio-based mixture of chemicals provided the control on selectivity of some MCRs of formaldehyde. Due to their hydrophilic nature, bio-based mixture could be separated after extraction of organic products and reused (n=4) with considerable activity. The combination of Glycerol with PS-PEG-supported sulfonic acid catalyst acted as green solvent for encouraging performance in several MCRs like Biginelli reaction and the reaction of amide, aldehydes and 2-naphthol.

11. Polyethylene glycol polymers (PEGs)

PEGs are nonvolatile, cheap, easily available, low toxicity, reusable, biodegradable and stable to acid, base and also to high temperature. PEGs are easily soluble in fairly polar solvents and reported insoluble in isopropanol and diethyl ether.This exceptional property facilitates the retrieval of PEGs by precipitation and filtration. Therefore, such solvents PEG and poly(propylene glycol) (PPG) have received great attention as novel solvents

Industrial Applications of Green Solvents I   Materials Research Forum LLC
Materials Research Foundations **50** (2019) 165-241   doi: https://doi.org/10.21741/9781644900239-7

for various catalytic processes. Both the solvents are comparatively cheap and freely available materials. PPG and PEG are well accepted for their use in beverages while PPG was used as a solvent in pharmaceutical and cosmetic industries. In the biphasic catalysis the combinations of PEG or PPG with, water or $scCO_2$ were found as important system. Polyoxometalate catalysed aerobic oxidation of benzylic alcohols, Wacker oxidation of propylene to acetone and Heck reactions using PEG-200 and/or PEG-400 (the number refers to the average molecular weight), have been reported by several researchers. [261-263].

During the last decade, several named reactions have been performed using PEG as solvent or catalyst. Biginelli reaction was promoted by PEG-400 and it was found capable to generate 3,4-dihydropyrimidinones in high yields under neutral conditions [264]. High yields of thiomorpholides were generated via condensation of aryl alkyl ketones, sulfur and morpholine using PEG-600 as solvent/catalyst [265].

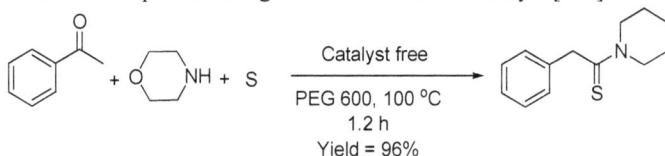

Fig 37 Willgeroldt Kindler reaction in PEG 600

Kouznetsov et al. [266], Polyfuncionalized 2,4-diaryl-1,2,3,4-tetrahydroquinoline derivatives via Povarov reaction have been synthesized in short span of time and small amount of solvent in PEG-400. High yield of polyfunctionalized 2-pyridones from the reaction of acetophenone, aldehyde, ethyl cyanoacetate and ammonium acetate in PEG-600 have been reported [267]. PEG-400 was reported as low-cost-effective and reusable medium for the one-pot synthesis of N-substituted decahydroacridine-1,8-diones, making the process potentially viable [268]. Certain substituted acridines, pyrazole, Hantzsch 1,4-dihydropyridines and pyridine were synthesized in PEG-400with or without using microwave irradiation ([269-272].

PEG has been showed to be an efficient medium for one-pot synthesis of N-substituted azepines from dialkylacetylene dicarboxylate, aniline and 2,5-dimethoxytetrahydrofuran and polysubstituted pyrrole derivatives under mild and catalyst-free conditions [273-274]. PEG-400 was found to provide high yield as compared to other solventslike acetonitrile, water, DMF and THF) under the similar reaction conditions, justifying the role of PEG solvent in execution of the reaction towards completion [275]

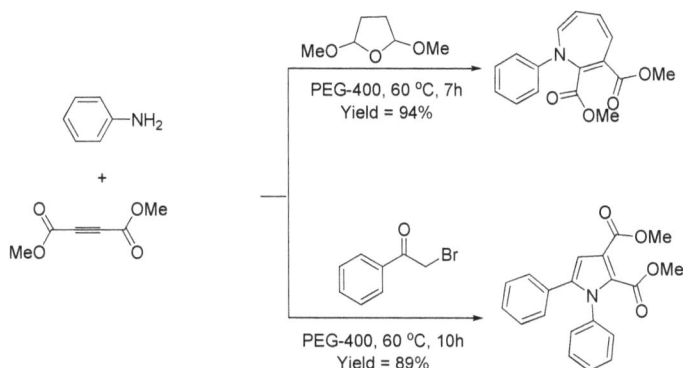

Fig 38. Three component reactions of aniline and dialkyl acetylene-dicarboxylate in PEG-400

Reactions of aromatic aldehydes, cyclohexanones and aniline (Mannich-type: one-pot three-component) was carried out in PEG-400 in conjunction with 2,4,6-trichloro-1,3,5-triazine as catalyst. Several organic carbamates were synthesized employing an efficient and green method using PEG-400 as both solvent and catalyst under ambient reaction conditions. Especially, the application of PEG could also reduce the alkylation of amine and the carbamate, resulted the improved selectivity toward the target carbamate. Kidwai et al. [268], reported synthesis of polysubstituted-tetrahydropyrimidines in PEG-400 within 45 min under mild reaction conditions. Similarly, high yields of thiazolidinone, imidazoles, and coumarins derivatives were generated in PEG-400 with or without assistance of microwave irradiation [276-279]. A mixture of 2-chloro-3-formylquinoline, piperidine and 2,4-thiazolidinedione yielded 5-[(2-(piperidin-1-yl)quinolin-3-yl)methylene]-2,4-thiazolidinedione under microwave irradiation in PEG-400 [280]. In this reaction, excess amount of piperidine was added so that it could act as base to neutralize the HCl (formed during the reaction).

Good to excellent yields several pyrazolophthalazinyl spirooxindoles has been established by using $NiCl_2$ as catalyst in PEG-600 from one-pot three-component reaction of isatin, malononitrile or cyanoacetic ester, and phthalhydrazide under ordinary reaction conditions [281]. Reaction of isocyanides, dialkyl acetylenedicarboxylates and α,β-unsaturated aldehydes goes well in PEG-400, yielding the corresponding styrylfuran derivatives. A multicomponent reaction (Ugi-type) of heterocyclic amidines with aldehydes and isocyanides catalyzed by ZrCl4 in PEG-400 was also described [282].

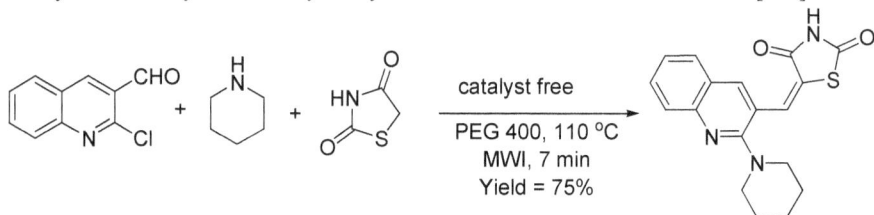

Fig 39. Three component reaction of 2-chloro-3-formyl quinoline, piperidine and 2,4-thiazolidinedione in PEG-400

12. Use of green solvents in nanomaterials synthesis

In order to get rid of hazardous solvents/reducing agents like sodium borohydrides, hydrazine, and dimethylformamide), green methodology employing environmentally benign and renewable materials like water, ionic liquids, aqueous plants extract/surfactant are preferred. It is a more reliable, sustainable and bioinspired bottom-up approach. The use of solvents is to provide a medium for the dissolution of precursors in the medium, transferring of heat and reactants and dispersion of resulting nanoparticles. Moreover, industrial-scale generation of nanoparticles demands the assortment and design of green solvent substitutes to decrease and eradicate ecological risks. The benefits of green synthesis of nanomaterials comprise (1) Economical (2) Eco-friendliness and safety (3) reusability (4) easy altering of the grain size, morphology, and surface functionality (5) generation of relative stability (6) remarkable biocompatibility and (7) biodegradability. [283-285]. Chettri et al. [285] reported the green synthesis of Ag-RGO Nanohybrids using Psidium guajava leaves to extract as a surfactant and the resultant nanomaterial showed enhanced capacity in the detection of methylene blue (MB). Prosopis cineraria leaf extracts was used to synthesized Ag-Cu NHs which showed high antibacterial activity against microbial pathogen and cytotoxicity for human breast cancer cell line (MCF-7), in comparison to the individual Ag and Cu NPs [286]. Kolya et al. [287] reported a green synthesis of Ag nanoparticles using Amaranthus gangeticus Linn (Chinese spinach) leaf extract. The synthesized Ag NPs showed decent catalytic efficiency (more than 50% within 15 minutes) in the degradation of hazardous Congo red dye. Green synthesis of ZnO, CuO and ZnO/GO nanoparticles exhibited high photocatalytic activities [288-289]. Green synthesis of several nanomaterials such as oxides of Zn, Cu, and Ni; ferrites of Ni–Cu–Mg, Ni-Cu-Zn and magnetic copper) was reported using gel, gum or plants. The catalytic efficiency of these NPs was evaluated for organic synthesis and in photocatalysis [290-293].

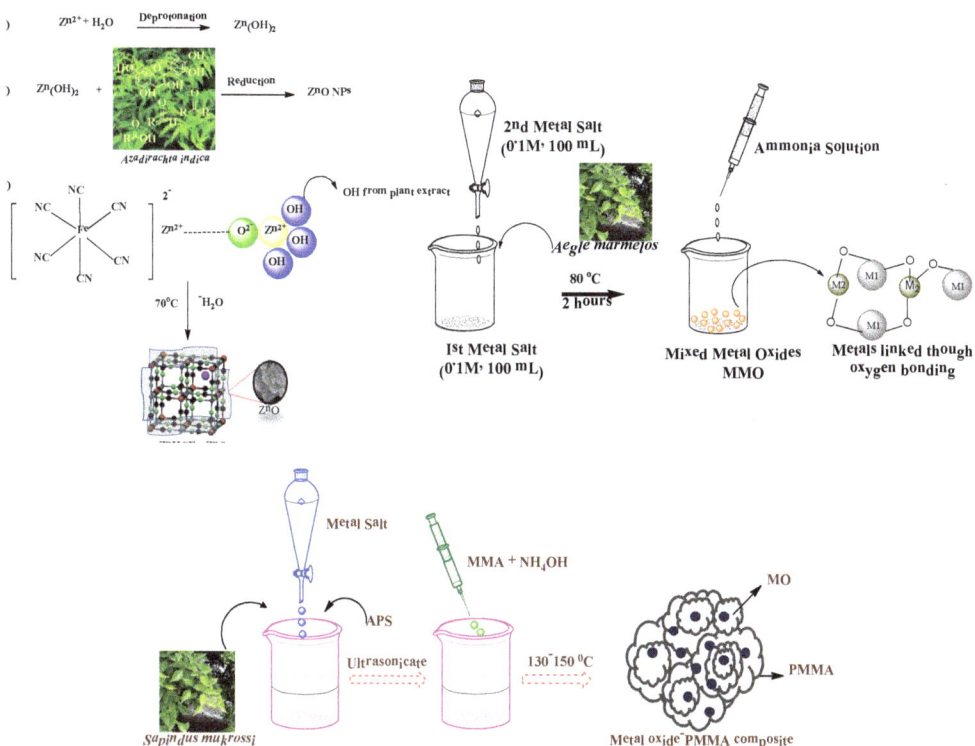

Fig. 40. Various green strategy for the synthesis of doped metal hexacynoferrates, bimetallic oxides simple and doped with PMMA

Shanker and his research group [294-296, 298-317] synthesized nanomaterials of metal hexacyanometallates and metal oxides using aqueous extracts of natural surfactants like sapindus mucrossi, and Aegle marmelos and used them for the degradation of organic pollutants. Green synthesized double metal Fe/Pd NPs were reported being superior than bared ones [318-319]. Green synthesized bimetallic oxides (BMO) with grain size 50 nm like NiCuO nanorods, $CuCr_2O_4$ nanoflowers and $NiCrO_3$ nanospheres were produced using Aegle marmelos leaf extract. Figure 40 showed synthetic strategy of BMO with A. marmelos used in different fields (medicinal, nutritional, commercial and environmental). Several natural products present in the leaves of A. marmelos are alkaloids, terpenoids and phenylpropanoids. Photodegradation of toxic phenols (1×10^{-4} M) from water using mixed BMO nanostructures (15 g) was carried out at neutral pH. Crystalline nanocubes of Fe_2O_3@ZnHCF and ZnO@ZnHCF nanocomposite were synthesized using water and plant extract of Azadirachta indica, a commonly found plant in India. A. indica consists of phytochemicals (benzoquinones) that reduce the interfacial tension to control the particle growth. Overall, green synthesized nanocomposites are cheap, reusable (n=10) with properties of greater active sites, high surface activity, low band gap with charge separation and semiconducting nature.

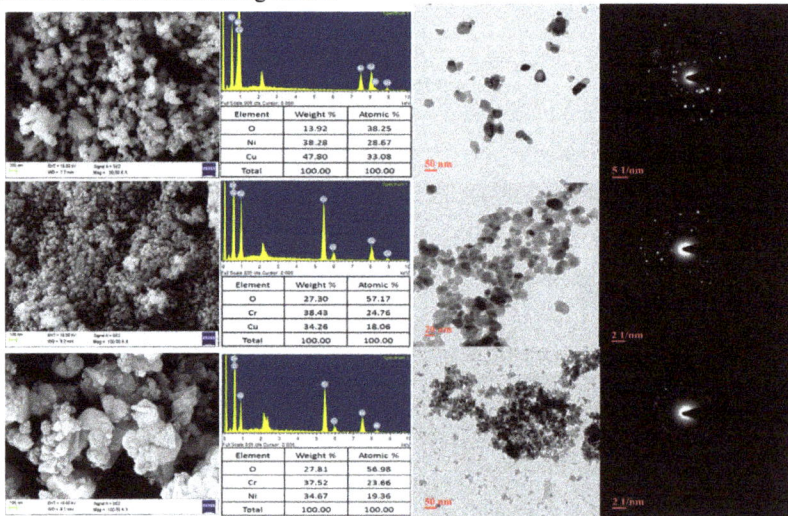

Fig. 41FE-SEM and TEM images with SAED patterns of (a) Ni-CuO (b) $CuCr_2O_4$ (c) $NiCrO_3$ nanoparticles

Shahwan et al. [320] observed that green tea-iron iron nanoparticles were a better catalyst than Fe nanoparticles produced by borohydride reduction. Silver nanoparticles synthesized with Morinda tinctoria leaf extract degraded 100% of dye within thirteen minutes. Highly crystalline sharp potassium zinc hexacyanoferrate nanocubes of ~100 nm and hexagonal, rod and spherical shaped iron hexacyanoferrate nanoparticles, size range: 10-60 nm were used for treatment of toxic PAHs (84-93%) at neutral pH under sunlight exposure.

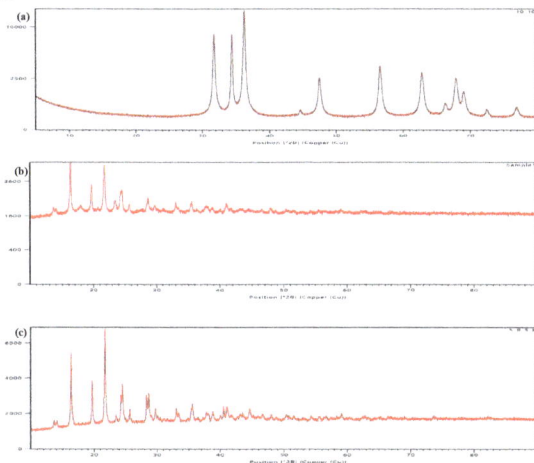

Fig 42. PXRD pattern of (a) ZnO, (b) ZnHCF and (c) ZnO@ZnHCF

Fig.43FE-SEM image and EDS pattern of (a)ZnO, (b) ZnHCF and (c)ZnO@ZnHCF

Sabbaghan et al. [321] synthesized distinct morphologies of ZnO NPs in water and imidazolium-based ionic liquids where strong hydrogen bonds formed between the hydrogen atom at position 2 of the imidazole ring and the oxygen atoms of O-Zn. Possible mechanism of generation showed that ZnO crystals are polar and their positive polar plane is rich in zinc, and the negative one is rich of O [322]. ILs have considerable impact on the morphology and structures of ZnO based on hydrogen bonds, π-π stack interactions, self-assembled mechanism and electrostatic attraction [323]. Bulky alkyl chain at position-1 of imidazole ring or using dicationic ionic liquid with a definite concentration results the more width of nano sheet (due to increase π-π stack interactions) [324].

No effect on solvent (water, alcohol) was observed but 2D ZnO nanostructures can only be yielded by use of excess NaOH. This results in the change of interaction between ZnO surface and ionic liquids i.e. weakened the interaction between hydrogen and carbon atoms at position 2 of the imidazole ring. In contrast, the hydrogen bond formed between the hydrogen atom at position 2 of the imidazole ring and the oxygen atoms of O-Zn were enhanced, playing a crucial rule in the morphology of ZnO nanostructures [324].

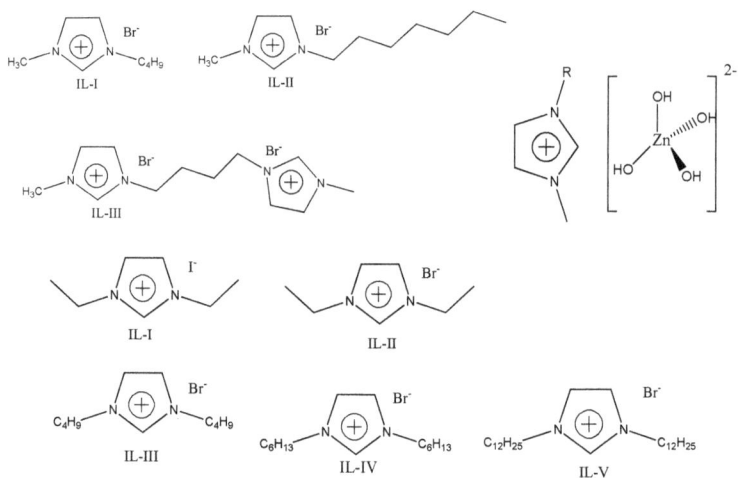

Fig. 44 Chemical structure of ionic liquids. Imidazolium combination with [Zn(OH)]$^{-2}$ through the electrostatic attraction

Industrial Applications of Green Solvents I   Materials Research Forum LLC
Materials Research Foundations **50** (2019) 165-241   doi: https://doi.org/10.21741/9781644900239-7

Sabbaghana et al. [325] synthesized ZnO nanostructures of distinct shape like nanocoral, spherical and nanosheet by the use of green synthesized imidazolium-based ILs and concluded that anion and cation of ionic liquids may affect the band gap and morphology of the zinc oxide NPs. bulkier alkyl chain at positions 1 and 3 of imidazole ring ionic liquid resulted in the nanosheet morphology. The ILs used in such reactions, were 1,3-Dihexylimidazolium bromide [DHIm][Br] (IL1), 1,3-Diethylimidazolium iodid [DEIm][Br] (IL2), 1,3-Dibutylimidazolium bromide [DBIm][Br] (IL3), 1,3-Diethylimidazolium bromide [DEIm][I] (IL4),

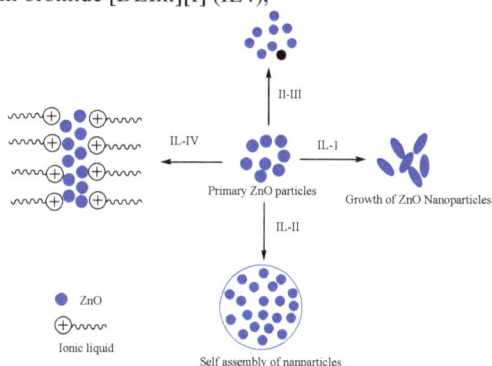

Fig. 45 Chemical structure of ionic liquids and Schematic drawing of the possible mechanism for the formation of products

Zhao et al. [326] performed photochemical synthesis of ZnO/Ag_2O heterostructures by ILs-assistnace two ionic liquids named 1-dodecyl-3-methylimidazolium and bromide anions ILs $[C12mim]^+Br^-$ (IL1) and triethylamine acetate (IL2) with enhanced visible light photocatalytic activity.

Well-ordered mesoporous SBA-15/TiO_2 nanocomposites (~0.1 nm) in varying ionic liquids ($CMITf_2N$ and $CMIBF_4$) with and without isopropyl alcohol. $CMIBF_4$ significantly favors TiO_2 anatase crystalline phase without alcohol while $CMITf_2N$ slightly favored TiO_2 rutile crystalline phase. This ordered mesoporous structure was lost once the concentration of ionic liquids increases. Narrow pore size distribution with large surface areas up to 680 m^2/g, pore volumes of up to 1.8 cm3/g and mean pore diameter of up to 10.4 nm was revealed by BET analysis.

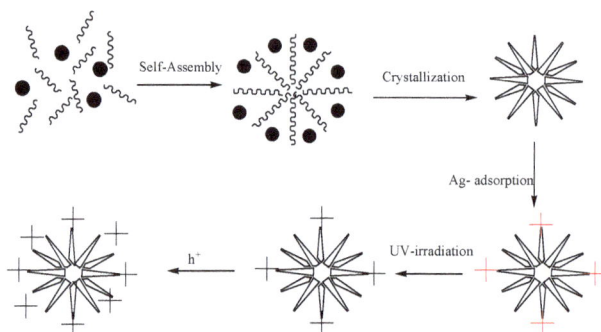

Fig.46 Structures of ionic liquids and Schematic illustration of the formation process for flower-like Z1/Ag2O heterostructure

13. Green solvents in analytical chemistry

Green analytical chemistry (GAC) [327-328] proposes many possibilities to avoid or apply alternative, more innocuous solvents in analytical extraction and chromatographic separation processes (mobile phase). The solventless processes (gas extraction and solid-phase microextraction) and alternative extracting solvents such as ionic liquids [329] supramolecular solvents [330], deep eutectic solvents [331] can be the basis of GAC. Supercritical fluids apart from contributing to greener extractions, make the chromatographic separations less environmentally problematic than liquid chromatography usually offers [332]. In spite of those advantages, organic solvents are still used because of convenience, less-substitute, analysts' habits or are applied in standard procedures. Though SSG focus on novel solvents (less toxic), usually ethers, esters or alcohols but not developed specifically for analytical chemists due to different waste solvents disposal practices and potentially increased occupational exposure during sample handling. LC (normal or reversed phase) uses large amounts of high-purity organic solvents. To make the process green, reversed phase LC (isopropyl alcohol in heptane or mixture of ethyl acetate with ethanol in heptane) [333] should be preferred than normal LC containing nonpolar and toxic solvents (dichloromethane). Ethanol, acetone and ethyl acetate, propylene carbonate or its mixture with ethanol, are preferentially applied over acetonitrile and methanol (more toxic) [334-336]. Ethanol or ethanol-water mixtures has drawback of relatively high viscosity [19] that can be removed at high temperature. Much effort is also done to increase the content of water in the mobile phase. Hexamethyldisiloxane, cyclopentyl methyl ether, 2-methyltetrahydrofuran and isopentyl acetate, D-limonene were preffered over chloroform and aliphatic hydrocarbons in normal LC used in determination of nonpolar and nonvolatile lipids [337]. However, miniaturization of columns dimensions is required for reduce amount of mobile phase.

The greenness of solvents in extraction processes increase in the order— from chlorinated solvents, aromatic and aliphatic hydrocarbon, terpenes, alcohol and esters to water. Liquid phase microextraction like hollow fibre microextraction, dispersive liquid-liquid microextraction or single drop microextraction usually require only sub- milliliter amounts of organic solvents [338-339]. Bio-based organic solvents e.g., (bio)ethanol, alcohols, esters, ethers and ketones [340-341] should be used, instead of petrochemistry derived ones. Ethyl lactate, 2-methyltetrahydrofuran and cyclopentylmethyl ether have been used in extraction of total petroleum hydrocarbons in soil samples and membrane proteins, respectively [342-343]. D-limonene (derived from citrus waste) is gaining much attention in replacing more toxic and petroleum-derived toluene, n-hexane in microwave-assisted Soxhlet determination of fats and lipids[344-346].

Another application is extraction simvastatin from human blood plasma and large volume injection (100 mL) to HPLC [347]. Terpenes can be treated as environmentally problematic as they are characterized by high potential for tropospheric ozone formation and are relatively toxic to aquatic organisms [348]. More solutions have been developed in the area of bioactive substances extraction from plant material. These solutions need some more research to be applied in analytical sciences for assurance that solvent is of adequate purity and to characterize the extraction efficiency. Such solvents not yet widely applied in analytical extractions are ethyl lactate [349], glycerol, furfural, furans or chyrene and others [350]. On the other hand, to some applications bio-based reagents applied in chemical analyses do not have to be refined. Similarly solvents for certain applications do not have to be ultrapure, under the condition that solvent contaminants are not analytes or interferences.

Conclusion

Green solvents in one-pot chemical reactions (in-organic, organic, natural products, pharmaceutical and agrochemical or nanomaterials synthesis) are used to protect the environmental from toxicity from conventional organic solvents and provide the ideal basis for a sustainable chemical industry. Obtained results are similar and sometimes even better than these, resulting from conventional syntheses in organic solvents.The solvent-free processes as well as more efficient recycling protocols have some limitations. In environmentally benign solvent alternatives, water, fluorous solvents, ionic liquids, organic carbonates, supercritical carbon dioxide, as well as biosolvents have been employed. Water, as a cheap, abundantly available, nontoxic and nonflammable solvent represents an ideal reaction medium for many processes, being mostly established in organometallic catalysis, hydroformylation processes and oxidations. Fluorous solvents and ionic liquids are attractive alternatives for performing reactions, which are not accomplishable in water or supercritical carbon dioxide. Organic carbonates, mostly used for extraction purposes and pharmaceutical and medical applications, feature characteristics like low (eco)toxicity, complete biodegradability as well as inexpensiveness. Supercritical carbon dioxide also exhibits outstanding characteristics for the utilization in Green Chemistry, such as the possibility to separate it from the resulting product by simple pressure release. Reaction rates are very high in scCO2, due to its intermediate properties, between gas and liquid state. Biosolvents, being produced from renewable sources are already widely used in cosmetics, cleaning agents, paint, inks, and agricultural chemicals and became to play an important role as an alternative to conventional solvents. A wide variety of heterocycles systems (different sizes and ring) and highly functionalized organic molecules can be readily synthesized through a combination of MCR strategies and unconventional green solvents. Using water and ionic liquids in metal-catalyzed MCRs also facilitates the recovery of metal catalysts. Some unique properties of unconventional solvents also allow the use of assisting technique, such as microwave irradiation.

There is a need to propose such a tool that would include assessment of solvents that are applied in analytical laboratories and are usually covered by pharmaceutical SSGs. The applications of bio-based solvents in strictly analytical applications are still limited.

References

[1] K. Tanaka, F. Toda, Solvent-Free Organic Synthesis, Chem. Rev. 100 (2000) 1025-1074.

[2] P Anastas, N. Eghbali, Green chemistry: principles and practice. Chem. Soc. Rev. 39 (2010) 301-12

[3] P. Koteswararao, S. L. Tulasi, Y. Pavani, Impact of solvents on environmental pollution. National Seminar on Impact of Toxic Metals, Minerals and Solvents leading to Environmental Pollution. Journal of Chemical and Pharmaceutical Sciences, 3, (2014), 132-135

[4] F.D. Dick, Solvent neurotoxicity. Occup. Environ Med., 63(2006) 221-226

[5] N. Sanni Babu, S. Mutta Reddy, Impact of solvents leading to environmental pollution. National Seminar on Impact of Toxic Metals, Minerals and Solvents leading to Environmental Pollution, Journal of Chemical and Pharmaceutical Sciences, 3, 2014, 49-52

[6] G. Malaguarnera, E. Cataudella, M. Giordano, G. Nunnari, G. Chisari, M. Malaguarnera, Toxic hepatitis in occupational exposure to solvents. World J Gastroenterol. 18(2012) 2756–2766.

[7] R. Lauwerys, A. Bernard, C. Viau, J. P. Buchet, Kidney disorders and hematotoxicity from organic solvent exposure. Scand J Work Environ Health 11(1985) 83-90

[8] T. Welton, Solvents and sustainable chemistry. Proc Math Phys Eng Sci., 471(2015) 20150502 -18

[9] H.C. Hailes, Reaction Solvent Selection: The Potential of Water as a Solvent for Organic Transformations Org. Process Res. Dev. 11 (2007) 114-120.

[10] C. J. Li, L. Chen, Organic chemistry in water Chem. Soc. Rev. 35 (2006) 68-82.

[11] M.O. Simon, C.J. Li, Green chemistry oriented organic synthesis in water Chem. Soc. Rev. 41 (2012) 1415-1427.

[12] D.C. Rideout, R. Breslow, Hydrophobic acceleration of Diels-Alder reactions J. Am. Chem. Soc. 102 (1980) 7816-7817.

[13] M.A. Hill Bembenic, C.E. Burgess Clifford,Subcritical Water Reactions of a Hardwood Derived Organosolv Lignin with Nitrogen, Hydrogen, Carbon Monoxide, and Carbon Dioxide Gases Energy Fuels 26 (2012) 4540-4549.

[14] M.A. Hill Bembenic, C.E. Burgess Clifford, Subcritical Water Reactions of Lignin-Related Model Compounds with Nitrogen, Hydrogen, Carbon Monoxide, and Carbon Dioxide Gases Energy Fuels 27 (2013) 6681-6694.

[15] M. Mooller, P. Nilges, F. Harnisch, U. Schroder, Subcritical water as reaction environment: fundamentals of hydrothermal biomass transformation. ChemSusChem 4 (2011) 566-579.

[16] S. Avola, F. Goettmann, M. Antonietti, W. Kunz, Organic reactivity of alcohols in superheated aqueous salt solutions: an overview New J. Chem. 36 (2012) 1568-1577.

[17] B. Smutek, W. Kunz, F. Goettmann, C. R.Hydrothermal alkylation of phenols with alcohols in diluted acids, Chimie 15 (2012) 96-101.

[18] J. A. Branch, P. N., Bartlett Electrochemistry in supercritical fluid. Philos Trans A Math Eng Sci,; 373(2015) 20150007-16

[19] S. P. Nalawade, F. Picchioni, L. P. B. M. Janssen, Supercritical carbon dioxide as a green solvent for processing polymer melts: Processing aspects and applications. Progress in polymer science 31(2006) 19-43

[20] M. Herrero, A. Cifuentes, E. Ibanez, Sub- and supercritical fluid extraction of functional ingredients from different natural sources: Plants, food-by-products, algae and microalgae: A review Food Chem. 98 (2006) 136-148.

[21] J. H. Clark, S. J. Taverner, Alternative solvents: shades of green. Org. Process Res. Dev., 11 (2007) 149–155.

[22] D. Zhao, Y. Liao, Z. Zhang, Toxicity of Ionic Liquids. Clean 35 (2007) 42-48.

[23] A. Romero, J. Santos, A. Tojo, Rodríguez, Toxicity and biodegradability of imidazolium ionic liquids J. Hazard. Mater. 151 (2008) 268-273.

[24] T.P.T. Pham, C.-W. Cho, Y.-S. Yun, Environmental fate and toxicity of ionic liquids: a review Water Res. 44 (2010) 352-372.

[25] Q. Yang, Y. Zhang, F. Fei, P. Zhou, Y. Wang, P. Deng, Biodegradable betaine-based aprotic task-specific ionic liquids and their application in efficient SO_2 absorption Green Chem. 17 (2015) 3798-3805.

[26] G. H. Tao, L. He, W.-S. Liu, L. Xu, W. Xiong, T. Wang, Y. Kou,Preparation, characterization and application of amino acid-based green ionic liquids Green Chem. 8 (2006) 639-646.

[27] B. L. Gadilohar, G.S. Shankarling, Choline based ionic liquids and their applications in organic transformation J. Mol. Liq. 227 (2017) 234-261.

[28] Y. Fukaya, Y. Iizuka, K. Sekikawa, H. Ohno, Bio ionic liquids: room temperature ionic liquids composed wholly of biomaterials Green Chem. 9 (2007) 1155.

[29] H. Garcia, R. Ferreira, M. Petkovic, J.L. Ferguson, M.C. Leit~ao, H.Q.N. Gunaratne, K.R. Seddon, L.P.N. Rebelo, C. Silva Pereira, Dissolution of cork biopolymers in biocompatible ionic liquids Green Chem. 12 (2010) 367.

[30] M. Zech, S. Kellermeier, E. Thomaier, R. Maurer, C. Klein, W. Schreiner, F. Kunz, Alkali Metal Oligoether Carboxylates—A New Class of Ionic Liquids Chemistry 15 (2009) 1341-1345.

[31] Y. Zhang, B.R. Bakshi, E.S. Demessie, Life Cycle Assessment of an Ionic Liquid versus Molecular Solvents and Their Applications Environ. Sci. Technol. 42 (2008) 1724-1730.

[32] W. Kunz, K. Hackl,The hype with ionic liquids as solvents Chem. Phys. Lett. 661 (2016) 6-12.

[33] A. P. Abbott, G. Capper, D.L. Davies, R.K. Rasheed, V. Tambyrajah,Novel solvent properties of choline chloride/urea mixtures Chem. Commun. (2003) 70-71.

[34] M. Hayyan, Y.P. Mbous, C.Y. Looi, W.F. Wong, A. Hayyan, Z. Salleh, O. Mohd-Ali, Natural deep eutectic solvents: cytotoxic profile SpringerPlus 5 (2016) 913-922.

[35] I. Juneidi, M. Hayyan, M.A. Hashim,Evaluation of toxicity and biodegradability for cholinium-based deep eutectic solvents RSC Adv. 5 (2015) 83636-83647.

[36] Q. Wen, J.-X. Chen, Y.-L. Tang, J. Wang, Z. Yang,Assessing the toxicity and biodegradability of deep eutectic solvents Chemosphere 132 (2015) 63-69.

[37] V. Fischer, D. Touraud, W. Kunz, Eco-friendly one pot synthesis of caffeic acid phenethyl ester (CAPE) via an in-situ formed deep eutectic solvent Sustain. Chem. Pharm. 4 (2016) 40-45.

[38] S. Chandrasekhar, C. Narsihmulu, S.S Sultana, N. R. Reddy, Poly(ethylene glycol) (PEG) as a reusable solvent medium for organic synthesis. Application in the Heck reaction. Org. Lett. 4(2002) 4399–4401.

[39] N. F. Leininger, R. Clontz, J. L. Gainer, D. J. Kirwan, Polyethylene glycol-water and polypropylene glycol-water solutions as benign reaction solvents. Chem. Eng. Commun., 190 (2003) 431–444.

[40] D. J. Heldebrant, H. N. Witt, S. M. Walsh, T. Ellis, J. Rauscher, P.G. Jessop Liquid polymers as solvents for catalytic reductions. Green Chem. 8 (2006) 807–815.

[41] K. Verschueren, Handbook of environmental data on organic chemicals, 4th edition, Wiley, New York, 2001

[42] J. Ulbricht, R. Jordan, R. Luxenhofer, On the biodegradability of polyethylene glycol, polypeptoids and poly(2-oxazoline)s. Biomaterials, 35 (2014) 4848 -61

[43] L. Lomba, B. Giner, I. Bandrés, C. Lafuenteb and M. R. Pinoa, Physicochemical properties of green solvents derived from biomass., Green Chem., 13 (2011) 2062-2070.

[44] B. Giner, C. Lafuente, A. Villares, M. Haro and M. C. Lopez,Volumetric and refractive properties of binary mixtures containing 1,4-dioxane and chloroalkanes J. Chem. Thermodyn., 2007, 39, 148–157.

[45] J. A. Riddick, W. B. Bunger, T. Sakano, and A.Weissberger, Organic solvents: physical properties and methods of purification, Wiley, New York, 1986.

[46] K.J. Kim, U. M. Diwekar, Efficient combinatorial optimization under uncertainty. Part II. Application to stochastic solvent selection. Ind Eng Chem Res 41 (2002) 1285–1296

[47] L. Chao-Jun and B. M. Trost, Green chemistry for chemical synthesis PNAS, 105 (2008) 13197-13202.

[48] C. Reichardt, Solvents and Solvent Effects: An Introduction, Org. Process Res. Dev., 11 (2007) 105–113.

[49] W. C. Berkeley, J. Zhang, Green process chemistry in the pharmaceutical industry. Green Chem. Lett. Rev., 2(2009) 193-211

[50] V. I. Parvulescu, C. Hardacre, Catalysis in ionic liquids. Chem. Rev., 107(2007) 2615–2665

[51] T. Agata, Green solvents. J. Edu., Health and Sport.;7(2017) 224-232

[52] K. Hackl, W. Kunz* Some aspects of green solvents C. R. Chimie, 21 (2018) 572-580.

[53] K. Shanab, C. Neudorfer, E. Schirmer, H. Spreitze, Current Organic Green Solvents in Organic Synthesis: An Overview Chemistry, 17 (2013)1179-1187.

[54] Y. Gu, Multicomponent reactions in unconventional solvents: state of the art. Green Chem., 14 (2012) 2091-2128.

[55] M. Tobiszewsk, J. Namiesnik, Greener organic solvents in analytical chemistry.Curr. Opi Green Sust Chem 5 (2017)1–4

[56] Z. Li, Z. Jia, Y. Luan, Tiancheng, Mud Ionic liquids for synthesis of inorganic nanomaterials , Curr Opin Solid St M 12 (2008) 1–8.

[57] Y. Lu, S. Ozcan, Green nanomaterials: On track for a sustainable future, Nano Today 10 (2015) 417—420

[58] R. A. Sheldon, Green solvents for sustainable organic synthesis: state of the art Green Chem., 7 (2005) 267-268

[59] H. Duan, D. Wang and Y. Li, Green chemistry for nanoparticle synthesis. Chem. Soc. Rev., 44 (2015) 5778-5792

[60] J. M. DeSimone, Practical approaches to green solvents, Science, 297(2002) 799–803.

[61] P.G. Jessop, Searching for green solvents Green Chem. 13 (2011) 1391.

[62] B. Cornils and E. Wiebus, Paradigms in Green Chemistry and Technology Recl. Trav. Chim. Pays-Bas, 115 (1996) 211–215.

[63] E.G. Kuntz, Fr. Patent 2314910 (1975).

[64] F. Joo, Z. Toth, Catalysis by water-soluble phosphine complexes of transition metal ions in aqueous and two-phase media. J. Mol. Catal., 8 (1980) 369-383

[65] S. Takashi, J. Dakka, R.A. Sheldon, Titanium-substituted zeolite beta(Ti- l- □)-catalysed epoxidation of oct-1-ene with tert-butyl hydroperoxide(TBHP). J. Chem. Soc., Chem. Commun., 8 (1994) 1887-1888.

[66] L. M. Papadogianakis, R. A. Sheldon, Catalytic conversions in water. Part 5: Carbonylation of 1-(4-iso-butylphenyl) ethanol to ibuprofen catalyzed by water-soluble palladium-phosphine complexes in a two-phase system. J. Chem. Technol. Biotechnol., 70 (1997) 83-91.

[67] G. J. Ten Brink, I. W. Arends, Sheldon, R.A. Green, catalytic oxidation of alcohols in water. Science, 287(2000) 1636-1639

[68] M. A. Wegman, H. A. Janssen, F. van Rantwijk, R.A. Sheldon, Towards biocatalytic synthesis of β-lactam antibiotics. Adv. Synth. Catal., 343 (2001) 559-576

[69] K. Oyama, In: Chirality in Industry; Collins, Sheldrake, Crosby, Eds.; Wile & Sons: NY, (1992) pp. 237-248.

[70] J. E. Gavagan, S. K. Fager, J. E. Seip, M. S. Payne, D. L. Anton, R. DiCosimo, Glyoxylic Acid Production Using Microbial Transformant Catalysts. J. Org. Chem., 60 (1995) 3957-3963

[71] Y. F. Han and M. Xia, Multicomponent Synthesis of Cyclic Frameworks on Knoevenagel-Initiated Domino Reactions Curr. Org. Chem., 14 (2010) 379–413

[72] B. Jiang and S.-J. Tu, Chimia, Active Methylene-based Multicomponent Reactions under Microwave Heating 65 (2011) 925–931.

[73] Y. Gu, R. De Sousa, G. Frapper, C. Bachmann, J. Barrault and F. Jérôme, Catalyst-free aqueous multicomponent domino reactions from formaldehyde and 1,3-dicarbonyl derivatives Green Chem., 11 (2009) 1968–1972

[74] M. Li, C. Tang, J. Yang and Y. Gu,Ring-opening reactions of 2-aryl-3, 4-dihydropyrans with nucleophiles Chem. Commun., 47 (2011) 4529–4531;

[75] M. Li, H. Li and Y. Gu, Ring-Opening Reactions of 2-Alkoxy-3,4-dihydropyrans with Thiols or Thiophenols Org. Lett., 13 (2011) 1064–1067

[76] M. Li, J. Yang and Y. Gu, Manganese Chloride as an Efficient Catalyst for Selective Transformations of Indoles in the Presence of a Keto Carbonyl Group Adv. Synth. Catal., 353 (2011)1551–1564;

[77] M. Li and Y. Gu, 2-Aryl-3,4-dihydropyrans as building blocks for organic synthesis: ring-opening reactions with nucleophiles Tetrahedron, 67 (2011) 8314–8320

[78] G. Frapper, C. Bachmann, Y. Gu, R. C. De Sousa and F. Jérôme, Mechanisms of the Knoevenagel hetero Diels–Alder sequence in multicomponent reactions to dihydropyrans: experimental and theoretical investigations into the role of water, Phys. Chem. Chem. Phys., 13 (2011) 628–636

[79] Kumar, A.; Maurya, R. A.'An unusual Mannich type reaction of tertiary aromatic amines in aqueous micelles'. Tetrahedron Lett., 49 (2008) 5471-5481

[80] L. Li, B. Liu, and X. Lin, Catalyst-free Multicomponent Synthesis of β-Mercapto Diketones in Water Chin. J. Chem., 29(2011) 1856–1862.

[81] A. Solhy, E. Tahir, A. R. Karkouri, M. Larzek, M. Bousmina, M. Zahouily, Clean chemical synthesis of 2-aminochromenes in water catalyzed by nanostructured diphosphate Na2CaP2O7 Green'. Chem., 12 (2010) 2261.

[82] X. Liu, J. Ma, W. Zhang, Applications of ionic liquids (ILs) in the convenient synthesis of nanomaterials, Rev. Adv. Mater. Sci. 27 (2011) 43–51.

[83] K. Ablajan and H. Xiamuxi, Efficient One-Pot Synthesis of β-Unsaturated Isoxazol-5-ones and Pyrazol-5-ones Under Ultrasonic Irradiation, Synth. Commun., 42 (2012) 1128–1136

[84] Q. Liu and R.-T. Wu, Facile synthesis of 3-methyl-4-arylmethylene-isoxazol-5(4H)-ones catalysed by sodium silicate in an aqueous medium J. Chem. Res., 35 (2011) 35, 598–599.

[85] K. Kumaravel, G. Vasuki, 'Multi-Component Reactions in Water'.Curr. Org. Chem. 13 (2009) 1820-1825.

[86] P. Gunasekaran, S. Perumal, P. Yogeeswari and D. Sriram, A facile four-component sequential protocol in the expedient synthesis of novel 2-aryl-5-methyl-2,3-dihydro-1H-3-pyrazolones in water and their antitubercular evaluation. Eur. J. Med. Chem., 46 (2011) 4530–4536.

[87] P. V. Shinde, S. S. Sonar, B. B. Shingate and M. S. Shingare, Boric acid catalyzed convenient synthesis of 2-amino-3,5-dicarbonitrile-6-thio-pyridines in aqueous media Tetrahedron Lett., 51 (2010) 1309–1312.

[88] P. V. Shinde, B. B. Shingate and M. S. Shingare, Aqueous Suspension of Basic Alumina: An Efficient Catalytic System for the Synthesis of Poly Functionalized Pyridines Bull. Korean Chem. Soc., 32 (2011) 459–462.

[89] Z. Q. Wang, Z.-M. Ge, T.-M. Cheng and R.-T. Li, Synthesis of Highly Substituted Pyridines via a One-Pot, Three-Component Cascade Reaction of Malononitrile with Aldehydes and S-Alkylisothiouronium Salts in Water Synlett, 22 (2009) 2020–2022.

[90] Y. Zhou, J. Wang, R. Du, G. Zhang, W. Wang and C. Guo, Microwave-Assisted One-Pot Synthesis of 3-Amino-1-aryl-8-bromo-2,4-dicyano-9H-fluorenes in Water Synth. Commun., 2011, 41, 3169–3176

[91] C. Mukhopadhyay and S. Ray, Synthesis of 2-amino-5-alkylidenethiazol-4-ones from ketones, rhodanine, and amines with the aid of re-usable heterogeneous silica-pyridine based catalyst Tetrahedron Lett., 67 (2011) 7936–7945.

[92] G.P. Lu, L.Y. Zeng and C. Cai, An efficient synthesis of dihydrothiophene ureidoformamides by domino reactions of 1,3-thiazolidinedione under catalyst-free conditions Green Chem., 13 (2011) 998–1003

[93] M. Hadjebi, M. S. Hashtroudi, H. R. Bijanzadeh and S. Balalaie, Novel Four-Component Approach for the Synthesis of Polyfunctionalized 1,4-Dihydropyridines in Aqueous Media Helv. Chim. Acta, 94 (2011) 382–388.

[94] S. M. Rajesh, B. D. Bala, S. Perumal and J. C. Menéndez, L-Proline-catalysed sequential four-component "on water" protocol for the synthesis of structurally complex heterocyclic ortho-quinones Green Chem., 13 (2011) 3248–3254.

[95] Y. Zou, H. Wu, Y. Hu, H. Liu, X. Zhao, H. Ji and D. Shi, A novel and environment-friendly method for preparing dihydropyrano[2,3-c]pyrazoles in water under ultrasound irradiation Ultrason. Sonochem., 18 (2011) 708–712

[96] S. J. Chai, Y. F. Lai, J. C. Xu, H. Zheng, Q. Zhu and P. F. Zhang, One-Pot Synthesis of Spirooxindole Derivatives Catalyzed by Lipase in the Presence of Water Adv. Synth. Catal., 353 (2011) 371–375.

[97] L. M. Wang, N. Jiao, J. Qiu, J. J. Yu, J. Q. Liu, F. L. Guo and Y. Liu, Sodium stearate-catalyzed multicomponent reactions for efficient synthesis of spirooxindoles in aqueous micellar media Tetrahedron, 66(2010) 339–343.

[98] Y. Li, H. Chen, C. Shi, D. Shi and S. Ji, Efficient One-Pot Synthesis of Spirooxindole Derivatives Catalyzed by l-Proline in Aqueous Medium J. Comb. Chem., 12 (2010) 231–237.

[99] R. Ghahremanzadeh, T. Amanpour and A. Bazgir, Clean synthesis of spiro[indole-3,8'-phenaleno[1,2-b]pyran]-9'-carbonitriles and spiro[indole-3,4'-pyrano[4,3-b]pyran]-3'-carbonitriles by one-pot, three-component reactions J. Heterocycl. Chem., 47 (2010) 46–49.

[100] V. Dandia, A. K. Jain and K. S. Rathore, Step-economic, efficient, ZnS nanoparticle-catalyzed synthesis of spirooxindole derivatives in aqueous medium via Knoevenagel condensation followed by Michael addition Green Chem., 13 (2011) 2135–2145.

[101] M. Rimaz and J. Khalafy, ARKIVOC, A novel one-pot, three-component synthesis of alkyl 6-aryl-3-methylpyridazine-4-carboxylates in water, ARKIVOC, 11 (2010) 110–117.

[102] M. Wang, T. T. Zhang, Y. Liang and J. J. Gao, Strontium chloride-catalyzed one-pot synthesis of 2, 3-dihydroquinazolin-4(1H)-ones in protic media Chin. Chem. Lett., 22(2011) 1423–1426.

[103] Z. H. Zhang, H. Y. Lü, S. H. Yang and J. W. Gao, Synthesis of 2,3-Dihydroquinazolin-4(1H)-ones by Three-Component Coupling of Isatoic Anhydride, Amines, and Aldehydes Catalyzed by Magnetic Fe3O4 Nanoparticles in Water J. Comb. Chem., 12 (2010) 643–646.

[104] M. Narasimhulu and Y. R. Lee, Ethylenediamine diacetate-catalyzed three-component reaction for the synthesis of 2, 3-dihydroquinazolin-4 (1H)-ones and their spirooxindole derivatives Tetrahedron, 67 (2011) 9627–9634.

[105] J. M. Khurana and S. Kumar, An efficient, catalyst free synthesis of 3-(2″-benzothiazolyl)-2,3-dihydroquinazolin-4(1H)-ones in aqueous medium Green Chem. Lett. Rev., 4 (2011) 321– 325.

[106] A. Alizadeh and N. Zohreh, A unique approach to catalyst-free, one-pot synthesis of spirooxindole-pyrazolines Synlett, 23 (2012) 428–432.

[107] F. Cruz-Acosta, P. de Armas and F. García-Tellado, A Metal-Free, Three-Component Manifold for the C2-Functionalization of 1-Substituted Imidazoles Operating 'On Water Synlett, 16 (2010) 2421– 2424.

[108] K. Tanaka and R. Shiraishi, Clean and efficient condensation reactions of aldehydes and amines in a water suspension medium Green Chem., 2 (2000) 272–273.

[109] N. Azizi and A. Davoudpour, Highly Efficient One-Pot Three-Component Mannich Reaction in Water Catalyzed by Heteropoly Acids Org. Lett., 141 (2011) 1506–1510.

[110] G. Li, H. Wu, Z. Wang and X. Wang, One-pot three-component Mannich-type reaction catalyzed by trifluoromethanesulfonic acid in water, Kinet. Catal., 52 (2011) 89–93.

[111] G. Zhang, Z. Huang and J. Zou, Ga(OTf)$_3$-catalyzed Three-component Mannich Reaction in Water Promoted by Ultrasound Irradiation, Chin. J. Chem., 27 (2009) 1967–1974

[112] H. Wu, X.-M. Chen, Y. Wan, L. Ye, H.-Q. Xin, H.-H. Xu, C.-H. Yue, L.-L. Pang, R. Ma and D.-Q. Shi, Stereoselective Mannich reactions catalyzed by Tröger's base derivatives in aqueous media Tetrahedron Lett., 50 (2009) 1062–1065.

[113] X.-S. Wang, J. Zhou, K. Yang and Q. Li, Facile and Green Method for the Synthesis of β-Aminoketone Derivatives in Aqueous Media, Synth. Commun., 40 (2010) 964–972.

[114] D. Hong, Y.-Y. Yang, Y.-G. Wang and X.-F. Lin, A Yb (OTf)$_3$/PEG-supported quaternary ammonium salt catalyst system for a three-component Mannich-type reaction in aqueous media, Synlett, 7 (2009) 1107–1110.

[115] Z. Xie, G. Li, G. Zhao and J. Wang, Three-component Synthesis of Homoallylic Amines Catalyzed by Phosphomolybdic Acid in Water Chin. J. Chem., 27 (2009) 925–929.

[116] Y. Yuan, F. Chen and D. Zhao, Iron-catalyzed efficient three-component allylation of imine in aqueous media Appl. Organomet. Chem., 23(2009) 485–491.

[117] K. Ando and T. Egami, Facile synthesis of α-amino phosphonates in water by Kabachnik–Fields reaction using magnesium dodecyl sulfate Heteroat. Chem., 22(2011) 358–362.

[118] B. Karmakar and J. Banerji, K$_2$PdCl$_4$ catalyzed efficient multicomponent synthesis of α-aminonitriles in aqueous mediaTetrahedron Lett., 51 (2010) 2748–2750.

[119] D. Fang, C. Jiao and C. Ni, SO3H-functionalized ionic liquids catalyzed the synthesis of α-aminophosphonates in aqueous media Heteroat. Chem., 21(2010) 546–550

[120] S. Sobhani and A. Vafaee Micellar solution of sodium dodecyl sulfate (SDS) catalyzes Kabacknik-Fields reaction in aqueous media, Synthesis, 11 (2009) 1909–1915

[121] M. N. Sefat, D. Saberi and K. Niknam, Preparation of silica-based ionic liquid an efficient and recyclable catalyst for one-pot synthesis of α-aminonitriles Catal. Lett., 141 (2011) 1713–1720.

Industrial Applications of Green Solvents I   Materials Research Forum LLC
Materials Research Foundations **50** (2019) 165-241   doi: https://doi.org/10.21741/9781644900239-7

[122] G. Brahmachari and S. Das, Bismuth nitrate-catalyzed multicomponent reaction for efficient and one-pot synthesis of densely functionalized piperidine scaffolds at room temperature Tetrahedron Lett., 53(2012) 1479–1484

[123] B. Das, D. B. Shinde, B. S. Kanth and G. Satyalakshmi, An efficient multicomponent synthesis of polysubstituted pyrrolidines and tetrahydropyrimidines starting directly from nitro compounds in water Synthesis, 16 (2010) 2823–2827.

[124] F. García-Tellado, F. Cruz-Acosta, A. Santos-Expósito, P. de Armas, Lewis base-catalyzed three-component Strecker reaction on water. An efficient manifold for the direct α-cyanoamination of ketones and aldehydes Chem. Commun., 0(2009) 6839–6841.

[125] P. Galletti, M. Pori and D. Giacomini, Catalyst-Free Strecker Reaction in Water: A Simple and Efficient Protocol Using Acetone Cyanohydrin as Cyanide Source Eur. J. Org. Chem.,11(2011) 3896–3903.

[126] B. P. Mathew and M. Nath, ne-pot three-component synthesis of dihydrobenzo- and naphtho[e]-1,3-oxazines in water J. Heterocycl. Chem., 46 (2009) 1003–1006.

[127] A. Sartori, L. Dell'Amico, C. Curti, L. Battistini, G. Pelosi, G. Rassu, G. Casiraghi and F. Zanardi, Aqueous and Solvent-Free Uncatalyzed Three-Component Vinylogous Mukaiyama–Mannich Reactions of Pyrrole-Based Silyl Dienolates Adv. Synth. Catal., 353 (2011) 3278–3284.

[128] D.N. Zhang, J.-T. Li, Y.-L. Song, H.-M. Liu and H.-Y. Li, Efficient one-pot three-component synthesis of N-(4-arylthiazol-2-yl) hydrazones in water under ultrasound irradiation. Ultrason. Sonochem., 19 (2012) 475–478.

[129] Z. Halimehjani, K. Marjani and A. Ashouri, Synthesis of dithiocarbamate by Markovnikov addition reaction in aqueous medium Green Chem., 12 (2010) 1306–1310.

[130] B. Karmakar and J. Banerji, K2PdCl4 catalyzed efficient multicomponent synthesis of α-aminonitriles in aqueous media. Tetrahedron Lett., 51 (2010) 2748–2750

[131] B. C. Ranu, T. ChatterjeeandS. Bhadra Transition metal-free procedure for the synthesis of S-aryl dithiocarbamates using aryl diazonium fluoroborate in water at room temperature Green Chem., 13 (2011) 1837–1842.

[132] S. Santra and P. R. Andreana, A Bioinspired Ugi/Michael/Aza-Michael Cascade Reaction in Aqueous Media: Natural-Product-like Molecular Diversity Angew. Chem., Int. Ed., 50 (2011) 9418–9422.

[133] A. Shaabani, A. Sarvary, S. Ghasemi, A. H. Rezayan, R. Ghadari and S. W. Ng, Green Chem., An environmentally benign approach for the synthesis of bifunctional sulfonamide-amide compounds viaisocyanide-based multicomponent reactions 13 (2011) 582–585.

[134] A. Ramazani, A. Rezaei, A. T. Mahyari, M. Rouhani and M. Khoobi, Three-Component Reaction of an Isocyanide and a Dialkyl Acetylenedicarboxylate with a Phenacyl Halide in the Presence of Water: An Efficient Method for the One-Pot Synthesis of γ-Iminolactone Derivatives Helv. Chim. Acta, 93 (2010) 2033–2036.

[135] R. Sarma, M. M. Sarmah, K. C. Lekhok and D. Prajapati, organic reactions in water: an efficient synthesis of pyranocoumarin derivatives Synlett, 12 (2010) 2847–2852.

[136] X. Zhu, X.-P. Xu, C. Sun, H.-Y. Wang, K. Zhao and S.-J. Ji, J. Comb. Direct construction of imino-pyrrolidine-thione scaffold via isocyanide-based Multicomponent reaction Chem., 12 (2010) 822–828.

[137] S. Pirrung, O. Kreye, and M. A. R. Ansgar, Tunable Polymers Obtained from Passerini Multicomponent Reaction Derived Acrylate Monomers Macromolecules 46 (2013) 6031-6037

[138] F. Marinelli, Cu-Mediated Organic Transformations in Water Curr. Org. Synth., 9 (2012) 2–16.

[139] J. Le Bras and J. Muzart, From Metal-Catalyzed Reactions with Hydrosoluble Ligands to Reactions in and on Water Curr. Org. Synth., 8 (2011) 330–334

[140] W. Wang, J. Wu, C. Xia and F. Li, Reusable ammonium salt-tagged NHC–Cu (I) complexes: preparation and catalytic application in the three component click reaction Green Chem.,13 (2011) 3440–3445

[141] M. Liu and O. Reiser, A copper (I) isonitrile complex as a heterogeneous catalyst for azide– alkyne cycloaddition in water Org. Lett., 13 (2011) 1102–1105.

[142] Y. Wang, J. Liu and C. Xia, Insights into supported copper(II)-catalyzed azide–alkyne cycloaddition in wate Adv. Synth. Catal., 353 (2011) 1534–1542

[143] R. B. Nasir Baig and R. S. Varma, A highly active magnetically recoverable nano ferrite-glutathione-copper (nano-FGT-Cu) catalyst for Huisgen 1, 3-dipolar cycloadditions Green Chem., 14 (2012) 625–632

[144] F. Alonso, Y. Moglie, G. Radivoy and M. Yus, Multicomponent Synthesis of 1, 2, 3-Triazoles in Water Catalyzed by Copper Nanoparticles on Activated Carbon Adv. Synth. Catal., 352 (2010) 3208–3214.

[145] M. Lei, W.-Z. Song, Z.-J. Zhan, S.-L. Cui and F.-R. Zhong, Multicomponent Reactions Stereo- and Regioselective Three-Component Reaction in Water: Synthesis of Triazole Substituted β-Lactams Via Click Chemistry Org. Chem. Lett., 8 (2011) 163–169.

[146] M. Dabiri, P. Salehi, M. Bahramnejad, F. Sherafat and M. Bararjanian, Facile and highly efficient procedure for the synthesis of triazolyl methoxyphenyl 1, 8-dioxo-decahydroacridines via one-pot, pseudo-five-component reaction Synth. Commun., 42 (2012) 3117-3127,

[147] S. R. Adapa, E., Ramu, R., Varala, N. Sreelatha, Zn(OAc)$_2$·2H$_2$O: a versatile catalyst for the onepot synthesis of propargylamines. Tetrahedron Lett., 48(2007). 7184-7190.

[148] D. Kumar, G. Patel and V. B. Reddy, Greener and expeditious synthesis of 1, 4-disubstituted 1, 2, 3-triazoles from terminal acetylenes and in situ generated α-azido ketones Synlett, 13 (2009) 399–402.

[149] I. T. Horvath and J. Rabai, Facile catalyst separation without water: fluorous biphase hydroformylation of olefins Science, 266 (1994) 72–75

[150] I. T. Horva´ th, Fluorous biphase chemistry Acc. Chem. Res., 31(1998) 641–650.

[151] J. A. Gladysz, D. P. Curran and I. T. Horvath, Wiley VCH, Handbook of Fluorous Chemistry, ed. Weinheim, 2004

[152] Orha, L.; Akien, G. R.; Horvath, I. T. In: Handbook of Green Chemistry; Wiley: 2012; Vol 7, pp. 93-120

[153] Carreira, M.; Contel, M. In: Fluorous Chemistry; Horváth, Ed.; Springer Berlin: Berlin, Heidelberg, 2012; Vol. 308, pp. 247-274.

[154] Yi, W.-B.; Cai, C.; Wang, X. A novel ytterbium/perfluoroalkylated-pyridine catalyst for Baylis-Hillman reactions in a fluorous biphasic system. J. Fluorine Chem., 128 (2007) 919-924.

[155] Yi, W.-B.; Cai, C.; Wang, X. Mannich-type reactions of aromatic aldehydes, anilines, and methyl ketones in fluorous biphase systems created by rare earth(III) perfluorooctanesulfonate catalysts in fluorous media. J. Fluorine Chem., 127 (2006) 1515-1521

[156] A. P. Dobbs, M. R. Kimberley, Fluorous phase chemistry: a new industrial technology. J. Fluorine Chem., 118 (2002) 3-17.

[157] A. Heydari, S. Khaksar, M. Tajbakhsh, Trifluoroethanol as a metal-free, homogeneous and recyclable medium for the efficient one-pot synthesis o Î±-amino nitriles and Î±-amino phosphonates. Tetrahedron Lett., 50 (2009) 77-80.

[158] A. Aghmiz, C. Claver, A. M. Masdeu-Bultó, D. Maillard, D. Sinou, Hydroformylation of 1-octene with rhodium catalysts in fluorous systems. J. Mol. Catal. A: Chem., 208 (2004) 97-101.

[159] A. Caballero, P. J. Perez, M.A. Fuentes, M. Etienne, B. K. Munoz, Abstracts of Papers, 244th ACS National Meeting & Exposition: Philadelphia, PA, United States, August 19-23, 2012 , pp. INOR-464

[160] K. P. Johnston and P. S. Shah, Materials science. Making nanoscale materials with supercritical fluids. Science, 303 (2004) 482–483.

[161] P. Pollet, E. A. Davey, E. E. Uren~a-Benavides, C. A. Eckert and C. L. Liotta, Solvents for sustainable chemical processes Green Chem., 16 (2014) 1034-1039

[162] T. Adschiri, Y.-W. Lee, M. Goto and S. Takami, Green materials synthesis with supercritical water Green Chem., 13 (2011) 1380-1386.

[163] K. Sue, T. Adschiri and K. Arai, Predictive Model for Equilibrium Constants of Aqueous Inorganic Species at Subcritical and Supercritical Conditions Ind. Eng. Chem. Res.,41 (2002) 3298–3306.

[164] R. Sui and P. Charpentier, Synthesis of Metal Oxide Nanostructures by Direct Sol–Gel Chemistry in Supercritical Fluids Chem. Rev., 112 (2012) 3057–3082

[165] R. A. Pai, R. Humayun, M. T. Schulberg, A. Sengupta, J. N. Sun and J. J. Watkins, Mesoporous silicates prepared using preorganized templates in supercritical fluids. Science, 303 (2004) 507–510.

[166] Ren, W.; Rutz, B.; Scurto, A. M. High-pressure phase equilibrium for the hydroformylation of 1-octene to nonanal in compressed CO_2. J. of Supercritical Fluids, 51 (2009) 142-147.

[167] A. C. J. Koeken, M. C. A. van Vliet, L. J. P. van den Broeke, B. J. Deelman, J. T. F. Keurentjes, Selectivity of rhodium-catalyzed hydroformylation of 1-octene during batch and semi-batch reaction using trifluoromethyl substituted ligands. Adv. Synth. Catal., 350 (2008) 179-188.

[168] A. C. Frisch,; Webb, P. B.; Zhao, G.; Muldoon, M. J.; Pogorzelec, P. J.; Cole-Hamilton, D. J. Emerging Strategies in Catalysis. Dalton Trans., 18 (2007) 5531-5538.

[169] B. Schäffner, J. Holz, S. P. Verevkin, A. Börner, Rhodium-catalyzed asymmetric hydrogenation with self-assembling catalysts in propylene carbonate. Tetrahedron Lett., 49(2008) 768-771.

[170] C. Li, B.M. Trost, Green chemistry for chemical synthesis. PNAS, 2008, 105, 13197-13202.

[171] CI Melo, R. Bogel-Kukasik, M. Gomes da Silva, E. Bogel-ukasik, Advantageous heterogeneously catalysed hydrogenation of carvone with supercritical carbon dioxide. Green Chem., 13 (2011) 2825-2830

[172] S. D. Endalkachew, G. D. Venu, A. H. Ashraf, Hydrogenation of Anthracene in Supercritical Carbon Dioxide Solvent Using Ni Supported on H2- Zeolite Catalyst. Catalysts, 2(2012) 85-100

[173] Y. Chen, Y. Wu, Y. Zhang, L. Long, L. Tao, M. Yang, M. Tang, Epoxidation of propylene to propylene oxide catalyzed by large-grain TS-1 in supercritical CO_2, J. Mol. Catal A: Chem, 352 (2012) 102-109

[174] H. F. Jiang, Q.X. Xu and A.Z. Wang, Stereoselective synthesis of tetrasubstituted olefins via palladium-catalyzed three-component coupling of aryl iodides, internal alkynes, and arylboronic acids in supercritical carbon dioxide J. Supercrit. Fluids, 49 (2009) 377–384

[175] F. Haviv, J.D. Ratajczyk, R.W. DeNet, F A Kerdesky, R L Walters, S.P Schmidt, J.H. Holms, P R Young, GW Carter, 3-[I-(2-Benzoxazolyl) hydrazinolpropanenitrile Derivatives: Inhibitors of Immune Complex Induced Inflammation. J. Med. Chem., 31 31(1988) 1719-1728.

[176] K. Tsuji, H. Ishikawa, Synthesis and anti-pseudomonal activity of new 2-isocephems with a dihydroxypyridone moiety at C-7. Bioorg. Med. Chem. Lett., 4 (1994) 1601-1606.

[177] F W. Bell, A.S. Cantrell, et al., . Phenethylthiazolethiourea (PEW) Compounds, a New Class of HIV- 1 Reverse Transcriptase Inhibitors. 1. Synthesis and Basic Structure-Activity Relationship Studies of PEW Analogs. J. Med. Chem., 38 (1995) 4929-4936.

[178] H. Zang, M. Wang. B.W. Cheng, J. Song, Ultrasound-promoted synthesis of oximes catalyzed by a basic ionic liquid [bmIm]OH. Ultrasonics Sonochem., 16 (2009) 301-303.

[179] X. Yannai, S. Ge, G. Rennert, N. Gruener, F.A. Fares, 3,3'- Diindolylmethane induces apoptosis in human cancer cells. Biochem. Biophys. Res. Comm., 228 (1996) 153-158

[180] X. H. Yuan, M. Chen, Q.X. Dai, X. N. Cheng, Friedel-Crafts acylation of anthracene with oxalyl chloride catalyzed by ionic liquid of [bmim]Cl/AlCl3. Chem. Eng. J., 146 (2009) 266-269

[181] H. G. Bonacorso, A. P. Wentz, et al., Trifluoromethyl-containing pyrazolinyl (p-tolyl) sulfones: The synthesis and structure of promising antimicrobial agents. J. Fluorine. Chem., 127 (2006) 1066-1072

[182] K. Gong, D. Fang, H.L. Wang, et al., The one-pot synthesis of 14 alkyl- or aryl-14H-dibenzo[a,j]xanthenes catalyzed by taskspecific ionic liquid. Dyes Pigments, 80 (2009) 80, 30-33.

[183] K. Jadidi, R. Ghahremanzadeh, D. Asgari, P. Eslami, H. Arvin-Nezhad, Eco-friendly synthesis of 1,4-benzodiazepine-2,5-diones in the ionic liquid [bmim]Br. Monatsh. Chem., 139 (2008) 1229-1232.

[184] D. N. Moreira, C.P. Frizzo, K. Longhi, N. Zanatta, H. G. Bonacorso, M. A. P. Martins, An efficient synthesis of 1-cyanoacetyl-5-halomethyl-4,5- dihydro-1H-pyrazoles in ionic liquid. Monatsh. Chem., 139 (2008) 1049-1054.

[185] H. Wakamatsu, Y. Saito, M. Masabuchi, R. Fujita, Synthesis of Imidazolium-Tagged Ruthenium Carbene Complex: Remarkable Activity and Reusability in Regard to Olefin Metathesis in Ionic Liquids. Synlett, 12 (2008) 1805-1808.

[186] S Kumar; J. S. Sandhu, Knoevenagel reaction: alum-mediated efficient green condensation of active methylene compounds with arylaldehydes. Green Chem. Lett. Rev., 2 (2009) 189-192.

[187] S. Kumar, J. S. Sandhu, An efficient green protocol for the production of 1,8-dioxo-octahydroxanthenes in triethylammonium acetate (TEAA), a recyclable inexpensive ionic liquid. J. Chem., 2, (2009) 937-940.

[188] S. Kumar, J. S. Sandhu Bismuth (III) chloride-mediated, efficient, solvent-free, MWI-enhanced Doebner condensation for the synthesis of (E)- cinnamic acids. Synth. Commun., 40 (2010) 1915-1919.

[189] Y. Zhang, C. Xia, Magnetic hydroxyapatite-encapsulated α- Fe_2O_3 nanoparticles functionalized with basic ionic liquids for aqueous Knoevenagel condensation. Appl. Cata. A: Gen., 366 (2009) 141-147.

[190] W. J. Wang, W.P. Cheng, L.L. Shao, C.H. Liu, J. G. Yang, Henry and Knoevenagel reactions catalyzed by methoxyl propylamine acetate ionic liquid. Kinet. Cata., 50 (2009) 186-191.

[191] C. Carrignon, P. Makowski, M. Antonietti, F. Goettmann, Chloride ion pairs as catalysts for the alkylation of aldehydes and ketones with C-H acidic compounds. Tetrahedron Lett., 50 (2009) 4833-4837.

[192] L.D.S. Yadav, S. Singha, V. K. Rai, A one-pot [Bmim]OH-mediated synthesis of 3-benzamidocoumarins. Tetrahedron Lett., 50 (2009) 2208-2212.

[193] Z. Zhou, J. Yuan, R. Yang Efficient Knoevenagel condensation catalyzed by 2-hydroxyethylammonium acetate under solvent-free conditions at room temperature Synth. Commun., 39 (2009) 2001-2007.

[194] D. Jiang, Y. Y. Wang, Y. N. Xu, L. Y. Dai, Doebner condensation in ionic liquids [Bmim]BF4 and [Bpy]BF4 to synthesize α,β-unsaturated carboxylic acid. Chin. Chem. Lett., 20 (2009) 279-282.

[195] B. Ni, Q. Zhang, K. Dhungana, A. D. Headley, Ionic Liquid-Supported (ILS) (S)-Pyrrolidine Sulfonamide, a Recyclable Organocatalyst for the highly enantioselective Michael Addition to Nitroolefins. Org. Lett., 11 (2009) 1037-1040.

[196] P. Singh, K. Kumari, A. Katyal, R Kalra, R. Chandra Copper Nanoparticles in Ionic Liquid: An Easy and Efficient Catalyst for Selective Carba- Michael Addition Reaction. Catal. Lett., 127 (2009) 119-125.

[197] M. A. Kolosov, V. D. Orlov, D.A. Beloborodov, V. V. Dotsenko, A chemical placebo. NaCl as an effective, cheapest, non-acidic and greener catalyst for Biginelli-type 3,4-dihydropyrimidin-2(1H)-ones (-thiones) synthesis. Mol. Divers., 13 (2009) 5-25.

[198] A. Suresh, A. Saini D. Kumar, J. S. Sandhu, Multicomponent eco-friendly synthesis of 3,4-dihydropyrimidine-2-(1H)-ones using an organocatalyst Lactic acid. Green Chem. Lett. Rev., 2, (2009) 29-33.

[199] S. S. Chavan, Y. O. Sharma, M. S. Degani, Cost-effective ionic liquid for environmentally friendly synthesis of 3,4-dihydropyrimidin-2(1H)-ones. Green Chem. Lett. Rev., 2, (2009) 175-179.

[200] S. Kumar, J. S. Sandhu, New efficient protocol for the production of Hantzsch 1,4-dihydropyridines using RuCl3. Synth. Commun., 39 (2009) 1957-1965.

[201] J. S. Sandhu, Recent advances in ionic liquids. Green unconventional solvents of this century. Part II. Green Chem. Lett. Rev., 4, (2011) 311-320.

[202] M. R. P. Heravi and F. Fakhr, Ultrasound-promoted synthesis of 2-amino-6-(arylthio)-4-arylpyridine-3,5-dicarbonitriles using ZrOCl2·8H2O/NaNH2 as the catalyst in the ionic liquid [bmim]BF4 at room temperature Tetrahedron Lett., 52 (2011) 6779–6782.

[203] A. K. Gupta, K. Kumari, N. Singh, D. S. Raghuvanshi and K. N. Singh, An eco-safe approach to benzopyranopyrimidines and 4H-chromenes in ionic liquid at room temperature Tetrahedron Lett., 53, (2012) 650–653

[204] G. Gupta, A. Kumar, S. Srivastava, Functional ionic liquid mediated synthesis (FILMS) of dihydrothiophenes and tacrine derivatives Green Chem., 13 (2011) 2459– 2463.

[205] V. K. Rao, B. S. Chhikara, R. Tiwari, A. N. Shirazi, K. Parang and A. Kumar, One-pot regioselective synthesis of tetrahydroindazolones and evaluation of their antiproliferative and Src kinase inhibitory activities Bioorg. Med. Chem. Lett., 22 (2012) 410–414.

[206] A. Kumar, M. S. Rao, I. Ahmad and B. Khungar, An Efficient and Simple One-Pot Synthesis of β-Acetamido Ketones Catalyzed by Ytterbium Triflate in Ionic Liquid Aust. J. Chem., 62 (2009) 322–327.

[207] S. S. Chavan, M. S. Degani, 'Ionic liquid mediated one-pot synthesis of 6-aminouracils'.Green Chem., 14(2012) 296-302.

[208] A. Hasaninejad, A. Zare, M. Shekouhy and J. Ameri Rad, Catalyst-Free One-Pot Four Component Synthesis of Polysubstituted Imidazoles in Neutral Ionic Liquid 1-Butyl-3-methylimidazolium Bromide J. Comb. Chem., 12, (2010) 844–849.

[209] Z. Xiao, M. Lei and L. Hu, An unexpected multi-component reaction to synthesis of 3-(5-amino-3-methyl-1H-pyrazol-4-yl)-3-arylpropanoic acids in ionic liquid Tetrahedron Lett., 52 (2011) 7099–7102.

[210] X. Zhang, X. Li, X. Fan, X. Wang, D. Li, G. Qu, J. Wang, Ionic liquid promoted preparation of 4H-thiopyran and pyrimidine nucleoside-thiopyran hybrids through one-pot multi-component reaction of thioamide'. Mol. Diversity., 13 (2009) 57-62.

[211] D. Q. Shi and F. Yang, An efficient synthesis of pyrazolo [3,4-b]quinolin-5(6H)-one derivatives in ionic liquids J. Heterocycl. Chem., 48(2011) 308–311

Industrial Applications of Green Solvents I   Materials Research Forum LLC
Materials Research Foundations **50** (2019) 165-241   doi: https://doi.org/10.21741/9781644900239-7

[212] S. Ramesh and R. Nagarajan, Efficient One-Pot Multicomponent Synthesis of (Carbazolylamino)furan-2(5H)-one and Carbazolyltetrahydropyrimidine Derivatives Synthesis, 7 (2011) 3307–3317.

[213] K. Rad-Moghadam and L. Youseftabar-Miri, Ambient synthesis of spiro[4H-pyran-oxindole] derivatives under [BMIm]BF4 catalysis Tetrahedron, 67 (2011) 5693–5699

[214] H. Y. Guo and Y. Yu, One-pot synthesis of 7-aryl-11,12-dihydrobenzo[h]pyrimido-[4,5-b]quinoline-8,10(7H,9H)-diones via three-component reaction in ionic liquid Chin. Chem. Lett., 21 (2010) 1435–1438.

[215] A. K. Yadav, G. R. Sharma, P. Dhakad and T. Yadav, A novel ionic liquid mediated synthesis of 4 (1H)-quinolones, 5H-thiazolo [3, 2-a] pyrimidin-5-one and 4H-pyrimido [2, 1-b] benzothiazol-4-ones Tetrahedron Lett., 53 (2012) 859–862.

[216] M. Shekouhy and A. Hasaninejad, Ultrasound-promoted catalyst-free one-pot four component synthesis of 2H-indazolo[2,1-b]phthalazine-triones in neutral ionic liquid 1-butyl-3-methylimidazolium bromide Ultrason. Sonochem., 19 (2012) 307–313.

[217] Z. E. Asri, Y. Génisson, F. Guillen, O. Baslé, N. Isambert, M. M. S. Duque, S. Ladeira, J. Rodriguez, T. Constantieux and J.-C. Plaquevent, Multicomponent reactions in ionic liquids: convenient and ecocompatible access to the 2,6-DABCO core Green Chem., 13 (2011) 2549–2552.

[218] M. Dabiri, M. Bahramnejad and S. Bashiribod, [Hmim] TFA catalyzed multicomponent reaction: direct, mild, and efficient procedure for the synthesis of 1,2-dihydroquinazoline derivatives. Mol. Diversity, 14 (2010) 507–512.

[219] E. Soleimani, M. M. Khodaei and A. T. K. Koshvandi, Three-Component, One-Pot Synthesis of Benzo[b][1,4]oxazines in Ionic Liquid 1-Butyl-3-methylimidazolium Bromide Synth. Commun., 42 (2012) 1367–1371

[220] O. B. Pawar, F. R. Chavan, S. S. Sakate and N. D. Shinde, Ultrasound Promoted and Ionic Liquid Catalyzed Cyclocondensation Reaction for the Synthesis of 4(3H)-Quinazolinones Chin. J. Chem., 28 (2010) 69–71

[221] L.J. Xu, L.C. Feng, Y.W. Sun, W.J. Tang, K.L. Lam, Z. Zhou and A. S. C. Chan, Highly efficient chemoselective construction of 2,2-dimethyl-6-substituted 4-piperidones via multi-component tandem Mannich reaction in ionic liquids Green Chem., 12 (2010) 949–952

[222] B. Wang, S. Zhou, Y. Sun, F. Xu and R. Sun, Salt-type organic acids: a class of green acidic catalysts in organic transformations Curr. Org. Chem., 15 (2011) 1392–1422

[223] D. Fang, J. Yang and C. Jiao, Thermal-regulated PEG 1000-based ionic liquid/PM for one-pot three-component synthesis of 2, 4, 5-trisubstituted imidazoles Catal. Sci. Technol.,1 (2011) 243–245

[224] Y. Yu, H. Guo and X. Li, An improved procedure for the three-component synthesis of benzo[g]chromene derivatives using basic ionic liquid J. Heterocycl. Chem., 48 (2011) 1264– 1268.

[225] J. M. Khurana, D. Magoo and A. Chaudhary, Efficient and Green Approaches for the Synthesis of 4H-Benzo[g]chromenes in Water, Under Neat Conditions, and Using Task-Specific Ionic Liquid Synth. Commun., 42 (2012) 3211-3219

[226] M. Zakeri, M. M. Heravi, M. Saeedi, N. Karimi, H. A. Oskooie and N. Tavakoli-Hoseini, One-pot Green Procedure for Synthesis of Tetrahydrobenzo[a]xanthene-11-one Catalyzed by Brønsted Ionic Liquids under Solvent-free Conditions Chin. J. Chem., 29 (2011) 1441–1445

[227] D. Kundu, A. Majee and A. Hajra, Task-specific ionic liquid catalyzed efficient microwave-assisted synthesis of 12-alkyl or aryl-8,9,10,12-tetrahydrobenzo[a]xanthen-11-ones under solvent-free conditions Green Chem. Lett. Rev., 4 (2011) 205–209.

[228] A. Chaskar, H. Shaikh, V. Padalkar, K. Phatangare and H. Deokar, IBX promoted one-pot condensation of β-naphthol, aldehydes, and 1,3-dicarbonyl compounds Green Chem. Lett. Rev., 4 (2011) 171–175.

[229] J. M. Khurana, B. Nand and Sneha, An efficient and convenient approach for the synthesis of novel 2-hydroxy-12-aryl-8,9,10,12-tetrahydrobenzo[a]xanthene-11-ones using p-toluenesulfonic acid in ethanol and ionic liquid J. Heterocycl. Chem., 48 (2011) 1388–1392.

[230] J. Luo and Q. Zhang, A one-pot multicomponent reaction for synthesis of 1-amidoalkyl-2-naphthols catalyzed by PEG-based dicationic acidic ionic liquids under solvent-free conditions Monatsh. Chem., 142 (2011) 923–930

[231] D. A. Kotadia and S. S. Soni, Silica gel supported–SO3H functionalized benzimidazolium based ionic liquid as a mild and effective catalyst for rapid synthesis of 1-amidoalkyl naphthols J. Mol. Catal. A: Chem., 353 (2012) 44–49.

[232] N. Tavakoli-Hoseini, M. M. Heravi, F. F. Bamoharram and A. Davoodnia, Brønsted acidic ionic liquids as efficient catalysts for clean synthesis of carbamatoalkyl naphthols Bull. Korean Chem. Soc., 32 (2011) 787–792

[233] M. M. Heravi, N. Tavakoli-Hoseini and F. F. Bamoharram, Brønsted acidic ionic liquids as efficient catalysts for the synthesis of amidoalkyl naphthols Synth. Commun., 41 (2011) 298–306.

[234] A. R. Hajipour, Y. Ghayeb, N. Sheikhan and A. E. Ruoho, Brønsted Acidic Ionic Liquid as an Efficient and Reusable Catalyst for One-Pot, Three-Component Synthesis of Pyrimidinone Derivatives via Biginelli-Type Reaction Under Solvent-Free Conditions Synth. Commun.,41 (2011) 2226–2233

[235] J. M. Khurana and S. Kumar, Ionic liquid: an efficient and recyclable medium for the synthesis of octahydroquinazolinone and biscoumarin derivatives Monatsh. Chem., 141 (2010) 561–564.

[236] M. Mirzai and H. Valizadeh, Microwave-promoted synthesis of 3, 4-dihydropyrimidin-2 (1H)-(thio) ones using IL-ONO as recyclable base catalyst under solvent-free conditions Synth. Commun., 42 (2012) 1268–1277.

[237] H. Valizadeh and A. Shockravi, Imidazolium-based phosphinite ionic liquid as reusable catalyst and solvent for one-pot synthesis of 3,4-dihydropyrimidin-2(1H)-(thio)ones Heteroat. Chem.,20 (2009) 284–288

[238] T. J. Li, C. S. Yao, C. X. Yu, X. S. Wang and S. J. Tu, Ionic Liquid–Mediated One-Pot Synthesis of 5-(Trifluoromethyl)-4,7-dihydrotetrazolo[1,5-a]pyrimidine Derivatives Synth. Commun.,42 (2012) 2728-2738.

[239] A. K. Chakraborti, S. R. Roy, P. S. Jadhavar, K. Seth, K. K. Sharma Organocatalytic Application of Ionic Liquids: [bmim][MeSO4] as a Recyclable Organocatalyst in the Multicomponent Reaction for the Preparation of Dihydropyrimidinones and -thiones Synthesis, 7 (2011) 2261–2267.

[240] D. Fang, J. Yang and C. Ni, Dicationic ionic liquids as recyclable catalysts for one-pot solvent-free synthesis of α-aminophosphonates Heteroat. Chem., 22 (2011) 5–10

[241] A. Davoodnia, M. Bakavoli, R. Moloudi, N. Tavakoli-Hoseini and M. Khashi, Highly efficient, one-pot, solvent-free synthesis of 2,4,6-triarylpyridines using a Brønsted-acidic ionic liquid as reusable catalyst Monatsh. Chem., 141 (2010) 867–870.

[242] H. Behmadi, S. Naderipour, S. M. Saadati, M. Barghamadi and M. Shaker, J. Solvent-free synthesis of new 2, 4, 6-triarylpyridines catalyzed by a Brønsted acidic ionic liquid as a green and reusable catalyst, Heterocycl. Chem.,48 (2011) 1117–1121.

[243] D. Fang, K. Gong, D.-Z. Zhang and Z.-L. Liu, One-pot, three-component Mannich-type reaction catalyzed by functionalized ionic liquid Monatsh. Chem., 140 (2009) 1325–1329.

[244] F.-F. Yong and Y.-C. Teo, Recyclable siloxy serine organocatalyst for the direct asymmetric mannich reactions in ionic liquids Synth. Commun., 41 (2011) 1293–1300.

[245] A. Paiva, R. Craveiro, I. Aroso, M. Martins, R.L. Reis, A.R.C. Duarte, Natural Deep Eutectic Solvents – Solvents for the 21st Century ACS Sustain. Chem. Eng. 2 (2014) 1063-1071.

[246] Y. Dai, J. van Spronsen, G.-J. Witkamp, R. Verpoorte, Y.H. Choi, Natural deep eutectic solvents as new potential media for green technology Anal. Chim. Acta 766 (2013) 61-68.

[247] H. Zhang, B. Tang, K. Row, Extraction of catechin compounds from green tea with a new green solvent Chem. Res. Chin. Univ. 30 (2014) 37-41.

[248] S. Bajkacz, J. Adamek, Evaluation of new natural deep eutectic solvents for the extraction of isoflavones from soy products Talanta 168 (2017) 329-335.

[249] I. Adeyemi, MRM Abu-Zahra, I. Alnashef, Novel green solvents for CO2 capture Energy Procedia, 114 (2017) 2552-2560

[250] C.A. Neuberg, Pro memoria Carl Neuberg Biochem. Z. 76 (1916) 107.

[251] P.G. Jessop, J. R. Vanderveen, J. Durelle, Design and evaluation of switchable hydrophilicity solvents Green Chem. 16 (2014) 1187-1197

[252] Y. Kohno, S. Saita, K. Murata, N. Nakamura, H. Ohno, Extraction of proteins with temperature sensitive and reversible phase change of ionic liquid/water mixture Polym. Chem. 2 (2011) 862-870

[253] D.E. Bergbreiter, P.L. Osburn, T. Smith, C. Li, J.D. Frels, Using Soluble Polymers in Latent Biphasic Systems J. Am. Chem. Soc. 125 (2003) 6254-6260.

[254] Z. Ludmer, T. Golan, E. Ermolenko, N. Brauner, A. Ullmann, Simultaneous Removal of Heavy Metals and Organic Pollutants from Contaminated Sediments and Sludges by a Novel Technology, Sediments Remediation Phase Transition Extraction Environ. Eng. Sci. 26 (2009) 419-430.

[255] C. VollmerandC. Janiak, Naked metal nanoparticles from metal carbonyls in ionic liquids: Easy synthesis and stabilization , Coord. Chem.Rev., 2011, 255, 2039–2057

[256] B. Schäffner, S. P. Verevkin, A. Börner, Green solvents for synthesis and catalysis. Organic carbonates, Chem. Unserer Zeit, 43 (2012) 12-21.

[257] M. Bandres, P. de Caro, S. Thiebaud-Roux, M.E. Borredon, Green syntheses of biobased solvents. Chimie, 14(2011) 636-646.

[258] Y. Gu Multicomponent reactions in unconventional solvents: state of the art, Green Chem., 14 (2012) 2091-2128.

[259] J. N. Tan, M. Li, Y. Gu, Multicomponent reactions of 1,3-disubstituted 5-pyrazolones and formaldehyde in environmentally benign solvent systems and their variations with more fundamental substrates. Green Chem., 12 (2010) 908-914.

[260] Z. J. Quan, R. G. Ren, Y.X. Da, Z. Zhang, X.C. Wang, Glycerol as an Alternative Green Reaction Medium for Multicomponent Reactions Using PS-PEG-OSO$_3$H as Catalyst. Synth. Comm., 41 (2011) 3106-3116.

[261] A. Haimov and R. Neumann, Polyethylene glycol as a non-ionic liquid solvent for polyoxometalate catalyzed aerobic oxidation Chem. Commun.,12 (2002) 876–877

[262] H. Alper, K. Januszkiewicz and D. J. H. Smith, Palladium chloride and polyethylene glycol promoted oxidation of terminal and internal olefins Tetrahedron Lett., 26 (1985) 2263–2264

[263] S. Chanrasekhar, Ch. Narsihmulu, S. S. Sultana and N. R. Reddy, Poly(ethylene glycol) (PEG) as a Reusable Solvent Medium for Organic Synthesis. Application in the Heck Reaction Org. Lett., 4 (2002) 4399–4401.

[264] S. L. Jain, S. Singhal and B. Sain, PEG-assisted solvent and catalyst free synthesis of 3,4-dihydropyrimidinones under mild reaction conditions Green Chem., 9 (2007) 740–741.

[265] S. S. Gawande, B. P. Bandgar, P. D. Kadam and S. S. Sable, Uncatalyzed synthesis of thiomorpholide using polyethylene glycol as green reaction media Green Chem. Lett. Rev., 3 (2010) 315–318.

[266] V. V. Kouznetsov, D. R. M. Arenas and A. R. R. Bohórquez, PEG-400 as green reaction medium for Lewis acid-promoted cycloaddition reactions with isoeugenol and anethole Tetrahedron Lett., 49 (2008) 3097–3100 .

[267] S. V. Nalage, A. P. Nikum, M. B. Kalyankar, V. S. Patil, U. D. Patil, K. R. Desale, S. L. Patil and S. V. Bhosale, One-Pot Four Component Synthesis of 4, 6-Disubstituted 3-Cyano-2- Pyridones in Polyethylene Glycol Lett. Org. Chem., 7 (2010) 406–410.

[268] M. Kidwai and D. Bhatnagar, Ceric ammonium nitrate (CAN) catalyzed synthesis of N-substituted decahydroacridine-1,8-diones in PEG Tetrahedron Lett., 51 (2010) 2700–2703.

[269] S. S. Chobe, G. G. Mandawad, O. S. Yemul, S. S. Kinkar and B. S. Dawane, An efficient one-pot synthesis of substituted pyrazolo [3,4 b:4',3'e]pyridine derivatives via the hantzch three component condensation using bleaching earth catalyst and their Invitro antimicrobial evaluation Int. J. ChemTech Res., 3 (2011) 938–943

[270] B. M. Shaikh, S. G. Konda, A. V. Mehare, G. G. Mandawad, S. S. Chobe and B. S. Dawane, One-pot multicomponent synthesis and antibacterial evaluation of some novel acridine derivatives Pharma Chem., 2 (2010) 25–29.

[271] X. Wang, H. Gong, Z. Quan, L. Li and H. Ye, One-Pot, Three-Component Synthesis of 1,4-Dihydropyridines in PEG-400 Synth. Commun., 41 (2011) 3251–3258

[272] A. Manvar, D. Karia, V. Trangadia, N. Vekariya and A. Shah, PEG-400 mediated and microwave assisted one pot three-component coupling reactions: Expedient and rapid synthesis of Hantzsch 1,4-dihydropyridines devoid of use of catalyst Org. Chem. Indian J., 3 (2007) 166–169

[273] R. Mallepalli, L. Yeramanchi, R. Bantu and L. Nagarapu, Polyethylene Glycol (PEG-400) as an Efficient and Recyclable Reaction Medium for the One-Pot

Synthesis of N-Substituted Azepines under Catalyst-Free Conditions Synlett, 9 (2011) 2730–2732.

[274] L. Nagarapu, R. Mallepalli, L. Yeramanchi and R. Bantu, Polyethylene glycol (PEG-400) as an efficient and recyclable reaction medium for one-pot synthesis of polysubstituted pyrroles under catalyst-free conditions Tetrahedron Lett., 52 (2011) 3401–3404

[275] P. V. Shinde, A. H. Kategaonkar, B. B. Shingate and M. S. Shingare, Polyethylene glycol (PEG) mediated expeditious synthetic route to 1,3-oxazine derivatives Chin. Chem. Lett., 22 (2011) 915–918

[276] S.-L. Wang, W.-J. Hao, S.-J. Tu, X.-H. Zhang, X.-D. Cao, S. Yan, S.-S. Wu, Z.-G. Han and F. Shi, Poly(ethyleneglycol): A versatile and recyclable reaction medium in gaining access to benzo[4,5]imidazo[1,2-a]pyrimidines under microwave heating J. Heterocycl. Chem., 46 (2009) 664–668.

[277] S. V. Nalage, M. B. Kalyankar, V. S. Patil, S. V. Bhosale, S. U. Deshmukh and R. P. Pawar, An efficient noncatalytic protocol for the synthesis of trisubstituted imidazole in polyethylene glycol using microwaves Open Catal. J., 3 (2010) 58–61.

[278] B. S. Dawane, S. G. Konda, R. G. Bodade and R. B. Bhosale, An efficient one-pot synthesis of some new 2,4-diaryl pyrido[3,2-c]coumarins as potent antimicrobial agents J. Heterocycl. Chem., 47 (2010) 237–241.

[279] A. A. Mulay and R. A. Mane, Polyethylene glycol mediated one-pot three-component synthesis of new 4-thiazolidinones Heteroat. Chem., 23 (2011) 166–170.

[280] J. R. Mali, M. R. Bhosle, S. R. Mahalle and R. A. Mane, One-Pot Multicomponent Synthetic Route for New Quinolidinyl 2,4-Thiazolidinediones. Bull. Korean Chem. Soc., 31 (2010) 1859–1862.

[281] X. N. Zhang, Y. X. Li and Z. H. Zhang, Nickel chloride-catalyzed one-pot three-component synthesis of pyrazolophthalazinyl spirooxindoles Tetrahedron, 67 (2011) 7426– 7430

[282] S. K. Guchhait and C. Madaan, Towards molecular diversity: dealkylation of tert-butyl amine in Ugi-type multicomponent reaction product establishes tert-butyl isocyanide as a useful convertible isonitrile Synlett, 12 (2009) 628–632

[283] V. V. Makarov, A. J. Love, O. V. Sinitsyna, S. S. Makarova, I. V. Yaminsky, M. E. Taliansky, and N. O. Kalinina "Green" Nanotechnologies: Synthesis of Metal Nanoparticles Using Plants, Acta Naturae. 6(2014) 35–44.

[284] J. Virkutyte and R.S. Varma, Environmentally Friendly Preparation of Metal Nanoparticles Royal Soc. Chem. 7, (2013) 564-569.

[285] P. Chettri, V. S. Vendamani, A Tripathi M. K. Singh A. P. Pathak, A. Tiwari A., Green synthesis of silver nanoparticle-reduced graphene oxide using Psidium

guajava and its application in SERS for the detection of methylene blue, Appl. Surf. Sci. 406 (2017) 312-318

[286] U. Jinu, M. Gomathi, N. Geetha, G. Benelli, P. Venkatachalam, Green engineered biomolecule-capped silver and copper nanohybrids using Prosopis cineraria leaf extract: enhanced antibacterial activity against microbial pathogens of public health relevance and cytotoxicity on human breast cancer cells (MCF-7), Microb. Pathogenesis 105 (2017) 8695-9705.

[287] H. Kolya, P. Maiti, A. Pandey, T. Tripathy, Green synthesis of silver nanoparticles with antimicrobial and azo dye (Congo red) degradation properties using Amaranthusgangeticus Linn leaf extract. J Anal Sci Technol 6 (2015) 33-39.

[288] SSM Hassan, W. Azab, H.R. Ali, M. S. M. Mansour, Green synthesis and characterization of ZnO nanoparticles for photocatalytic degradation of anthracene. Adv Nat Sci Nanosci Nanotechnol 6 (2015) 1-10.

[289] K. Lellala, K. Namratha, K. Byrappa, Microwave assisted synthesis and characterization of nanostructure zinc oxide-graphene oxide and photo degradation of brilliant blue. Mater. Today Proc 3 (2016) 74-83.

[290] S. T. Fardood, A. Ramazani, S. M. Pegah, A. Asiabi, Green synthesis of zinc oxide nanoparticles using arabic gum and photocatalytic degradation of direct blue 129 dye under visible light, J Mater Sci: Mater Electro 28 (2017) 13596

[291] A. Ramazani, S. T. Fardood, Z. Hosseinzadeh, F. Sadri, S.W. Joo, Green synthesis of magnetic copper ferrite nanoparticles using tragacanth gum as a biotemplate and their catalytic activity for the oxidation of alcohols. Iran. J Catal 7(2017) 181-185

[292] S. T. Fardood, A. Ramazani, Z. Golfar, S. W. Joo, Green synthesis of Ni-Cu-Zn ferrite nanoparticles using tragacanth gum and their use as an efficient catalyst for the synthesis of polyhydroquinoline derivatives. Appl. Organomet. Chem., 31(2017) 3823-3870.

[293] M. Sorbiun, E. S. Mehr, A. Ramazani, S. T. Fardood, Green Synthesis of Zinc Oxide and Copper Oxide Nanoparticles Using Aqueous Extract of Oak Fruit Hull (Jaft) and Comparing Their Photocatalytic Degradation of Basic Violet 3, Int J Environ Res 12(2018) 29-37.

[294] M. Rani, Rachna, U. Shanker, Metal hexacyanoferrates nanoparticles mediated degradation of carcinogenic aromatic amines. Environ Nanotechnol Monit Manage 10 (2018) 36-45.

[295] M. Rani, U. Shanker, Removal of chlorpyrifos, thiamethoxam, and tebuconazole from water using green synthesized metal hexacyanoferrate nanoparticles. Environ Sci Pollut Res 25 (2018)10878-10888

[296] M. Rani, U. Shanker, Promoting sunlight-induced photocatalytic degradation of toxic phenols by efficient and stable double metal cyanide nanocubes. Environ Sci Pollut R 25 (2018) 23764–23779

[297] M. Rani, Studies on decay profiles of quinalphos and thiram pesticides. Ph.D Thesis, Indian Institute of Technology Roorkee, Roorkee, Uttarakhand, India, Chapter 1- 5(2012).

[298] M. Rani, U. Shanker, A. Chaurasia, Catalytic potential of laccase immobilized on transition metal oxides nanomaterials: degradation of alizarin red S dye. J. Env. Chem. Engg. 5(2017) 2730-2739.

[299] M. Rani, U. Shanker, Effective adsorption and enhanced degradation of various pesticides from aqueous solution by Prussian blue nanorods, J. Env. Chem. Engg. 6(2018) 1512-1519.

[300] M. Rani, U. Shanker, Removal of carcinogenic aromatic amines by metal hexacyanoferrates nanocubes synthesized via green process, J. Env. Chem. Engg. 5(2017)5298-5309.

[301] M. Rani, U. Shanker, Sun-light driven rapid photocatalytic degradation of methylene blue by poly (methyl methacrylate)/metal oxide nanocomposites, Colloids Surf A Physicochem Eng Asp. 559 (2018)136–147.

[302] M. Rani, U. Shanker, Insight in to the degradation of bisphenol A by doped ZnO@ZnHCF nanocubes: High photocatalytic performance, J. Colloid Interf. Sci. 530(2018) 16–28.

[303] M. Rani, U. Shanker, Photocatalytic degradation of toxic phenols from water using bimetallic metal oxide nanostructures, Colloids Surf A Physicochem Eng Asp. 553 (2018) 546–561

[304] M. Rani, U. Shanker , Advanced Treatment Technologies In: C. M. Hussain (ed.), Handbook of Environmental Materials Management, Springer International Publishing AG (2018) https://doi.org/10.1007/978-3-319-58538-3_33-1.

[305] M. Rani, U. Shanker Remediation of Polycyclic Aromatic Hydrocarbons Using nanomaterials , In eds: Green Adsorbents for Pollutant Removal, Springer International Publishing AG, part of Springer Nature, (2018) https://doi.org/10.1007/978-3-319-92111-2_10

[306] M. Rani, U. Shanker, Degradation of traditional and new emerging pesticides in water by nanomaterials: recent trends and future recommendations, Int. J. Environ. Sci. Technol. 15 (2018) 1347–1380

[307] M. Rani, U. Shanker, V. Jassal, Recent strategies for removal and degradation of persistent and toxic organochlorine pesticides using nanoparticles: a review. J. Environ. Manage. 190 (2017) 208

[308] Shanker U., Rani M., Jassal V., Degradation of hazardous organic dyes in water by nanomaterials. Environ Chem Lett 15 (2017) 623.

[309] Shanker U., Rani M., Jassal V., B. S. Kaith, Towards green synthesis of nanoparticles: From bio-assisted sources to benign solvents. A review, Int. J. Environ. Anal. Chem, 96 (2016) 801.

[310] Shanker U., Rani M., Jassal V., Green synthesis of iron hexacyanoferrate nanoparticles: Potential candidate for the degradation of toxic PAHs. J Env Chem Engg 5 (2017)4108-4120

[311] Shanker U., Rani M., Jassal V., Catalytic removal of organic colorants from water using some transition metal oxide nanoparticles synthesized under sunlight, RSc Adv. 6(2017) 94989-9499.

[312] U. Shanker, V. Jassal, M. Rani, Degradation of toxic PAHs in water and soil using potassium zinc hexacyanoferrate nanocubes, J. Environ. Manage. 204(2017) 337-345.

[313] V. Jassal U. Shanker S. Gahlot B.S. Kaith, Kamaluddin, M. A. Iqubal, P. Samuel, Sapindus mukorossi mediated green synthesis of some manganese oxide nanoparticles interaction with aromatic amines, Appl. Phys. A 122 (2016) 271-280.

[314] V. Jassal, U. Shanker, B. S. Kaith, Aegle marmelos mediated green synthesis of different nanostructured metal hexacyanoferrates: activity against photodegradation of harmful organic dyes, Scientifica 2016 (2016) 1-17.

[315] V. Jassal, U. Shanker, B. S. Kaith, S. Shankar, Green synthesis of potassium zinc hexacyanoferrate nanocubes and their potential application in photocatalytic degradation of organic dyes, RSC Adv. 5 (2015) 26141-26150.

[316] V. Jassal, U. Shanker S. Gahlot, Green synthesis of some iron oxide nanoparticles and their interaction with 2-Amino, 3-Amino and 4-Aminopyridines, Mater. Today. Proc. 3(2016) 1874-1880.

[317] Rachna, M. Rani, U. Shanker, Enhanced photocatalytic degradation of chrysene by Fe2O3@ZnHCF nanocubes. Chem Eng J, 348(2018) 754-760.

[318] V. Smuleac, R. Varma, S. Sikdar, D. Bhattacharyya, Green synthesis of Fe and Fe/Pd bimetallic nanoparticles in membranes for reductive degradation of chlorinated organics. J. Membr. Sci. 379(2011) 131-140.

[319] F. Luo, D. Yang, Z. Chen, M. Megharaj, R. Naidu, One-step green synthesis of bimetallic Fe/Pd nanoparticles used to degrade orange II. J Hazard Mater 303(2016) 145-151.

[320] T. Shahwan, S. Abu Sirriah, M. Nairat, E. Boyacı, A.E. Eroğlu, T. B. Scott, K. R. Hallam, Green synthesis of iron nanoparticles and their application as a fenton-like catalyst for the degradation of aqueous cationic and anionic dyes. Chem. Eng. J. 172(2011) 258-268.

[321] M. Sabbaghan, A. S. Shahvelayati, S. F. Bashtani, Synthesis and optical properties of ZnO nanostructures in imidazolium-based ionic liquids, Solid State Sciences 14 (2012) 1191-1195

[322] M. Sabbaghan, A. S. Shahvelayati, S. Banihashem, Green synthesis of symmetrical imidazolium based ionic liquids and their application in the preparation of ZnO nanostructures, Ceramics International 42 (2016) 3820–3825

[323] Q. Liu and Y.-N. Zhang, One-pot Synthesis of 3-Methyl-4-arylmethylene-isoxazol-5(4H)-ones Catalyzed by Sodium Benzoate in Aqueous Media: A Green Chemistry Strategy, Bull. Korean Chem. Soc., 32 (2011) 3559– 3560

[324] H. Olivier-Bourbigou, L. Magna, Ionic liquids: perspectives for organic and catalytic reactions Journal of Molecular Catalysis A: Chemical 182–183 (2002) 419–437

[325] M. Sabbaghana, J. Beheshtian, S.A.M. Mirsaeidi, Preparation of uniform 2D ZnO nanostructures by the ionic liquid-assisted sonochemical method and their optical properties, Ceram. Int. 40 (2014) 7769–7774.

[326] S. Zhao, Y. Zhang, Y. Zhou, C. Zhang, J. Fang, X. Sheng, Ionic liquid-assisted photochemical synthesis of ZnO/Ag$_2$O heterostructures with enhanced visible light photocatalytic activity, Appl. Surf. Sci. 410 (2017) 344–353

[327] M. Koel, Do we need green analytical chemistry? Green Chem. 18 (2016) 923–931.

[328] M. Tobiszewski and J. N.Snik Greener organic solvents in analytical chemistry Current Opinion in Green and Sustainable Chemistry 5 (2017) 1–4

[329] K. D. Clark, O. Nacham, J. A. Purslow, S. A. Pierson, J. L. Anderson, Magnetic ionic liquids in analytical chemistry: a review. Anal. Chim. Acta 934 (2016) 9–21.

[330] A. Ballesteros-Gómez, M. Dolores Sicilia, S. Rubio, Supramolecular solvents in the extraction of organic compounds. review. Anal. Chim. Acta 677 (2010) 108–130.

[331] M. Espino, M. de los Ángeles Fernández, F. J. V. Gomez, M. Fernanda Silva, Natural designer solvents for greening analytical chemistry. Trends Anal. Chem. 76(2016)126–136.

[332] V. Abrahamsson, N. Andersson, B. Nilsson, C. Turner, Method development in inverse modeling applied to supercritical fluid extraction of lipids. J. Supercrit. Fluids 111(2016) 111:14–27

[333] J. P. Taygerly, L. M. Miller, A. Yee, E. A. Peterson, A convenient guide to help select replacement solvents for dichloromethane in chromatography. Green Chem. 14 (2012) 3020–3025.

[334] S. Yao, B. Chen, T. A. van Beek, Alternative solvents can make preparative liquid chromatography greener. Green Chem. 17 (2015) 4073–4081.

[335] F. Tache, S. Udrescu, F. Albua, F. Micale, A. Medvedovici, Greening pharmaceutical applications of liquid chromatography through using propylene carbonate–ethanol mixtures instead of acetonitrile as organic modifier in the mobile phases. J. Pharm. Biomed. Anal. 75 (2013) 230–238

[336] C. S. Funari, R. L. Carneiro, M. M. Khandagale, A. J. Cavalheiro, E. F. Hilder EF: Acetone as a greener alternative to acetonitrile in liquid chromatographic fingerprinting. J. Sep. Sci. 38 (2015) 1458–1465.

[337] N. Prache, S. Abreu, P. Sassiat, D. Thiébaut, P. Chaminade, Alternative solvents for improving the greenness of normal phase liquid chromatography of lipid classes. J. Chromatogr. A 1464 (2016) 55–63.

[338] A. Spietelun, L. Marcinkowski, M. de la Guardia, J. N.Snik, Green aspects, developments and perspectives of liquid phase microextraction techniques. Talanta 119 (2014) 34–45.

[339] P. Bigus, N. J. Snik, M. Tobiszewski, Application of multicriteria decision analysis in solvent type optimization for chlorophenols determination with a dispersive liquid–liquid microextraction. J. Chromatogr. A 1446 (2016) 21–26.

[340] L. Wang, J. Littlewood, R. J. Murphy, An economic and environmental evaluation for bamboo-derived bioethanol. RSC Adv. 4 (2014) 29604–29611.

[341] F. Pena-Pereira, A. Kloskowski, N. J. Snik, Perspectives on the replacement of harmful organic solvents in analytical methodologies: a framework toward the implementation of generation of eco-friendly alternatives. Green Chem. 17 (2015) 3687–3705

[342] S.P.J. Ahmadkalaei, S Gan, H K. Ng, S. A. Talib, Investigation of ethyl lactate as a green solvent for desorption of total petroleum hydrocarbons (TPH) from contaminated soil. Environ. Sci. Pollut. Res. 23 (2016) 22008–22018.

[343] S. J. Tenne, J. Kinzel, M. Arlt, F. Sibilla, M. Bocola, U. Schwaneberg, 2-Methyltetrahydrofuran and cyclopentylmethylether: two green solvents for efficient purification of membrane proteins like FhuA. J. Chromatogr. B 937 (2013) 13–17.

[344] K. Sharma, N. Mahato, M. Hwan Cho, Y. Rok Lee, Converting citrus wastes into value-added products: economic and environmently friendly approaches. Nutrition, 34 (2017) 29–46.

[345] S. Veillet, V. Tomao, K. Ruiz, F. Chemat, Green procedure using limonene in the Dean–Stark apparatus for moisture determination in food products. Anal. Chim. Acta 674 (2010) 49–52.

[346] M. Virot, V. Tomao, C. Ginies, F. Visinoni, F. Chemat, Green procedure with a green solvent for fats and oils' determination Microwave-integrated Soxhlet using limonene followed by microwave Clevenger distillation. J. Chromatogr. A 1196–1197 (2008) 147–152.

[347] A. Medvedovici, S. Udrescu, V. David, Use of a green (bio) solvent – limonene – as extractant and immiscible diluent for large volume injection in the RPLC-tandem MS assay of statins and related metabolites in human plasma. Biomed. Chromatogr. 27 (2013) 48–57.

[348] M. Tobiszewski, N. J.Snik, F. Pena-Pereira, Environmental risk – based ranking of solvents by the combination of multimedia model and multi-criteria decision analysis. Green Chem. 19 (2017) 1034–1043.

[349] Y. Leng Kua, S. Gan, A. Morris, H. Kiat Ng, Ethyl lactate as a potential green solvent to extract hydrophilic (polar) and lipophilic (non-polar) phytonutrients simultaneously from fruit and vegetable by-products. Sustain. Chem. Pharm. 4 (2016) 21–31.

[350] Z. Li, K. H. Smith, G. W. Stevens, The use of environmentally sustainable bio-derived solvents in solvent extraction applications— a review. Chin. J. Chem. Eng 24 (2016) 215–220.

Industrial Applications of Green Solvents I
Materials Research Foundations **50** (2019) 242-268

Materials Research Forum LLC
doi: https://doi.org/10.21741/9781644900239-8

Chapter 8

Supercritical Carbon Dioxide in Esterification Reactions

Baithy Mallesham[1,*], A. Jeyanthi[1], Pothu Ramyakrishna[2], and Boddula Rajender[3,*]

[1]Department of Chemistry, University College of Science, Satavahana University, Telangana-505 002, India

[2]College of Chemistry and Chemical Engineering, Hunan University, Changsha 410082, PR China

[3]Chinese Academy of Sciences (CAS) Key Laboratory of Nanosystem and Hierarchy Fabrication, CAS Center for Excellence in Nanoscience, National Center for Nanoscience and Technology, Beijing 100190, People's Republic of China

baithy.m@gmail.com (Dr. Baithy Mallesham),

research.raaj@qq.com (Dr. Rajender Boddula)

Abstract

Green and sustainable solvents are gaining attention in the research institutes and industries owing to the minimal influence on the environment. The importance of supercritical fluids is associated with their *"tunable"* properties that could be easily altered by monitoring reaction parameters like pressure and temperature. The physical properties of supercritical fluids are in-between the gases and liquids. Therefore, $ScCO_2$ is used as an environmentally friendly solvent in various esterification reactions.

Keywords

Green Solvent, Supercritical CO_2, Esterification, Critical Temperature, Critical Pressure, Biocatalyst

Contents

1. Introduction

With the increase in environmental hazards throughout the world, the focus has been extended to control the excess of utilization and disposal of harmful compounds by the industries [1,2]. The research community and chemical industries have immense interests on the sustainable solvents due to the solvents impact on disposal of harmful compounds, usage of energy, pollution, and contributions to the change in the climate [2-4]. Hence, the researchers are constantly in search of new and clean alternatives to existing methodologies [5-7]. One of the most obvious targeted areas is solvent usage.

Solvents are a major segment of organic air pollution. A variety of new and green reusable solvents have been therefore developed during the last few decades [8-10]. The focus is mainly on the environmental impact the solvent makes and the sustainability from which it is made [11,12]. Suitable solvents such as supercritical fluids (particularly CO_2), eutectic solvents, switchable solvents, liquid polymers, reusable solvents, ionic liquids, perfluorinated hydrocarbons, and water are taken into account for replacing the conventional organic solvents, or alternatively reactions that could be performed without solvents [1,2,13-16].

Industrial Applications of Green Solvents I   Materials Research Forum LLC
Materials Research Foundations **50** (2019) 242-268   doi: https://doi.org/10.21741/9781644900239-8

Among these solvents, the supercritical carbon dioxide ($ScCO_2$) as a polymerization solvent has been utilized in the production of the polystyrene and polymethylmethacrylate in addition of its use by DuPont (USA) in a pilot plant for the fluoropolymers production [1,10,17,18]. However, most important large-scale promising applications of $ScCO_2$ are in spray painting and dry cleaning [1, 19-21].

Green chemistry is replacing more hazardous solvents or reagents with less harmful ones and which could be stated as well-designed chemistry owing to the following three factors [21].

- environmentally friendly

- chemically efficient and selective

- economically viable

Table 1 Critical temperatures and critical pressures of different fluids.

Fluids	Critical Temperature T_C (°C)	Critical Pressure P_C (MPa)	Critical Density (g/mL)
Carbon dioxide	31.10	7.40	0.448
Ammonia	132.4	11.25	0.235
Water	374.1	22.1	0.315
Nitrous oxide	36.5	71.7	0.45
Xenon	16.6	57.6	0.118
Methane	-82.1	45.8	0.2
Ethane	32.50	4.91	0.203
Propane	96.80	4.26	0.217
Pentane	196.6	33.3	0.232
Ethylene	9.21	49.7	0.218
Methanol	240.0	7.95	0.272
Ethanol	243.1	6.39	0.280
Isopropanol	235.6	5.37	0.270
Acetone	235.0	4.76	0.273

Industrial Applications of Green Solvents I   Materials Research Forum LLC
Materials Research Foundations **50** (2019) 242-268   doi: https://doi.org/10.21741/9781644900239-8

1.1. The concept of supercritical fluid (SCF)

The fluid containing only one phase above its critical temperature and pressure called supercritical fluid. For example, the $ScCO_2$ has become a most utilized solvent due to its moderate critical temperature of 31.1 °C and pressure of 73.8 bar, as shown in Figure 1. Fluid over its critical temperature and pressure exhibits good solvent power in most of the applications. Some of the supercritical fluids are listed in Table 1.

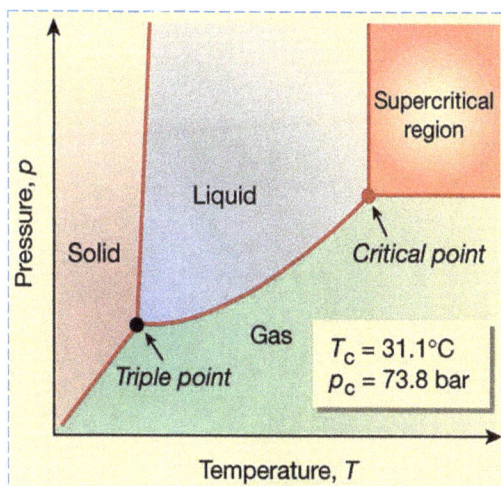

Figure 1   Phase diagram of CO_2

1.2 What is a supercritical fluid?

In Figure 2, two separate phases that are a liquid phase and gaseous phase are evident. An increase in the temperature of the liquid phase converts into the gaseous phase. Further, an increase in the temperature results in the complete conversion of the liquid phase into the gaseous phase. In this case, there is no formation of the supercritical liquid phase.

A supercritical fluid (SCF) is a phase where the matter is compressible and behaves like a gas over its critical temperature (T_c) and pressure (P_c), which is not the case when the fluid is in a liquid phase (Figure 3). However, a SCF has a typical density of a liquid and hence its characteristics dissolving power. The main importance of SCFs is due to its *"tunable"* properties that can be simply altered by examining pressure and temperature. The fluids have a very good solvent power at more densities, *i.e.,* temperature close to their critical temperature and pressure greater than their critical pressure but extremely poor solvent property at low densities i.e., a temperature equal or above their critical temperature, whereas the pressure at lower critical pressure of the solvent. That is why we cannot define the supercritical fluid as a liquid or a gas and this is a new phase of matter.

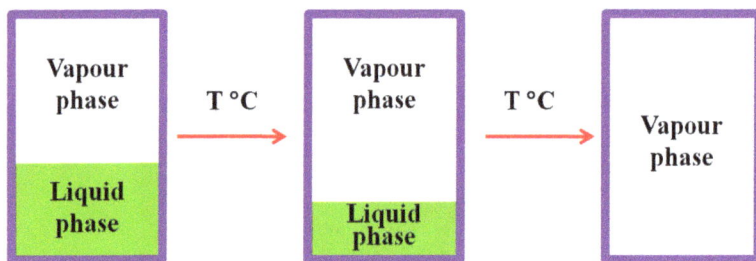

Figure 2   *Heating results obeying usual gas laws*

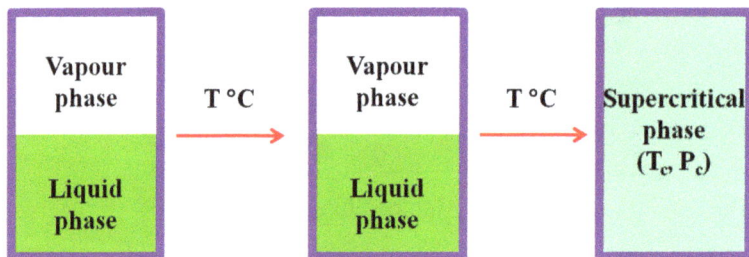

Figure 3   *The formation of a new phase at critical temperature and critical pressure*

In Figure 4, the separate carbon dioxide phases (liquid and gas phases) and the meniscuses are clearly visible (Figure 4 A). With an increase in temperature, the meniscus starts to reduce (Figure 4B). Further, an increase in the temperature results in the equalization of densities of gas and liquid. The meniscus is quite difficult to observe but still, it exists (Figure 4C). At critical temperature and pressure of the liquid and gas phases, the two separate phases of liquid and gas do not exist. Hence, the meniscus can no longer be seen, there exist only one homogeneous phase called the "*supercritical carbon dioxide*" phase (Figure 4D).

At standard temperature and pressure (STP), carbon dioxide is a gas when it is frozen it form solids known as dry ice. However, if the temperature and pressure increase from the STP to a tipping point higher than CO_2, it shows the nature between a liquid and a gas.

"The liquid state of carbon dioxide that exhibits above its critical temperature and critical pressure is called supercritical carbon dioxide (ScCO₂)".

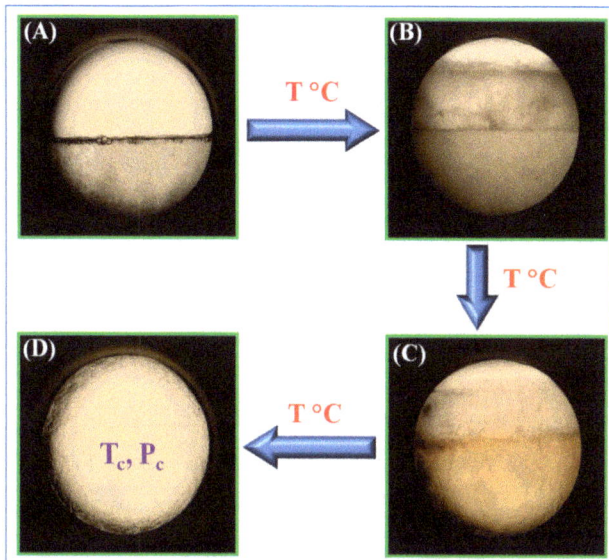

Figure 4   Visualization of liquid and gas phase transitions into the supercritical phase for carbon dioxide (CO₂) [25]

Supercritical CO_2, due to its stability in the chemical reactions, low toxicity, is widely used commercially as an industrial solvent. Due to its stability, it is used extensively in the extraction process where the compounds to be extracted are devoid of denaturing.

Therefore, the $ScCO_2$ is used in various organic transformations such as esterification, acetalization, hydrogenation, condensation and oxidation as a solvent [1,23,24]. In this chapter, we aimed to cover the esterification reactions in the supercritical carbon dioxides to produce esters.

1.3   Esterification reaction and its applications

The reaction in which a carboxylic acid reacts with an alcohol in the presence of a catalyst and solvent to form an ester is called esterification reaction. It is a reversible reaction. The esters formed are sweet smelling compounds and have a fruity odour.

For example acetic acid in excess ethanol and concentrated H_2SO_4, by removal of water results in ethyl acetate which is an ester, it represented in Scheme 1. Hence, this reaction is also called a dehydration reaction.

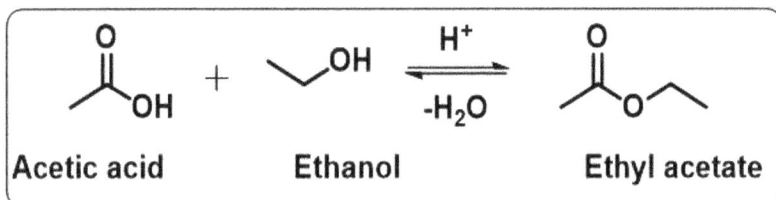

Scheme 1   Acid-catalyzed esterification of acetic acid with ethyl alcohol

The applications of esterification reactions are listed below:

i.   Esterification is used to test the nature of carboxylic acid and alcohols.

ii. This is used in fragrance and flavor industries and also used in the polymer industry

iii. The manufacturing of paints, dyes, medicines, soaps, varnishes, and synthetic rubber involves esterification reactions.

iv. Chemical likes chloroform, iodoform, and ethers are made by this reaction.

1.4   Experimental setup for esterification reaction in supercritical CO_2

Generally, the esterification reactions are carried out using the experimental setup which has the following components [Fig 5].

- Plunger pump 1
- Methanol reservoir...2
- Constant temperature water-bath 3
- High-pressure autoclave 4
- Power controller..5
- Sampling pipe 6
- High-performance liquid chromatography 7
- Vacuum flask 8
- Vacuum pump..9

- Mercury U-tube manometer..............................10
- CO_2 cylinders...11

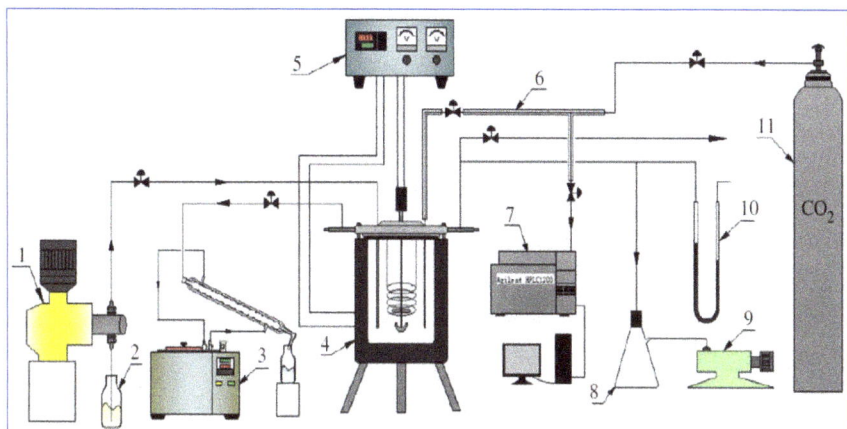

Figure 5   The experimental set-up [26]

2.   Esterification reactions in $ScCO_2$

2.1   Enzymatic esterification reactions

The effects of pressure and temperature can be employed on the density and transport properties of SCFs, including viscosity, thermal conductivity, diffusivity, etc. These properties affect the solubility and transport of reactant molecules and products to/from the enzyme in various esterification reactions [27]. The high-temperature condition results in the less mass transfer limitations in physical properties such as surface tension, viscosity, solvating power, etc. Although it is significant to specify that the most favorable temperature for the enzymatically catalyzed esterification reactions in SCFs depends on the activity of the enzyme, the enzyme activity was decreased because of the thermal deactivation takes place with rising reaction temperatures [28-30]. The most favorable temperature for the enzyme-catalyzed reactions in SCFs was also dependant on the pressure of the reactor. Hence, the solubility of the reactants and products are significantly dependent on the density of the solvent, it has controlled with changing the reaction temperature and pressures during the course of the reaction. The efficiency of the esterification reaction was indirectly affected by changing pressure. The interaction between solute and solvent was increased with increasing reaction pressures, resulting in a higher solvent capacity of SCFs [31]. In addition, the density of a solvent depending on physical properties (i.e., partition coefficient, dielectric constant and solubility parameters) was affected by changing the pressure of ScCO$_2$. These physical properties indirectly regulate the activity, specificity, and stability of the enzymes [32,33].

The dipole moment of various alcohols and carboxylic acids is a significant descriptor, which is much closer to the conversion of reactants to produce products. In supercritical CO$_2$, the dipole moment is highly correlated with the compound's solubility. The electrostatic interaction between the solute and solvent has a considerable effect on the solubility of reactants. As per the "*like-dissolves-like*" principle, the more polarity of the solute has lowered its solubility when compared to that of non-polar solvent i.e., CO$_2$ [34,35]. The esterification reactions are strongly affected by the solubility of the reactants. Therefore, the dipole moment is a key parameter in investigating factors which influence the percentage of conversions.

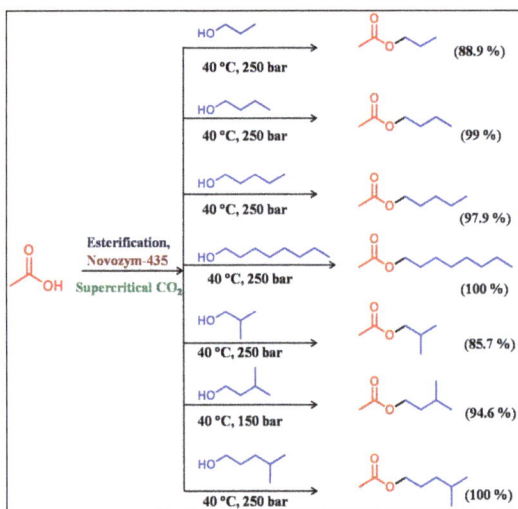

Scheme 2   Esterification of acetic acid with different alcohols in the presence of Novozym-435 biocatalyst

Scheme 2 shows the esterification of acetic acid with different alcohols at a constant temperature, 40°c and at constant pressure 250 bar in the presence of *Novozym-435* biocatalyst. It has been observed that in a straight chain, as the length of the chain increases, the percentage of conversion significantly increases. When the carbon chain of alcohols increases from 3 to 8, the conversions increased from 88.9 to 100 %, respectively. In addition, it has also been observed that in a branched carbon chain, as the number of carbon atoms in the branch increases, the percentage of conversions also considerably increased.

In Scheme 3, when isoamyl alcohol is esterified with different acids at 45 °C of temperature and 90 bar of pressure over the hydroperoxide lyase (HPL) as a catalyst, the corresponding esters are obtained. It is clearly observed that with the increase in the length of the chain, the percentage of conversions increases. In addition, when the carbon chain of acids increases from 3 to 8, the conversions were also increased from 60 to 77 % respectively.

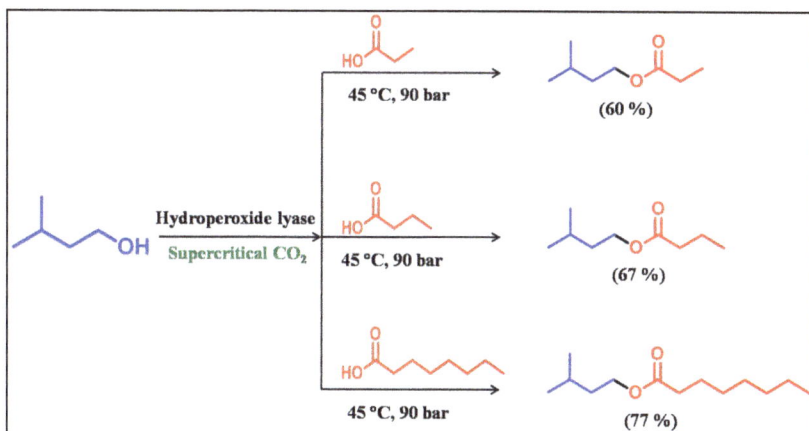

Scheme 3   *Esterification of different carboxylic acids with isoamyl alcohol in presence of Hydroperoxide lyase (HPL) biocatalyst*

Scheme 4 represents the esterification of oleic acid with oleyl alcohol at different pressures and temperatures. As the pressure or temperature is increased, the percentage of conversion is decreased. From Table 2, it is evident that the ideal temperature and pressure for maximum conversion are 50°C and 80 bars.

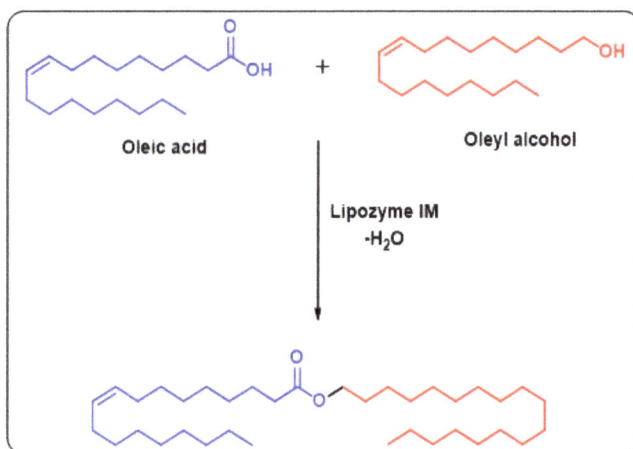

Scheme 4   Esterification of oleic acid with oleyl alcohol over lipozyme IM catalyst at different temperatures and pressures

Table 2 Effect of reaction pressure on the esterification of oleic acid with oleyl alcohol as a function of temperatures.

Pressure [bars]	Conversions [%] at different temperatures				
	40 °C	50 °C	60 °C	70 °C	80 °C
80	84	90	87	86	86
150	83	86	88	84	76
200	83	84	84	76	76
300	80	83	83	77	75
450	80	81	82	77	71

2.2   Mechanism of esterification reactions

Generally, the colophony is a complex and containing a different percentage of carboxylic acids such as abietic acid (40%), palustric acid (20%), neoabietic acid (30%) and levopimaric acid (10%). These acids are produced with varies by the location of two double bonds in the colophony structure, causing different concentrations of different acids to be present. The colophony compounds, dehydroabietic acids, acetic acid, and long chain carboxylic acids, and these compounds were able to undergo esterification with different alcohols to produces a variety of esters [36-39]. In the presence of acid catalyst, the esterification reaction of the carboxylic group of acid reacts with the hydroxyl group of an alcohol is significantly improved in the presence of $ScCO_2$ [26,40].

Industrial Applications of Green Solvents I   Materials Research Forum LLC
Materials Research Foundations **50** (2019) 242-268   doi: https://doi.org/10.21741/9781644900239-8

Scheme 5 showed that the mechanism for the production of esters from the reaction between the acid and alcohol over the acid catalyst is discussed in two steps as follows:

Step –I: A heterogeneous acid catalyst used in the esterification of carboxylic acids reacts with alcohols in supercritical CO_2. In this step-I, the carbon dioxide reacts with water to form H^+ ions and the supercritical fluid phase in a closed vessel.

Step –II: The produced H^+ ions react with a carboxylic acid to form protonated acid followed by dehydration to produce the acyl carbocation as one of the intermediate, which is represented in the pathway II(a). On the other hand, the carboxylic acid also produces acyl carbocation in the presence of an acid catalyst as shown in pathway II (b) of Scheme 5. Subsequently, the formed carbocation reacts with alcohol to generate an intermediate as protonated ester followed by removal of a proton to produce ester in the final step.

Step –I:

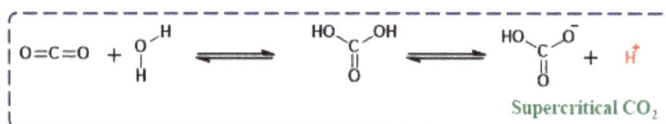

Step –II:

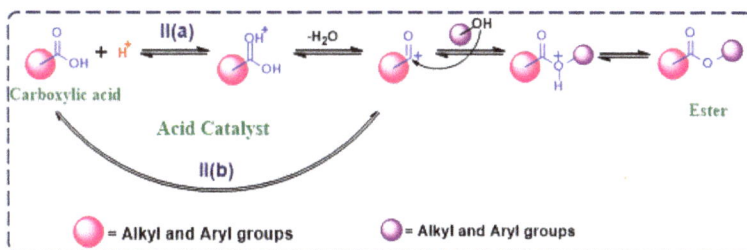

Scheme 5   *Mechanism of acid-catalyzed esterification reaction in the supercritical CO$_2$*

2.3   Effect of size of the carbon chain of alcohol in the esterification reactions

Different alcohols (short chain, long chain, and branched alcohols) have reactions with vinyl laurate to produce outstanding productivity of corresponding esters by using immobilized lipase biocatalyst under optimum reaction conditions. In addition, the lower straight chain alcohols (such as ethanol, propanol, and butanol) reacted very quickly with acid than that of long carbon chain alcohols to provide a better yield (Scheme 6). It could be explained by the mass transfer limitations of long carbon chain alcohols on the biocatalyst reactive sites [41,42].

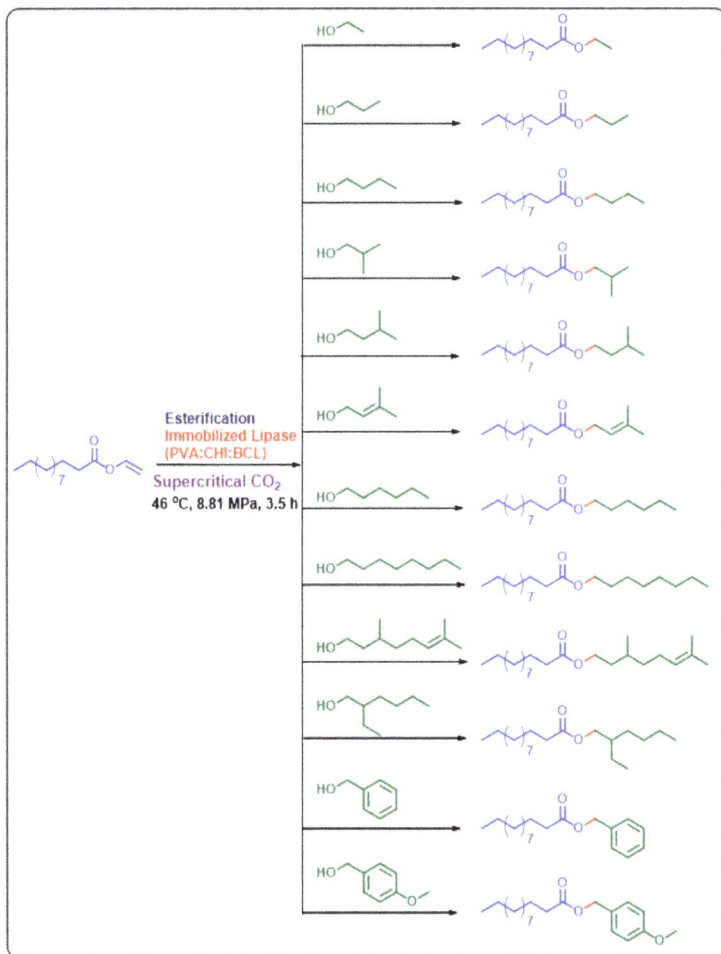

Scheme 6   *Synthesis of various valuable fatty acid esters by immobilized lipase biocatalyst*

Furthermore, the branched chain alcohols also affected by mass transfer limitations and give slightly lower yield than straight long chain alcohols. Among these, the aromatic benzyl alcohol derivatives gave excellent productivity in comparison with long carbon chain alcohol due to the superior nucleophilicity. All these laurate esters are also called as fatty acid esters, which are broadly applied in spin finishes in textiles, emulsifiers or oiling agents for foods, lubricant, paint or ink additives, surfactants, lubricants for plastics, etc. [41-45]. However, these compounds have also been applied as solvents, co-solvents and base materials in perfumery, flavor in food and pharma industries [42–45]. The reactivity order of the various alcohols in the esterification of vinyl laurate was given in the order: lower chain alcohols > branched chain alcohols > aromatic benzyl alcohol derivatives > straight long chain alcohols.

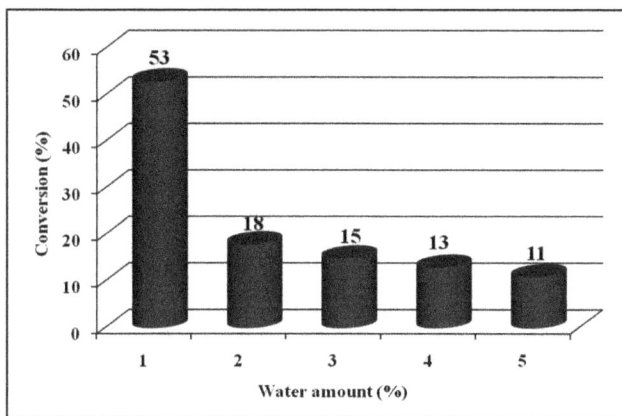

Figure 6   Effect of water on the esterification reactions over biocatalyst

2.4   Influence of water on the esterification reaction

The influence of water quantity on esterification reaction was studied using biocatalysts by adding different quantities of water to the reaction mixture as represented in Figure 6. Srivastava et al. [46] observed that the excess amount of water led to the sintering of the enzyme catalyst, and hence the availability of active sites on the biocatalyst surface area was decreased for the reaction. Sophie et al. [47] stated that the product formation in the esterification reactions, water can not only affect the activity of enzyme-catalyst but also the thermodynamic equilibrium of the esterification reactions. The high amount of water shifts the equilibrium of the esterification reaction towards backward direction by hydrolysis of the ester to form reactants, resulting in low yields of the ester products [46-

Industrial Applications of Green Solvents I | Materials Research Forum LLC
Materials Research Foundations **50** (2019) 242-268 | doi: https://doi.org/10.21741/9781644900239-8

53]. This could be probably for the reason that the excess water shifted into the heavier phase from the lighter phase after it is saturated by water dilution. Thus, the water amount in the lighter phase always remains constant to control the temperature and pressure in the reaction mixture. In the presence of excess water, the enzyme added at the reactor bottom in the esterification reactions, an enzyme in direct contact with the heavy phase [48]. Hence, the reverse rate of reaction was rapidly increased, while decreasing the forward reaction rate towards the formation of ester products. For example, the yield of terpinyl acetate was lowered rapidly when the water introduced into the reaction mixture increased to 5% [46-48].

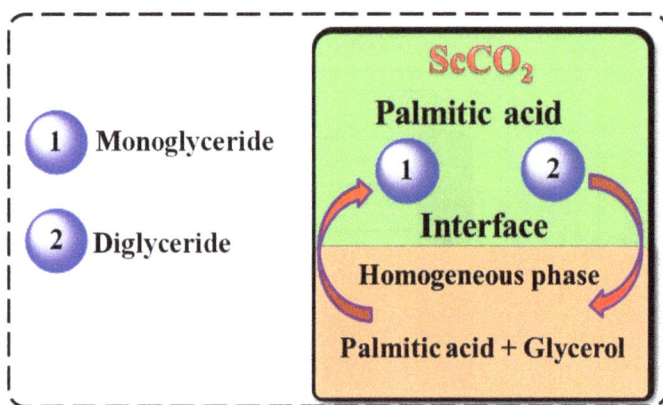

Figure 7   Phase transfer model of palmitin in ScCO$_2$

2.5   Phase transfer model of palmitin

The phase transfer representation was designed to show the high selective production of diglyceride in the ScCO$_2$ medium. The monoglyceride is formed as the starting compound, which instantly dissolved into the ScCO$_2$ phase and then it improves termination of palmitic acid in ScCO$_2$ solvent [54]. The main benefits of ScCO$_2$ include more diffusivity and low viscosity to produce diglyceride products [55,56]. Hence, the rate of formation of diglyceride from monoglyceride has significantly improved by preparation of the glycerol and palmitic acid homogeneous phase in the presence of ScCO$_2$. However, monoglyceride formed instantly disappears from the homogeneous phase of reaction mixture due to its unique properties of ScCO$_2$, and an interface forms between two-phase systems (Figure 7). This type of phase behavior is reduced the further reactivity of diglyceride with acid to form triglyceride and concurrently stops the reaction

[54-56]. Therefore, the superior selectivity towards diglyceride product was obtained as a result of the solubility changes between the different reaction products [54].

2.6   The influence of pressure and temperature on the phase behaviour system

Generally, every reaction is strongly dependant on the solubility of substrates in $ScCO_2$, which could be altered with little difference in pressure and/or temperature mainly close to the critical point of the compounds [57-59]. The solubility of all the reactant molecules increased with increase in the pressure as a consequence of the higher density of the supercritical fluid. In fact, the solvent power of supercritical CO_2 can be used to perform the esterification reactions. By increasing the temperature of the reaction, the substrate solubility in the presence of $ScCO_2$ can be enhanced. However high-temperature reactions could lead to enzyme deactivation [58-60]. Hence, these esterification reactions have to be performed at most favorable temperatures for the enzyme activity. [60].

For example, the solubility investigations of lactic acid in $ScCO_2$ have revealed that the solubility of lactic acid in $ScCO_2$ improved at a temperature of 55 °C and pressure of 200 bars, i.e., 17×10^8 [61,62]. Therefore, the effects of temperature and pressure have been studied with lipase-catalyzed esterification reactions in $ScCO_2$. The phase performance of lactic acid/n-butanol/Novozyme-435/$ScCO_2$ and lactic acid/n-butanol/Novozyme-435/n-hexane/$ScCO_2$ systems where n-hexane was acting as a co-solvent has been examined, at different temperatures and pressures in a reactor [61-63]. Finally, comparable phase performances of the esterification reaction mixtures were achieved at the most favorable pressures and temperatures in the reaction mixtures [63].

2.7   Comparison between the presence and absence of biocatalyst

The catalyzed esterification reaction has been performed with the equivalent quantity over immobilized lipase biocatalyst (PVA/CHI) and free lipase immobilized biocatalyst under optimal reaction parameters. It was observed that the reaction productivity was greatly improved in the presence of biocatalyst, while the reaction productivity was significantly lowered in the absence of biocatalyst as is shown in Figure 8. This could be explained as the absence of lipases straightforwardly expose to the $ScCO_2$ background, as PVA/CHI biocatalyst was better connected with the support matrix [64-66]. However, after lipase immobilization, the lipases have well dispersed on the surface of immobilization support and then, easily activated the diffusion of the reactant on active sites of biocatalyst surface [41,64,67]. Therefore, the PVA/CHI biocatalyst exhibited ~4-fold higher catalytic activity in comparison with the absence of biocatalyst in $ScCO_2$. It was observed that the yield of final product was significantly achieved in presence of biocatalyst in 3–5-folds greater than that of the absence of biocatalyst [41]. Finally, these results indicated that the esterification reaction could be significantly enhanced in the presence of lipase-based biocatalysts when compared to that of the absence of biocatalysts.

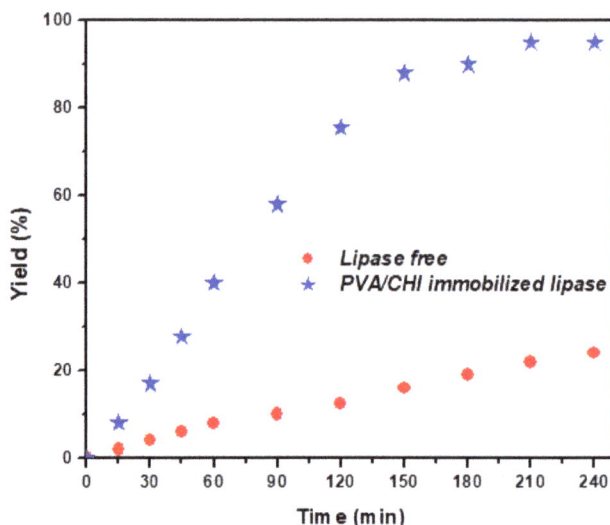

Figure 8   Comparison between the presence and absence of immobilized lipase biocatalyst as a function of time in the synthesis of citronellyl laurate in supercritical CO_2

2.8   Comparison of activity between $ScCO_2$ and organic solvents

In the last few decades, a lot of research work signifying that enzyme-catalyzed esterification reactions in supercritical CO_2 solvent gave better activity in comparison with conventional organic solvents or solvent-free conditions has been reported [68,69]. $ScCO_2$ has been undertaken as an alternative to usual organic solvents for the biocatalyzed production of different esters. As acids are soluble in the $ScCO_2$, they easily partition into the ester products from the reaction mixture and also enzyme may be easily obtained as the reaction is carried out in $ScCO_2$ solvent. The main drawback of the organic solvent is that it is applied as a solvent or co-solvent in esterification reactions, the separation of the solvent or co-solvent become more difficult than $ScCO_2$ solvent.

Examples: i) Yu et al. [70] reported the more rapid production of ethyl oleate catalyzed over *Candida Cylindracae Lipase* in supercritical CO_2 in comparison with organic solvents and Knez et al. [71,72] have reported the improved activities for the production

Industrial Applications of Green Solvents I   Materials Research Forum LLC
Materials Research Foundations **50** (2019) 242-268   doi: https://doi.org/10.21741/9781644900239-8

of corresponding ester catalyzed by *Lipozyme TL IM* in supercritical CO_2 in comparison to absence of solvent conditions, which are clearly shown in Scheme 7. In addition to the enhanced rate of reactions and activities, improved selectivity in $ScCO_2$ has been reported by several researchers [73-75,76-78], which is due to the specific properties of $ScCO_2$ including lower viscosity and higher diffusivity of the substrates and formation of carbamate on the catalyst surface [73,74,78].

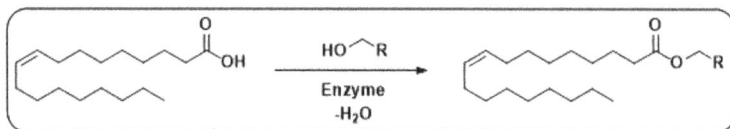

Scheme 7   Esterification of oleic acid with an alcohol

ii) Tewari et al. [79] explained the rate of reactions on the transesterification of benzyl alcohol and butylacetate by lyophilized *Candida Antarctica Lipase* biocatalyst. The rate of the reaction was enhanced in $ScCO_2$ than inorganic solvents such as hexane and toluene or in the absence of solvent reaction conditions.

Summary

The chapter focuses on the substitution of the conventional solvents with stable, non-toxic and environmentally friendly solvent, supercritical carbon dioxide. The unique properties of $ScCO_2$, it exhibits liquid-like density and gas-like diffusivity, surface tension and viscosity. Hence, it has widely used as a green solvent for esterification reactions in comparison with organic solvents. It also emphasizes the conversions towards the esters from the acids employing different reaction parameters such as temperatures, pressures, water dilution, and carbon chain of alcohols. In addition, a comparative study showed that the enzymatic esterification was much faster in $ScCO_2$ than in other organic solvents. The phase transfer model of palmitin was proposed to show highly selective production of diglycerides in $ScCO_2$ solvent. This type of phase behavior is reduced the further reactivity of diglyceride with acid to form triglyceride. Overall outlook gives the importance of supercritical carbon dioxide over the conventional solvent employing various parameters.

Industrial Applications of Green Solvents I   Materials Research Forum LLC
Materials Research Foundations **50** (2019) 242-268   doi: https://doi.org/10.21741/9781644900239-8

Table 3 Conversions of vinyl laureate with different alcohols in the presence and absence of biocatalyst (Lipase).

Substrate	Laurate yield (%)	
	Lipase	**Lipase free**
	99	31
	99	30
	98	31
	98	33
	98	34
	99	32
	97	29
	93	23
	94	23
	95	28
	96	27
	95	29

Industrial Applications of Green Solvents I   Materials Research Forum LLC
Materials Research Foundations **50** (2019) 242-268   doi: https://doi.org/10.21741/9781644900239-8

References

[1] C.J. Clarke, W.C. Tu, O. Levers, A. Brohl, J.P. Hallett, Green and sustainable solvents in chemical processes, Chem. Rev. 118 (2018) 747-800. https://doi.org/10.1021/acs.chemrev.7b00571

[2] R.S. Oakes, A.A. Clifford, C.M. Rayner, The use of supercritical fluids in synthetic organic chemistry, J. Chem. Soc., Perkin Trans. 1 (2001) 917-941. https://doi.org/10.1039/b101219n

[3] B. Mallesham, P. Sudarsanam, G. Raju, B.M. Reddy, Design of highly efficient Mo and W-promoted SnO_2 solid acids for heterogeneous catalysis: acetalization of bio-glycerol, Green Chem. 15(2013) 478-489. https://doi.org/10.1039/c2gc36152c

[4] R.T. Baker, W. Tumas, Toward greener chemistry, Sci. 284 (1999) 1477-1479.

[5] P.T. Anastas, T.C. Williamson, Green Chemistry: Frontiers in benign chemical syntheses and processes, Oxford University Press, Oxford, 1998.

[6] P.T. Anastas, J.C. Warner, Green Chemistry: Theory and practice, Oxford University Press, Oxford, 1998.

[7] T. Welton, Room-temperature ionic liquids. Solvents for synthesis and catalysis, Chem. Rev. 99 (1999) 2071-2084. https://doi.org/10.1021/cr980032t

[8] E. Buncel, R. Stairs, H. Wilson, The role of the solvent in chemical reactions, Oxford University Press, UK, 2003.

[9] S. Abou-Shehada, J.H. Clark, G. Paggiola, J. Sherwood, Tunable solvents: Shades of green, Chem. Eng. Process. 99 (2016) 88-96. https://doi.org/10.1016/j.cep.2015.07.005

[10] R.A. Sheldon, Green solvents for sustainable organic synthesis: State of the art, Green Chem. 7 (2005) 267-278. https://doi.org/10.1039/b418069k

[11] B. Mallesham, P. Sudarsanam, B.M. Reddy, Eco-friendly synthesis of bio-additive fuels from renewable glycerol using nanocrystallineSnO_2-based solid acids, Catal. Sci. Technol. 4 (2014) 803-813. https://doi.org/10.1039/c3cy00825h

[12] B. Mallesham, P. Sudarsanam, B.M. Reddy, Production of biofuel additives from esterification and acetalization of bioglycerol over SnO_2 based solid acids, Ind. Eng. Chem. Res. 53 (2014) 18775-18785. https://doi.org/10.1021/ie501133c

[13] I.T. Horvath, Fluorous biphase chemistry, Acc. Chem. Res. 31 (1998) 641-650. https://doi.org/10.1021/ar970342i

[14] B. Betzemeier, P. Knochel, Perfluorinated solvents a novel reaction medium in organic chemistry, Top. Curr. Chem. 206 (1999) 60-68. https://doi.org/10.1007/3-540-48664-x_3

[15] J.P. Genet, M. Savignac, Recent developments of palladium (0) catalyzed reactions in aqueous medium, J. Organomet. Chem. 576 (1999) 305-317. https://doi.org/10.1016/s0022-328x(98)01088-2

[16] S. Kobayashi, Scandium triflate in organic synthesis, Eur. J. Chem. 1999 (1999) 15-27.

[17] D.A. Canelas, D.E. Betts, J.M. DeSimone, M.Z. Yates, K.P. Johnson, Poly(vinyl acetate) and poly(vinyl acetate-co-ethylene) latexes via dispersion polymerizations in carbon dioxide, Macromolecules, 31 (1998) 6794-6805. https://doi.org/10.1021/ma980596z

[18] M. McCoy, Chem. Eng. News, June 14th, 1999, 11.

[19] M. McCoy, Chem. Eng. News, June 14th, 1999, 13; see also http://www.micell.com.

[20] M.D. Donohue, J.L. Geiger, A.A. Kiamos, K.A. Nielsen, Green chemistry, ACS Symp. Ser., Am. Chem. Soc. Washington, 626 (1996) 152-167.

[21] R. Scott Oakes, A.A. Clifford, C.M. Rayner, The use of supercritical fluids in synthetic organic chemistry, J. Chem. Soc. Perkin Trans. 1 (2001) 917-941. https://doi.org/10.1039/b101219n

[22] W. Leitner, Green chemistry: Designed to dissolve, Nature 405 (2000) 129-130.

[23] H.R. Hobbs, N.R. Thomas, Biocatalysis in supercritical fluids, in fluorous solvents, and under solvent-free conditions, Chem. Rev. 107 (2007) 2786-2820. https://doi.org/10.1021/cr0683820

[24] S. Sabeder, M. Habulin, Z. Knez, Comparison of the esterification of fructose and palmitic acid in organic solvent and in supercritical carbon dioxide, Ind. Eng. Chem. Res. 44 (2005) 9631-9635. https://doi.org/10.1021/ie050266k

[25] R.S. Oakes, A.A. Clifford, C.M. Rayner, The use of supercritical fluids in synthetic organic chemistry, J. Chem. Soc., Perkin Trans. 1 (2001) 917-941. https://doi.org/10.1039/b101219n

[26] X. Wang, L. Wang, X. Chen, D. Zhou, H. Xiao, X. Wei, J. Liang, Catalytic methyl esterification of colophony over ZnO/SFCCR with subcritical CO_2: Catalytic performance, reaction pathway and kinetics, R. Soc. open sci. 5 (2018) 172124-172139. https://doi.org/10.1098/rsos.172124

[27]K. Rezaei, F. Temellib, E. Jena, Effects of pressure and temperature on enzymatic reactions in supercritical fluids, Biotechnol. Adv. 25 (2007) 272-280.

[28] M. Esteki, M. Rezayat, H.S. Ghaziaskar, T. Khayamian, Application of QSPR for prediction of percent conversion of esterification reactions in supercritical carbon dioxide using least squares support vector regression, J. Supercrit. Fluids, 54 (2010) 222-230. https://doi.org/10.1016/j.supflu.2010.04.007

[29] O. Hemminger, A. Marteel, M.R. Mason, J.A. Davies, A.R. Tadd, M.A. Abraham, Hydroformylation of 1-hexene in supercritical carbon dioxide using a heterogeneous rhodium catalyst. 3. Evaluation of solvent effects, Green Chem. 4 (2002) 507-512. https://doi.org/10.1039/b204822c

[30] W. Leitner, Reactions in Supercritical Carbon Dioxide (ScCO$_2$). In: P. Knochel (eds.), Modern Solvents in Organic Synthesis, Springer, Top. Cur. Chem. 206 (1999) 107-132. https://doi.org/10.1007/3-540-48664-x_5

[31]Z. Knez, Enzymatic reactions in dense gases, J. Supercrit. Fluids, 47 (2009) 357-372.

[32] G.K. Nagesha, B. Manohar, K.U. Sankar, Enzymatic esterification of free fatty acids of hydrolyzed soy deodorizer distillate in supercritical carbon dioxide, J. Supercrit. Fluids, 32 (2004) 137-145. https://doi.org/10.1016/j.supflu.2004.02.001

[33]M. Habulin, M. Primozic, Z. Knez, Stability of proteinase form Carica papaya latex in dense gases, J. Supercrit. Fluids 33 (2005) 27-34.

[34]J. McHardy, S.P. Sawan (Eds.), Supercritical fluid cleaning: Fundamental, technology and applications, Noyes Publications, New Jersey, 1998.

[35] K. Fukui, T. Yonezawa, H. Shingu, A molecular orbital theory of reactivity in aromatic hydrocarbons, J. Chem. Phy. 20 (1952) 722-725. https://doi.org/10.1063/1.1700523

[36] Y. Huang, L. Wang, X. Chen, X. Wei, J. Liang, W. Li, Intrinsic kinetics study of rosin hydrogenation on a nickel catalyst supported on spent equilibrium catalyst. Rsc. Adv. 7 (2017) 780-788. https://doi.org/10.1039/c7ra03611f

[37] F. Ren, Y. Zheng, X. Liu, L. Ma, W. Li, An investigation of the oxidation mechanism of abietic acid using two-dimensional infrared correlation spectroscopy, J. Mole. Struct. 1084 (2015) 236-243. https://doi.org/10.1016/j.molstruc.2014.12.055

[38]M.D. Romero, L. Calvo, C. Alba, M. Luis, The production of flavor esters in supercritical carbon dioxide, Proceedings of the Sixth International Symposium on Supercrit. Fluids, 2 (2003) 1445-1450.

[39] M.D. Romero, L. Calvo, C. Alba, M. Habulin, M. Primozi, Z. Knez, Enzymatic synthesis of isoamyl acetate with immobilized Candida Antarctica lipase in supercritical carbon dioxide, J. Supercrit. Fluids, 33 (2005) 77-84. https://doi.org/10.1016/j.supflu.2004.05.004

[40] J.S. Brown, H.P. Lesutis, D.R. Lamb, D. Bush, K. Chandler, B.L. West, C.L. Liotta, C.A. Eckert, D. Schiraldi, J.S. Hurley, Supercritical fluid separation for selective quaternary ammonium salt promoted esterification of terephthalic, Ind. Eng. Chem. Res. 38 (1999) 3622-3627. https://doi.org/10.1021/ie990040f

[41] K.C. Badgujar, B.M. Bhanage, Synthesis of geranyl acetate in nonaqueous media using immobilized Pseudomonas cepacia lipase on biodegradable polymer film: Kinetic modelling and chain length effect study, Process Biochem. 49 (2014) 1304-1313. https://doi.org/10.1016/j.procbio.2014.04.014

[42] G.D. Yadav, P.S. Lathi, Synthesis of citronellol laurate in organic media catalyzed by immobilized lipases: kinetic studies, J. Mol. Catal. B: Enzym. 27 (2004) 113-119. https://doi.org/10.1016/j.molcatb.2003.10.004

[43] M. Habulin, S. Sabeder, M. Paljevac, M. Primozi, Z. Knez, Lipase-catalyzed esterification of citronellol with lauric acid in supercritical carbon dioxide/co-solvent media, J. Supercrit. Fluids, 43 (2007) 199-203. https://doi.org/10.1016/j.supflu.2007.05.001

[44] M. Habulin, S. Sabeder, M.A. Sampedro, Z. Knez, Enzymatic synthesis of citronellol laurate in organic media and supercritical carbon dioxide, Biochem. Eng. J. 42 (2008) 6-12. https://doi.org/10.1016/j.bej.2008.05.012

[45] A.Z. Abdullah, N.S. Sulaiman, A.H. Kamaruddin, Biocatalytic esterification of citronellol with lauric acid by immobilized lipase on aminopropyl-grafted mesoporous SBA-15, Biochem. Eng. J. 44 (2009) 263-270. https://doi.org/10.1016/j.bej.2009.01.007

[46] S. Srivastava, G. Madras, J.M. Modak, Esterification of myristic acid in supercritical carbon dioxide, J. Supercrit. Fluids, 27 (2003) 55-64. https://doi.org/10.1016/s0896-8446(02)00191-2

[47] R.J.T. Sophie Colombié, J.S. Condoret, A. Marty, Water activity control: Away to improve the efficiency of continuous lipase esterification, Biotechnol. Bioeng. 60 (1998) 362-368. https://doi.org/10.1002/(sici)1097-0290(19981105)60:3%3C362::aid-bit13%3E3.0.co;2-o

[48] K.J. Liu, Y.R. Huang, Lipase-catalyzed production of a bioactive terpene ester in supercritical carbon dioxide, J. Biotechnol. 146 (2010) 215-220. https://doi.org/10.1016/j.jbiotec.2010.02.017

[49] S. Srivastava, J. Modak, G. Madras, Enzymatic synthesis of flavors in supercritical carbon dioxide, Ind. Eng. Chem. Res. 41 (2002) 1940-1945. https://doi.org/10.1021/ie010651j

[50] J.S. Condoret, S. Vankan, X. Joulia, A. Marty, Prediction of water adsorption curves for heterogeneous biocatalysis in organic and supercritical solvents, Chem. Eng. Sci. 52 (1997) 213-220. https://doi.org/10.1016/s0009-2509(96)00413-7

[51] N. Fontes, J. Partridge, P.J. Halling, S. Barreiros, Zeolite molecular sieves have dramatic acid–base effects on enzymes in nonaqueous media, Biotechnol. Bioeng. 77 (2002) 296-305. https://doi.org/10.1002/bit.10138

[52] N. Harper, S. Barreiros, Enhancement of enzyme activity in supercritical carbon dioxide via changes in acid-base conditions, Biotechnol. Prog. 18 (2002) 1451-1454. https://doi.org/10.1021/bp025602w

[53] N. Fontes, N. Harper, P.J. Halling, S. Barreiros, Salt hydrates for in situ water activity control have acid-base effects on enzymes in nonaqueous media, Biotechnol. Bioeng. 82 (2003) 802-808. https://doi.org/10.1002/bit.10627

[54] M. Tao, Q. Li, J. Qu, M. Zhang, Enzymatic synthesis of dipalmitin in supercritical carbon dioxide and mechanism study, Ind. Eng. Chem. Res. 52 (2013) 13528-13535. https://doi.org/10.1021/ie4015364

[55] M.V. Oliveira, S.F. Rebocho, A.S. Ribeiro, E.A. Macedo, J.M. Loureiro, Kinetic modelling of decyl acetate synthesis by immobilized lipase-catalyzed transesterification of vinyl acetate with decanol in supercritical carbon dioxide, J. Supercrit. Fluids, 50 (2009) 138-145. https://doi.org/10.1016/j.supflu.2009.05.003

[56] C. Pereyra, D. Gordillo, E.J.M. De La Ossa, Supercritical fluid-solid phase equilibria calculations by cubic equations of state and empirical equations: application to the palmitic acid + carbon dioxide system, J. Chem. Eng. Data 49 (2004) 435-438. https://doi.org/10.1021/je0340598

[57] S. Sabeder, M. Habulin, Ž. Knez, Comparison of the esterification of fructose and palmitic acid in organic solvent and in supercritical carbon dioxide, Ind. Eng. Chem. Res. 44 (2005) 9631-9635. https://doi.org/10.1021/ie050266k

[58]D.W. Chung, M.H. Cho, A comparative study on the effect of commercialized immobilized lipases on the selective synthesis of 1,3-diglyceride, J. Korean Ind. Eng. Chem. 21 (2010) 452.

[59] Z. Guo, Y. Sun, Solvent-free production of 1,3-diglyceride of CLA: Strategy consideration and protocol design, Food Chem. 100 (2007) 1076-1084. https://doi.org/10.1016/j.foodchem.2005.11.011

[60] M. Habulin, S. Šabeder, M. Paljevac, M. Primozic, Z. Knez, Lipase-catalyzed esterification of citronellol with lauric acid in supercritical carbon dioxide/co-solvent media, J. Supercrit. Fluids, 43 (2007) 199-203. https://doi.org/10.1016/j.supflu.2007.05.001

[61] J. Gregorowicz, Solubilities of lactic acid and 2-hydroxyhexanoic acid in supercritical CO_2, Fluid Phase Equilibria, 166 (1999) 39-46. https://doi.org/10.1016/s0378-3812(99)00283-6

[62] J. Gregorowicz, P. Bernatowicz, Phase behavior of l-lactic acid based polymers of low molecular weight in supercritical carbon dioxide at high pressures, J. Supercrit. Fluids, 51 (2009) 270-277. https://doi.org/10.1016/j.supflu.2009.08.002

[63] Z. Knez, S. Kavcic, L. Gubicza, K. Belafi-Bako, G. Nemeth, M. Primozic, M. Habulin, Lipase-catalyzed esterification of lactic acid in supercritical carbon dioxide, J. Supercrit. Fluids, 66 (2012) 192- 197. https://doi.org/10.1016/j.supflu.2011.11.006

[64] K.P. Dhake, K.M. Deshmukh, Y.P. Patil, R.S. Singhal, B.M. Bhanage, Improved activity and stability of Rhizopus oryzae lipase via immobilization for citronellol ester synthesis in supercritical carbon dioxide, J. Biotechnol. 156 (2011) 46-51. https://doi.org/10.1016/j.jbiotec.2011.08.019

[65] D. Chen, C. Peng, H. Zhang, J. Xu, Y. Yan, Assessment of activities and con-formation of lipases treated with sub- and supercritical carbon dioxide, Appl. Biochem. Biotechnol. 169 (2013) 2189-2201. https://doi.org/10.1007/s12010-013-0132-3

[66] Y. Liu, D. Chen, X. Xu, Y. Yan, Evaluation of structure and hydrolysis activity of Candida rugosa Lip7 in presence of sub-/super-critical CO_2, Enzyme Microb. Technol. 51 (2012) 354-358. https://doi.org/10.1016/j.enzmictec.2012.08.003

[67] G. Fernandez-Lorente, Z. Cabrera, C. Godoy, R. Fernandez-Lafuente, J.M. Palomo, J.M. Guisan, Interfacially activated lipases against hydrophobic supports: Effect of the support nature on the biocatalytic properties, Process Biochem. 43 (2008)1061-1067. https://doi.org/10.1016/j.procbio.2008.05.009

[68] H.R. Hobbs, N.R. Thomas, Biocatalysis in supercritical fluids, in fluorous solvents, and under solvent-free conditions, Chem. Rev. 107 (2007) 2786-2820. https://doi.org/10.1021/cr0683820

[69] S. Sabeder, M. Habulin, Z. Knez, Comparison of the esterification of fructose and palmitic acid in organic solvent and in supercritical carbon dioxide, Ind. Eng. Chem. Res. 44 (2005) 9631-9635. https://doi.org/10.1021/ie050266k

[70] Z.R. Yu, S.S.H. Rizvi, J.A. Zollweg, Enzymic esterification of fatty acid mixtures from milk fat and anhydrous milk fat with canola oil in supercritical carbon dioxide, Biotechnol. Prog. 8 (1992) 508-513. https://doi.org/10.1021/bp00018a006

[71] Z. Knez, M. Habulin, Z. Knez, M. Habulin, Lipase-catalyzed esterification in supercritical carbon dioxide. biocatalysis in non-concentional media. J. Tramper, Elsevier Science Publishers: (1992) 401-407. https://doi.org/10.1016/b978-0-444-89046-7.50061-8

[72] Z. Knez, M, Habulin, Lipase catalysed esterification at high pressure, Biocatal. 9 (1994) 115-121. https://doi.org/10.3109/10242429408992113

[73] Y. Ikushima, N. Saito, M. Arai, H.W. Blanch, Activation of a lipase triggered by interactions with supercritical carbon dioxide in the near-critical region, J. Phys. Chem. 99 (1995) 8941-8944. https://doi.org/10.1021/j100022a001

[74] Y. Ikushima, Supercritical fluids: An interesting medium for chemical and biochemical processes, Adv. Colloid Interface Sci. 71-72 (1997) 259-280. https://doi.org/10.1016/s0001-8686(97)00021-3

[75] N. Mase, T. Sako, Y. Horikawa, K. Takabe, Novel strategic lipase-catalyzed asymmetrization of 1,3-propanediacetate in supercritical carbon dioxide, Tetrahedron Lett. 44 (2003) 5175-5178. https://doi.org/10.1016/s0040-4039(03)01266-8

[76] Y. Ikushima, N. Saito, T. Yokoyama, K. Hatakeda, S. Ito, M. Arai, H.W. Blanch, Solvent effects on an enzymatic ester synthesis in supercritical carbon dioxide, Chem. Lett. 22 (1993) 109-112. https://doi.org/10.1246/cl.1993.109

[77] E. Catoni, E. Cernia, C. Palocci, Different aspects of 'solvent engineering' in lipase biocatalysed esterifications, J. Mol. Catal. A: Chem. 105 (1996) 79-86. https://doi.org/10.1016/1381-1169(95)00153-0

[78] M. Rantakyla, M. Alkio, O. Aaltonen, Stereospecific hydrolysis of 3-(4-methoxyphenyl) glycidic ester in supercritical carbon dioxide by immobilized lipase, Biotechnol. Lett. 18 (1996) 1089-1094. https://doi.org/10.1007/bf00129737

Industrial Applications of Green Solvents I   Materials Research Forum LLC
Materials Research Foundations **50** (2019) 242-268   doi: https://doi.org/10.21741/9781644900239-8

[79] Y.B. Tewari, T. Hara, K.W. Phinney, M.P. Mayhew, A thermodynamic study of the lipase-catalyzed transesterification of benzyl alcohol and butyl acetate in supercritical carbon dioxide media, J. Mol. Catal. B: Enzym. 30 (2004) 131-136. https://doi.org/10.1016/j.molcatb.2004.04.005

Industrial Applications of Green Solvents I
Materials Research Foundations 50 (2019) 269-319

Materials Research Forum LLC
doi: https://doi.org/10.21741/9781644900239-9

Chapter 9

Multicomponent Synthesis of Biologically Relevant Spiroheterocycles in Water

Bubun Banerjee[*]

Department of Chemistry, Indus International University, Bathu, Una, Himachal Pradesh, 174301, India

banerjeebubun@gmail.com

Dedicated to my little Agnish

Abstract

This chapter deals with the up-to-date developments of one-pot multicomponent synthesis of biologically relevant spiroheterocycles in aqueous media. As the current topic is one of the challenging areas for today's organic chemists, therefore the present chapter will surely be a valuable document to boost the on-going developments in this direction.

Keywords

Spiroheterocycles, Multicomponent Reactions, Bioactivity, Aqueous Media, Green Synthesis

List of abbreviations

CNS: Central nervous system
PEG: Polyethylene glycol
SDS: sodium dodecyl sulfate
CSA: Camphor-10-sulfonic acid
CTAB: Cetyltrimethylammonium bromide
p-TSA: *p*-Toluene sulfuonic acid

EDDA: Ethylenediamine diacetate
β-CD: β-Cyclodextrin
TEBAC: Triethylbenzylammonium chloride
TBAB: Tetrabutylammonium bromide
TEBA: Triethylbenzylammonium chloride
DBSA: *p*-dodecylbenzenesulfonic acid

Contents

1.   Introduction

Heterocycles are the main building block of many organic compounds [1, 2]. Majority of organic compounds consist of diverse heterocyclic motifs. Spiroheterocycles where two rings attached through a common carbon atom are structurally interesting [3]. These significant classes of organic compounds are important as they possess a wide range of biological activities [4, 5]. Many naturally occurring organic compounds possess spiroheterocyclic skeleton that include horsfiline (anesthetics) [6], abamectin (insecticide and antihelmintic) [7], coerulescine (anesthetics) [8], ajugarin I (insect antifeedant) [9], calcimycin (antibiotic and antifungal) [10], chlorogenin (cytotoxic) [11], alstonisine (antitumour and antimicrobial agents) [12], convallamarogenin (anti-food-deteriorating agent) [13], coriamyrtin (investigative tool in neuroscience) [14], spirotryprostatin (inhibitors of microtubule assembly) [15], digitogenin (cardiac glycoside) [16], fredericamycin-A (cytotoxic) [17], fumagillin (antibiotic and anti protozoal) [18], isopteropodine (serotonin receptor modulators) [19], griseofulvin (antifungal), [20], gelsemine (CNS stimulant) [21], grindelic acid (HIV-1 reverse transcriptase inhibitor) [22] and hecogenin (cholesterol absorption inhibitor) [23] etc. Furthermore, many synthetic spiroheterocyclic scaffolds are being clinically used as potent antimicrobial [24], diuretic [25], antitumor [26], antipsychotic [27], antibroncho-constrictor [28], inhibitors of the human NK-1 receptor [29], antihypertensive [30], anxiolytic [31], antibiotic [32], anti-diabetic [33], anti-ulcer [34] and antihypertensive [35] agents.

Recently, a multi-component reaction (MCR) strategy is becoming one of the valuable tools to synthesize various structurally diverse organic scaffolds in a pot. MCR strategy offers a wide range of advantages including operational simplicity, reduction in the number of work-up steps by minimizing the purification processes, atom economic and energy efficiency [36-38]. As a result, both in academia as well as in industry there is a constant effort to design environmentally sustainable multi-component reactions in a pot.

On the other hand, it is interesting to note that the last decade has shown a tremendous outburst of aqueous mediated organic transformations to make them 'sustainable' for the betterment of our *Mother Nature* [39]. Now-a-days, to carry out organic reactions, water

Industrial Applications of Green Solvents I   Materials Research Forum LLC
Materials Research Foundations **50** (2019) 269-319   doi: https://doi.org/10.21741/9781644900239-9

has become the first choice as a solvent because of its environmental friendliness as well as it is non-flammable, cheap and abundantly available [40].

2.   Synthesis of *N*-containing spiroheterocycles

2.1   Synthesis of spiro[pyrimido[4,5-*b*]quinoline-5,5-pyrrolo[2,3-*d*]pyrimidine]-pentaone derivatives

Bazgir and his co-researchers [41] developed a simple, mild and efficient protocol for the synthesis of biologically promising spiro[pyrimido[4,5-*b*]quinoline-5,5-pyrrolo[2,3-*d*]pyrimidine]-pentaone derivatives (**3**) *via* one-pot pseudo three-component reactions of two equivalents of 6-amino-uracil derivatives (**1, 1a, 1b**) and one equivalent of substituted isatins (**2**) using *p*-toluene sulfuonic acid (*p*-TSA) as catalyst in aqueous media (Scheme 1). It was proposed that the reaction undergoes through the unusual ring opening of isatin moiety followed by recyclization (Scheme 2). Compound **3b** showed the highest antibacterial activity among the all synthesized compounds.

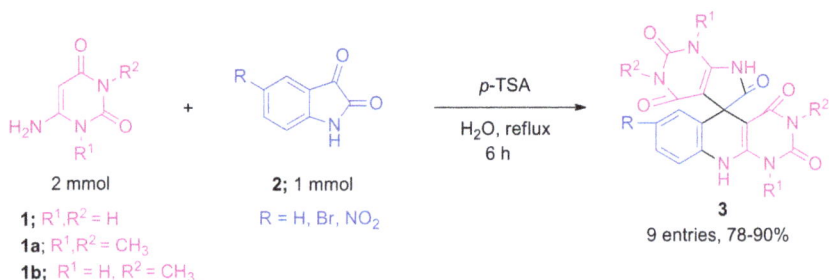

Scheme 1 *p-TSA catalyzed pseudo three-component synthesis of spiro[pyrimido[4,5-b]quinoline-5,5-pyrrolo[2,3-d]pyrimidine]-pentaone derivatives in aqueous medium*

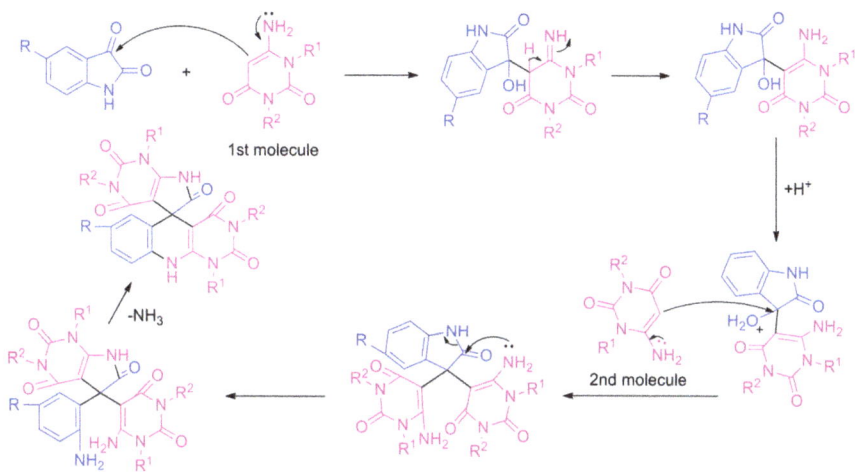

Scheme 2 *Plausible mechanism for the synthesis of spiro[pyrimido[4,5-b]quinoline-5,5-pyrrolo[2,3-d]pyrimidine]-pentaone derivatives in water*

2.2 Synthesis of spirooxindole-containing fused 1,4-dihydropyridine derivatives

Another *p*-toluene sulfonic acid (*p*-TSA) catalyzed protocol was developed by Alizadeh et al. [42] for the efficient synthesis of spirooxindole-containing fused 1,4-dihydropyridine derivatives (**7**) *via* one-pot pseudo five-component reactions between two equivalents of ammonia, one equivalent of 1,1-bis(methylthio)-2-nitroethylene (**4**), one equivalent of isatin or its derivatives (**2**) and one equivalent of 1,3-cyclohexanedione (**6**) in aqueous medium under heating conditions (Scheme 3). The reaction underwent through the formation of intermediate 1,1-bis(amino)-2-nitroethylene (**5**). Uses of aqueous media, metal free organocatalyst and high product yields have been the major benefits of this method.

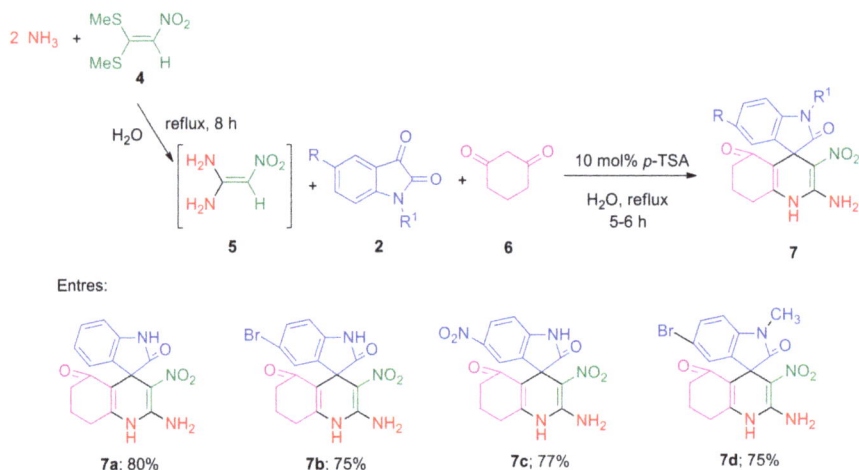

Scheme 3 *p-TSA catalyzed synthesis of spirooxindole-containing fused 1,4-dihydropyridine derivatives in an aqueous medium.*

2.3   Synthesis of spiro[indoline-3,5′-pyrimido[4,5-*b*]quinoline] derivatives

Bazgir and his group [43] have developed a simple and straightforward method for the one-pot three-component synthesis of a series of novel spiro[indoline-3,5′-pyrimido[4,5-*b*]quinoline] derivatives (**9**) *via* the condensation between isatins (**2**), 2,6-diaminopyrimidin-4(3*H*)-one (**8**) and dimedone (**6a**) using the same *p*-toluene sulfonic acid (*p*-TSA) as catalyst in water under reflux conditions (Scheme 4). Use of green solvent, operational simplicity, easy work-up procedure and good to excellent yields were some of the major advantages of this developed method.

2.4   Synthesis of spiro[indoline-3,4′-pyrazolo[3,4-*b*]pyridine]-2,3′ (7′*H*)-dione

A series of novel spiro[indoline-3,4'-pyrazolo[3,4-*b*]pyridine]-2,3'(7'*H*)-diones (**12**) was synthesized *via* one-pot three-component reactions of isatins (**2**), 3-amino-1-phenyl-1*H*-pyrazol-5(4*H*)-one (**10**) and 1,2-diphenylethan-1-one (**11**) using acetic acid as catalyst in water at 90 °C (Scheme 5) [44]. Excellent yields, wide scope substrates, mild reaction conditions, use of water as a solvent were some of the major advantages of this protocol.

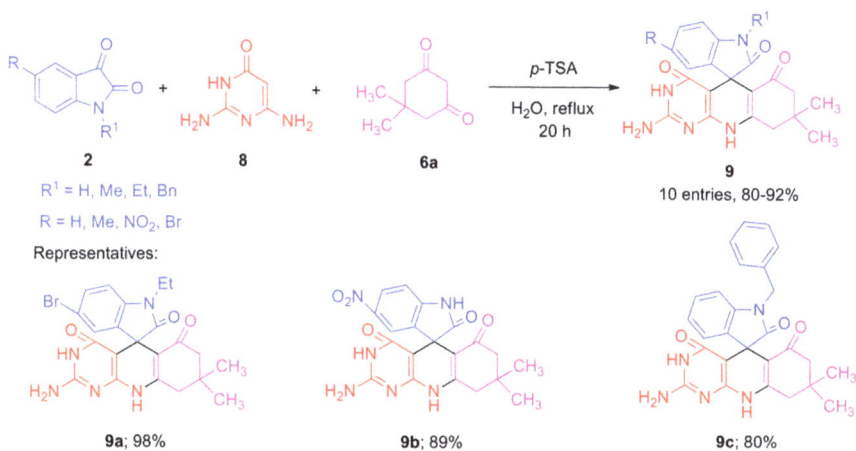

Scheme 4 *p-TSA catalyzed synthesis of spiro[indoline-3,5'-pyrimido[4,5-b]quinoline] derivatives in water*

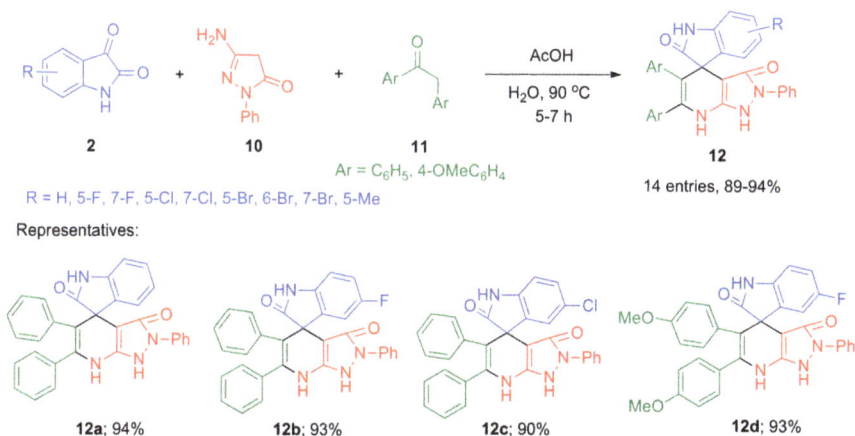

Scheme 5 *Acetic acid-catalyzed synthesis of spiro[indoline-3,4'-pyrazolo[3,4-b]pyridine]-2,3' (7'H)-dione in aqueous medium.*

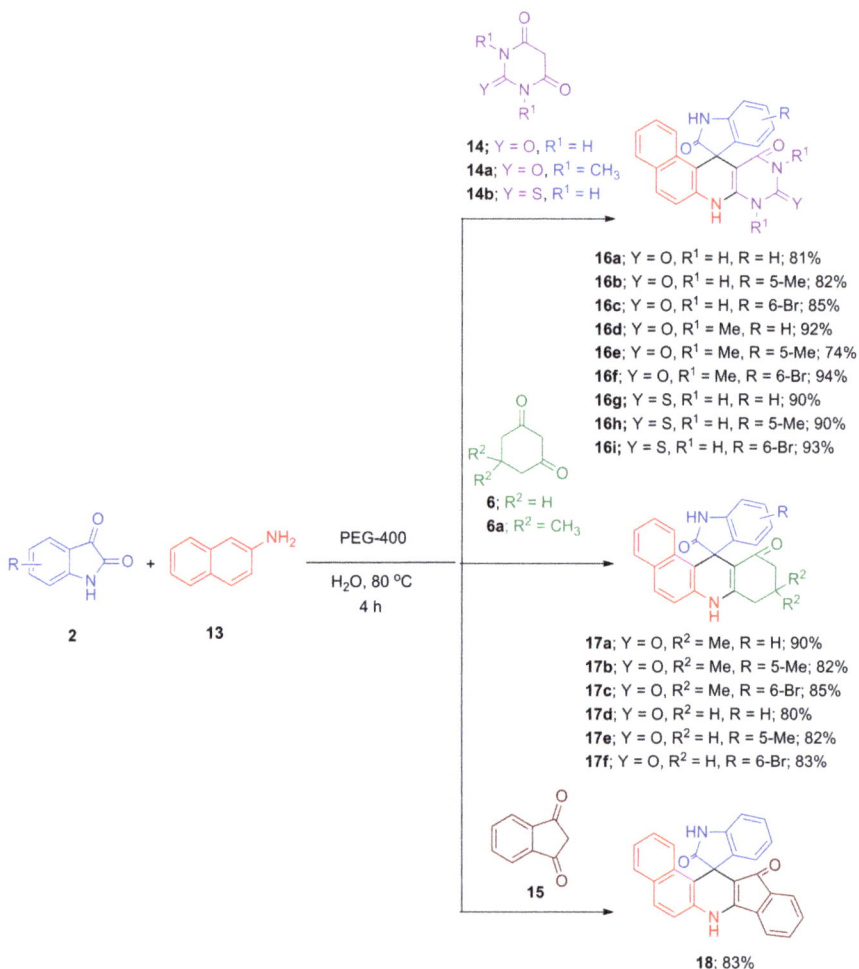

14; Y = O, R^1 = H
14a; Y = O, R^1 = CH$_3$
14b; Y = S, R^1 = H

16a; Y = O, R^1 = H, R = H; 81%
16b; Y = O, R^1 = H, R = 5-Me; 82%
16c; Y = O, R^1 = H, R = 6-Br; 85%
16d; Y = O, R^1 = Me, R = H; 92%
16e; Y = O, R^1 = Me, R = 5-Me; 74%
16f; Y = O, R^1 = Me, R = 6-Br; 94%
16g; Y = S, R^1 = H, R = H; 90%
16h; Y = S, R^1 = H, R = 5-Me; 90%
16i; Y = S, R^1 = H, R = 6-Br; 93%

6; R^2 = H
6a; R^2 = CH$_3$

2 + **13** → PEG-400, H$_2$O, 80 °C, 4 h

17a; Y = O, R^2 = Me, R = H; 90%
17b; Y = O, R^2 = Me, R = 5-Me; 82%
17c; Y = O, R^2 = Me, R = 6-Br; 85%
17d; Y = O, R^2 = H, R = H; 80%
17e; Y = O, R^2 = H, R = 5-Me; 82%
17f; Y = O, R^2 = H, R = 6-Br; 83%

15

18; 83%

Scheme 6: *PEG-400 mediated synthesis of spiro[dihydropyridine-oxindole] derivatives*

2.5   Synthesis of spiro[dihydropyridine-oxindole] derivatives

A large number of biologically promising structurally diverse spiro[dihydropyridine-oxindole] derivatives (**16a-16i,17a-17f,18**) were synthesized by Lu et al. [45] *via* one-pot three-component reactions of isatins (**2**), β-naphthylamine (**13**) and various 1,3-dicarbonyl compounds such as barbituric acids (**14,14a**) or thiobarbituric acid (**14b**) or 1,3-cyclohexadiones (**6,6a**) or indane-1,3-dione (**15**) using polyethylene glycol (PEG-400) in water as solvent at 80 °C (Scheme 6).

2.6   Synthesis of spirooxindolyl-dihydroquinazolinone derivatives

A catalytic amount of ethylenediamine diacetate (EDDA) was employed as an efficient organocatalyst for the synthesis of biologically interesting spirooxindolyl-dihydroquinazolinone derivatives (**21**) starting from isatins (**2**), isatoic anhydride (**19**) and various primary amines (**20**) in an aqueous medium under reflux conditions (Scheme 7) [46]. The proposed mechanism of this conversion is described in Scheme 8.

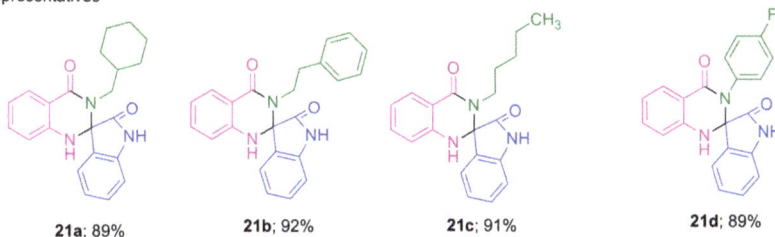

Scheme 7 EDDA*-catalyzed synthesis of spirooxindolyl-dihydroquinazolinone derivatives in water*

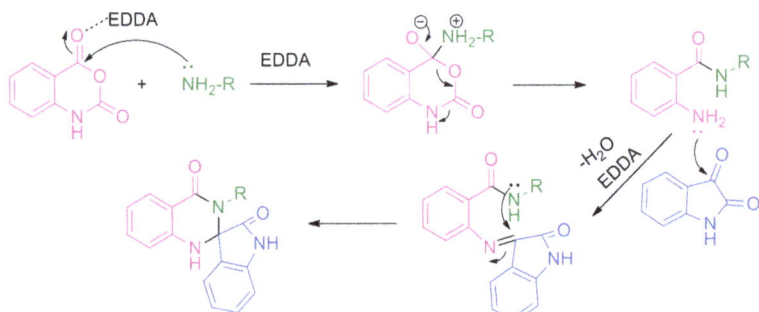

Scheme 8 *Plausible mechanism for the synthesis of spirooxindolyl-dihydroquinazolinone derivatives in water*

2.7 Synthesis of spiro[acridine-9,3′-indole]-2′,4,4′(1′*H*,5′*H*,10*H*)-trione derivatives

Chate et al. [47] reported a simple, efficient and environmentally benign protocol for the synthesis of a series of bioactive spiro[acridine-9,3′-indole]-2′,4,4′(1′*H*,5′*H*,10*H*)-trione derivatives (**22**) *via* one-pot four component condensation reactions of two equivalents of dimedone (**6a**), one equivalent of substituted anilines (**20**) and one equivalent of isatin (**2**) using a catalytic amount of β-cyclodextrin (β-CD) in aqueous media at 80 °C (Scheme 9). The catalyst was recovered quantitatively and reused three times without any significant loss in its catalytic activities. Among the all twenty synthesized compounds, **22c** showed the highest antibacterial activity against the tested bacterial strain.

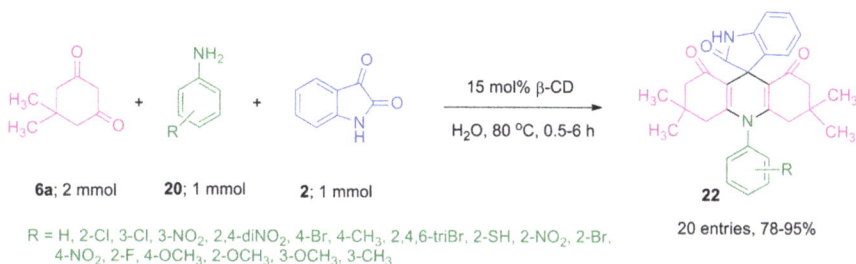

6a; 2 mmol   **20**; 1 mmol   **2**; 1 mmol   **22**

20 entries, 78-95%

R = H, 2-Cl, 3-Cl, 3-NO₂, 2,4-diNO₂, 4-Br, 4-CH₃, 2,4,6-triBr, 2-SH, 2-NO₂, 2-Br, 4-NO₂, 2-F, 4-OCH₃, 2-OCH₃, 3-OCH₃, 3-CH₃

Representatives:

22a; 91%   **22b**; 88%   **22c**; 81%

Scheme 9 *β-CD-catalyzed synthesis of spiro[acridine-9,3'-indole]-2',4,4'(1'H,5'H,1'H)-trione derivatives in aqueous medium*

2.8   Synthesis of 6-spiro-substituted pyrido[2,3-*d*]pyrimidines

Catalyst-free synthesis of a series of stereoselective novel 6-spirosubstituted pyrido[2,3-*d*]-pyrimidines (**24**) was achieved by the reactions of two equivalents of various aldehydes (**23**), one equivalent of 1,3-dimethyl barbituric acid (**14a**) and one equivalent of 2,6-diaminopyrimidine-4-one (**8**) in water under microwave-irradiated conditions at 100 °C (Scheme 10) [48]. High stereoselectivity (up to 99%), catalyst-free conditions, use of a green solvent, very short reaction times, operational simplicity and excellent yields were some of the benefits of this reported method.

23; 2 mmol **14a**; 1 mmol **8**; 1 mmol **24**

Ar = C₆H₅, 4-FC₆H₄, 4-ClC₆H₄, 4-BrC₆H₄, 4-MeC₆H₄, 4-OMeC₆H₄, 3,4,5-(OMe)₃C₆H₂ 7 entries, 80-89%

Representatives

24a, 85% **24b**, 89% **24c**, 87%

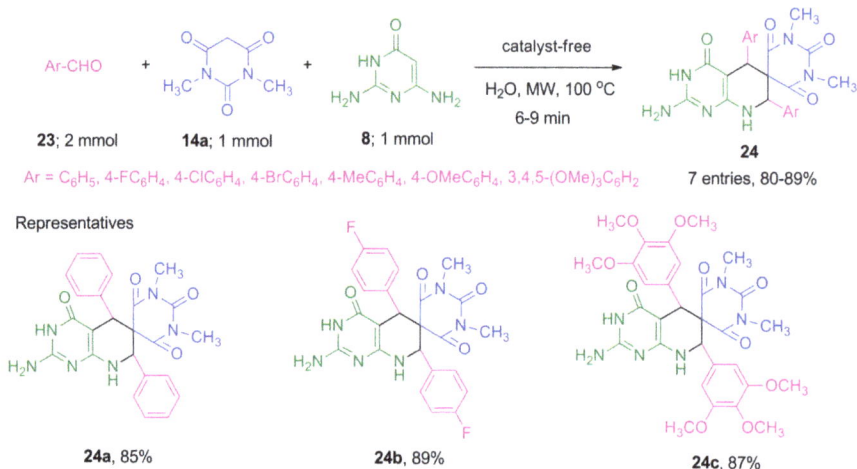

Scheme 10 *Microwave-assisted synthesis of 6-spiro-substituted pyrido[2,3-d]pyrimidines in water under catalyst-free conditions*

2.9 Synthesis of pyrimidine fused spiro-benzoquinolines

A simple, facile and environment friendly protocol was designed by Wang et al. [49] for the efficient synthesis of a series of pyrimidine fused spiro-benzoquinolines (**26**) starting from various aromatic aldehydes (**23**), 1,3-dimethyl barbituric acid (**14a**) and *N*-(arylidene)naphthalen-2-amine (**25**) using triethylbenzylammonium chloride (TEBAC) as catalyst in water at 100 °C (Scheme 11). Aromatic aldehydes with both electron-donating as well as electron-withdrawing substituents produced excellent yields of the desired products. It was proposed that the desired products were formed by following the hetero Diels-Alder reaction pathway (Scheme 12).

Industrial Applications of Green Solvents I   Materials Research Forum LLC
Materials Research Foundations **50** (2019) 269-319   doi: https://doi.org/10.21741/9781644900239-9

Ar = C_6H_5, 4-FC_6H_4, 4-ClC_6H_4, 3-ClC_6H_4, 4-BrC_6H_4, 4-$OMeC_6H_4$,
2-$NO_2C_6H_4$, 3-$NO_2C_6H_4$, 4-$NO_2C_6H_4$, 2,4-$diClC_6H_3$, 3,4-$diClC_6H_3$

11 entries, 72-84%

Representatives:

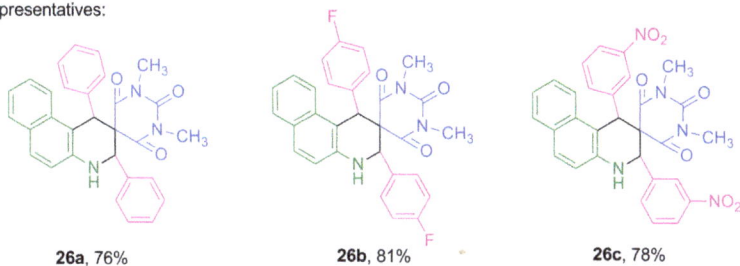

26a, 76%   **26b**, 81%   **26c**, 78%

Scheme 11 *TEBAC-catalyzed synthesis of pyrimidine fused spiro-benzoquinolines in water*

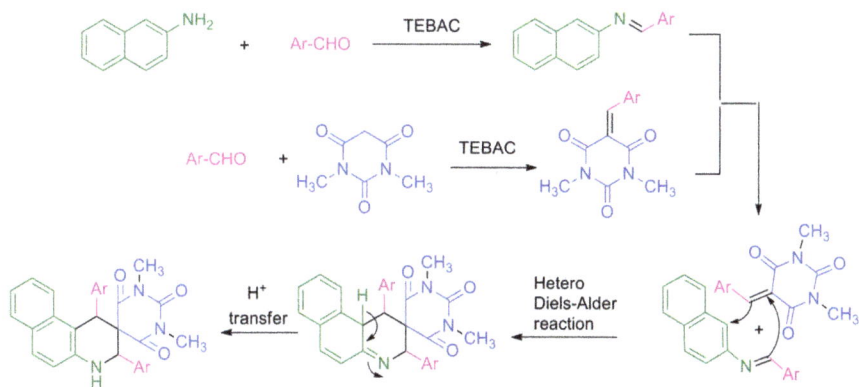

Scheme 12 *Plausible mechanism for the synthesis of pyrimidine fused spiro-benzoquinolines in water*

Industrial Applications of Green Solvents I   Materials Research Forum LLC
Materials Research Foundations **50** (2019) 269-319   doi: https://doi.org/10.21741/9781644900239-9

2.10 Synthesis of spirooxindolyl fused pyrazolopyridine derivatives

A variety of structurally diverse spiroheterocycles were synthesized involving 5-aminopyrazoles (**27,27a**). Kalita et al. [50] reported a novel and facile approach for the efficient synthesis of a series of spiro(indoline-3,4'-pyrazolo[4',3':5,6]pyrido[2,3-d]pyrimidine) derivatives (**28**) *via* one-pot three-component condensation of isatins (**2**), 5-aminopyrazoles (**27**) and 6-aminouracils (**1,1a,1b**) using 20 mol% of *p*-toluene sulfonic acid (*p*-TSA) as catalyst in aqueous medium under reflux conditions (Scheme 13). Schematic representation of the plausible mechanism of this conversion is shown in Scheme 14.

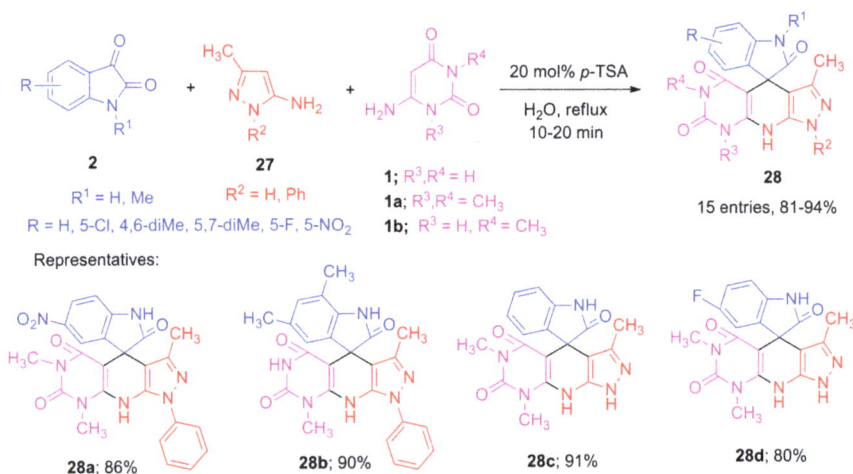

Scheme 13 *p-TSA catalyzed synthesis of spiro(indoline-3,4'-pyrazolo[4',3':5,6]pyrido[2,3-d]pyrimidine) derivatives in an aqueous medium*

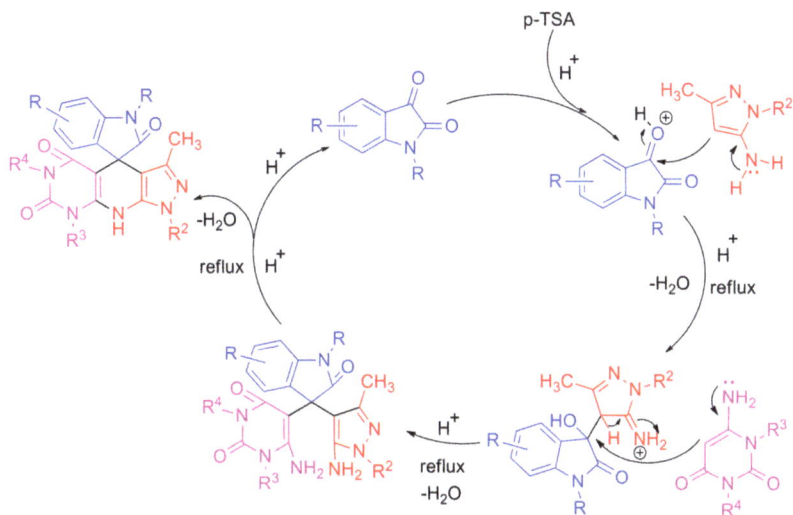

Scheme 14 *Plausible mechanism for the synthesis of spiro(indoline-3,4'-pyrazolo[4',3':5,6]pyrido[2,3-d]pyrimidine) derivatives in an aqueous medium*

Synthesis of spiro[indoline-pyrazolo[4',3':5,6]pyrido[2,3-*d*]pyrimidine]trione derivatives (**29**) was achieved by Ghahremanzadeh et al. [51] by employing barbituric acids (**14,14a**) or thiobarbituric acid derivatives (**14b,14c**) instead of 6-aminouracils under the above mentioned conditions (Scheme 15). Later on, in 2014, the same group also synthesized a number of spiro[indoline-pyrazolo[4',3':5,6]pyrido[2,3-*d*]pyrimidine]trione derivatives (**29**) using the inorganic-organic hybrid silica-based tin complex as a catalyst in water under reflux conditions [52]. Balamurugan et al. [53] achieved the synthesis of the same scaffolds (**29**) by the reactions of isatins (**2**), phenylhydrazine (**30**), 3-aminocrotononitrile (**31**) and barbituric acid (**14**)/thiobarbituric acid (**14b**) using camphor-10-sulfonic acid (CSA) as catalyst in water at 100 °C (Scheme 16). This domino reaction proceeded *via* the generation of 3-methyl-1-phenyl-1*H*-pyrazol-5-amine (**27**) (Scheme 17).

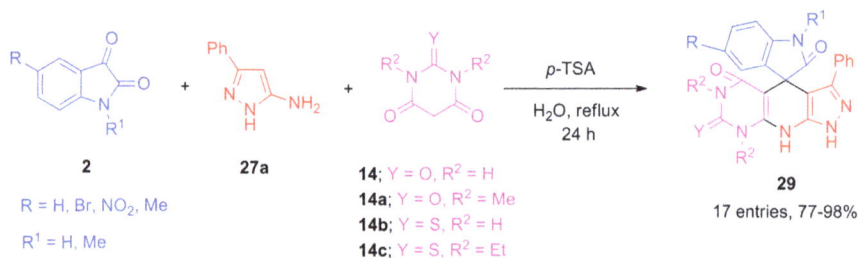

2

R = H, Br, NO$_2$, Me

R^1 = H, Me

27a

14; Y = O, R^2 = H
14a; Y = O, R^2 = Me
14b; Y = S, R^2 = H
14c; Y = S, R^2 = Et

p-TSA

H$_2$O, reflux
24 h

29

17 entries, 77-98%

Representatives:

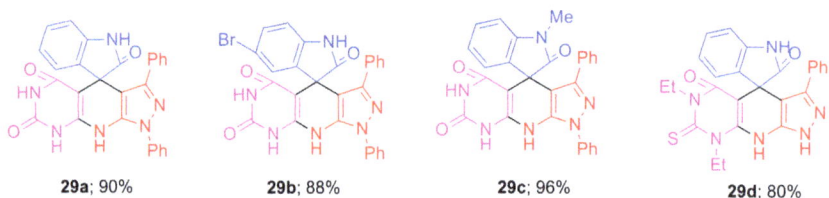

29a; 90%   **29b**; 88%   **29c**; 96%   **29d**; 80%

Scheme 15 *p-TSA catalyzed synthesis of novel spiro[indoline-pyrazolo[4',3':5,6]pyrido[2,3-d]pyrimidine]trione derivatives in an aqueous medium*

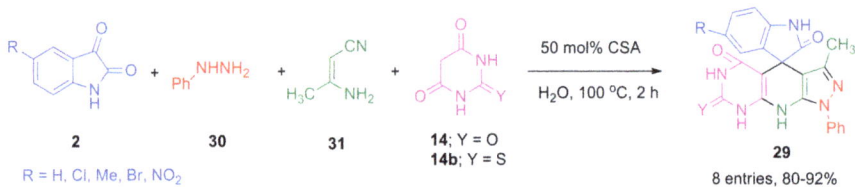

2   **30**   **31**

R = H, Cl, Me, Br, NO$_2$

14; Y = O
14b; Y = S

50 mol% CSA

H$_2$O, 100 °C, 2 h

29

8 entries, 80-92%

Representatives:

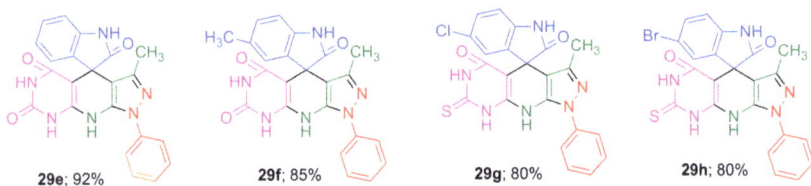

29e; 92%   **29f**; 85%   **29g**; 80%   **29h**; 80%

Scheme 16 *CSA catalyzed water-mediated synthesis of spiro[indoline/acenaphthylene-3,4'-pyrazolo[3,4-b]pyridine derivatives*

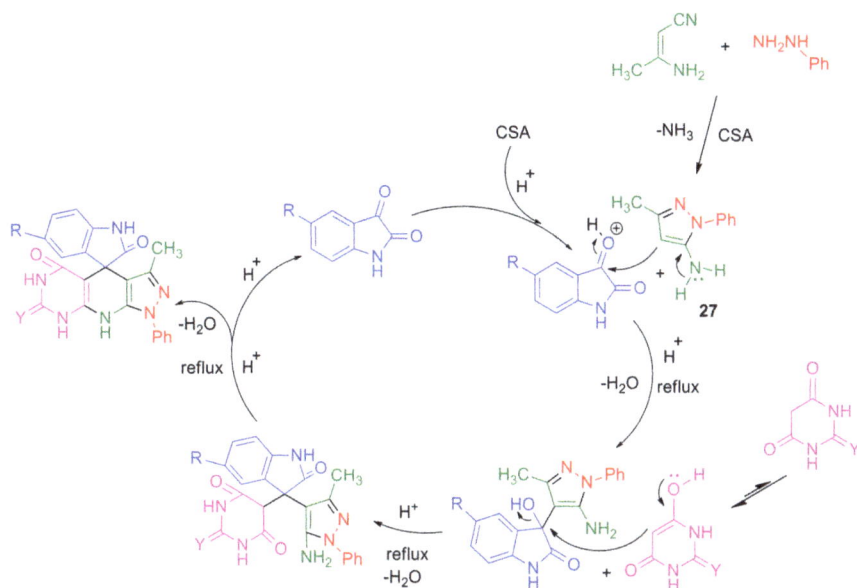

Scheme 17 *Plausible mechanism for the synthesis of spiro[indoline/acenaphthylene-3,4'-pyrazolo[3,4-b]pyridine derivatives in aqueous medium*

A catalytic amount of *p*-toluene sulfuonic acid (*p*-TSA) was also efficiently catalyzed the reactions of isatin (**2**), 3-methyl-1*H*-pyrazol-5-amine (**27**) and 2-hydroxy-1,4-naphthaquinone (**32**) which afforded the corresponding 3-methylspiro[benzo[g]pyrazolo[3,4-*b*]quinoline-4,3'-indoline]-2',5,10(1*H*,11*H*)-trione (**33**) in aqueous medium at 80 °C (Scheme 18) [54].

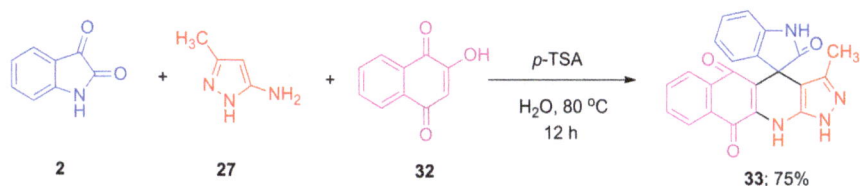

Scheme 18 *p-TSA catalyzed synthesis of 3-methylspiro[benzo[g]pyrazolo[3,4-b]quinoline-4,3'-indoline]-2',5,10(1H,11H)-trione in aqueous medium*

Industrial Applications of Green Solvents I   Materials Research Forum LLC
Materials Research Foundations **50** (2019) 269-319   doi: https://doi.org/10.21741/9781644900239-9

Kamal et al. [55] demonstrated an efficient, facile and environment friendly protocol for the one-pot three-component synthesis of another series of spirooxindole fused pyrazolopyridine derivatives (**35**) *via* the reactions among isatins (**2**), tetronic acid (**34**) and 3-phenyl-1*H*-pyrazol-5-amine (**27a**) involving sulfamic acid as an efficient organocatalyst in aqueous medium at 100 °C (Scheme 19). The catalyst was recovered quantitatively and recycled up to four successive runs without any significant loss in its activities. The mechanism involved in the transformation is described in Scheme 20. Cytotoxic activity of the synthesized compounds was screened against three human cancer cell lines. Compounds **35c** and **35d** showed remarkable cytotoxic activities. Dabiri et al. [56] also synthesized the same series of compounds employing *L*-proline as a catalyst in water under reflux conditions.

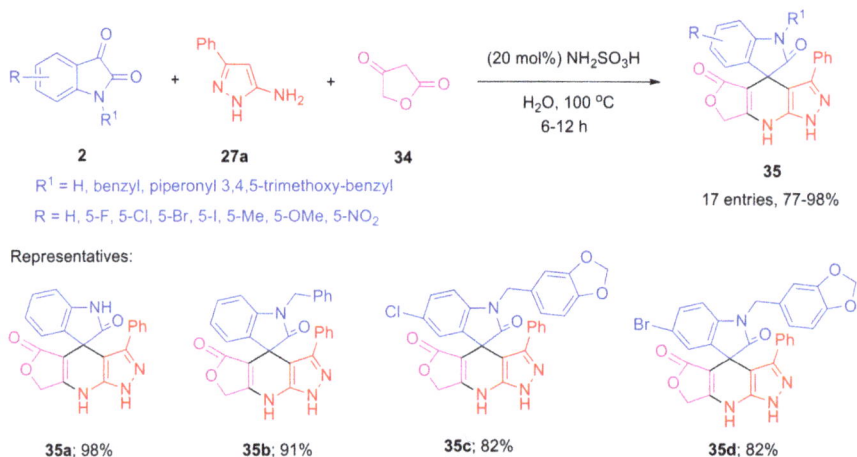

Scheme 19 *NH$_2$SO$_3$H catalyzed synthesis of spiro[furo[3,4-b]pyrazolo[4,3-e]pyridine-4,3'-indoline]-2',5(1H)-dione derivatives in water*

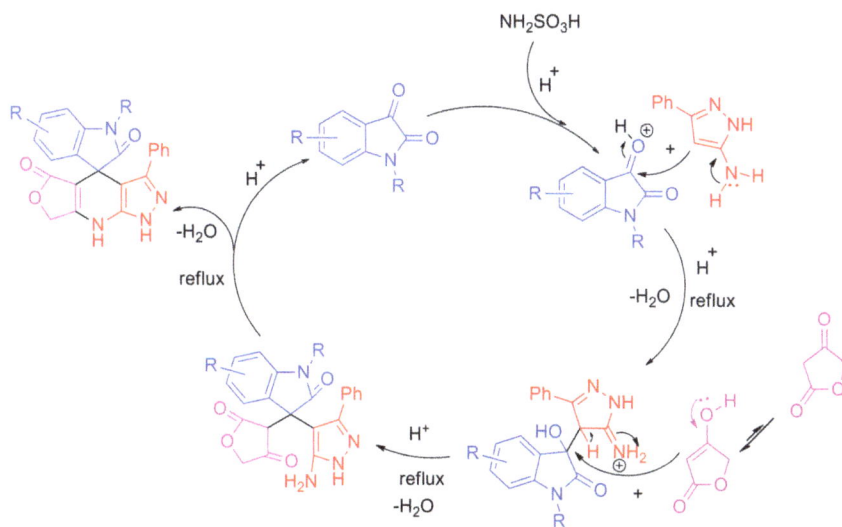

Scheme 20 *Plausible mechanism for the synthesis of spiro[furo[3,4-b]pyrazolo[4,3-e]pyridine-4,3'-indoline]-2',5(1H)-dione derivatives in aqueous medium*

Bazgir and his research group [57] developed another aqueous mediated *p*-toluene sulfonic acid (*p*-TSA) catalyzed protocol for the synthesis of spiro[chromenopyrazolo-pyridine-indoline]-diones (**37**) starting from isatins (**2**), 3-phenyl-1*H*-pyrazol-5-amine (**27a**) and 4-hydroxy-1-methylquinolin-2(1*H*)-one (**36**) (Scheme 21). Employing *p*-toluene sulfonic acid as catalyst, the same research group also carried out the reactions of isatins (**2**), 3-phenyl-1*H*-pyrazol-5-amine (**27a**) and 4-hydroxycoumarin (**38**) in aqueous media under ultrasonic irradiation at 60 °C. Under these reaction conditions, they obtained a series of unexpected spiro[indoline-3,4'-pyrazolo[3,4-*b*]pyridine]-2,6'(1'*H*)-dione derivatives (**39**) (Scheme 22) [58].

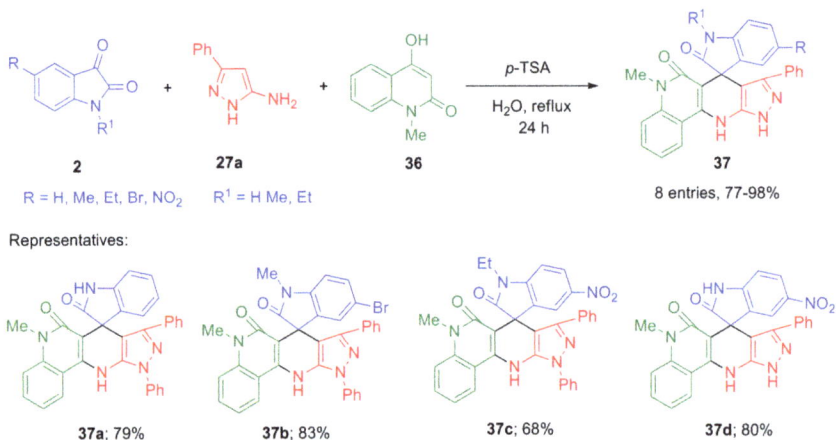

Scheme 21 *p-TSA catalyzed synthesis of spiro[chromenopyrazolo-pyridine-indoline]-diones in an aqueous medium*

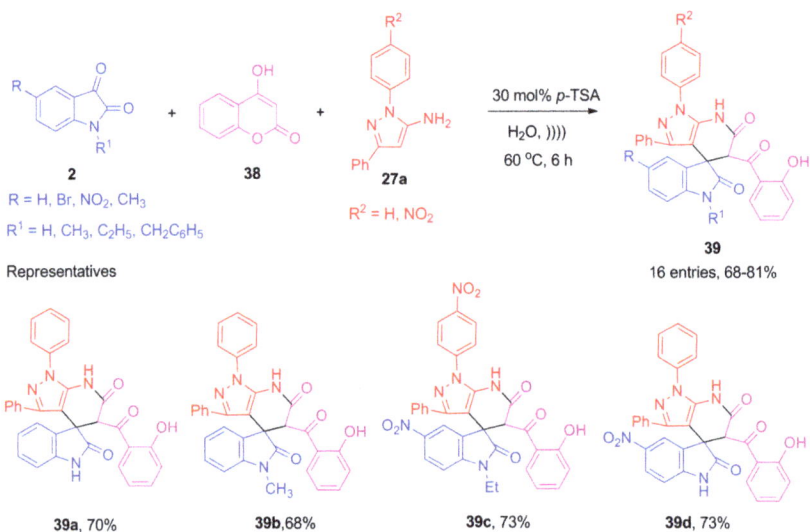

Scheme 22 *Ultrasound promoted aqueous mediated synthesis of spiro[indoline-3,4'-pyrazolo[3,4-b]pyridine]-2,6'(1'H)-dione derivatives*

Industrial Applications of Green Solvents I   Materials Research Forum LLC
Materials Research Foundations **50** (2019) 269-319   doi: https://doi.org/10.21741/9781644900239-9

2.11   Synthesis of pyrazolopyridinyl spirooxindoles

Dandia et al. [59] demonstrated a simple and efficient approach for the synthesis of a series of biologically interesting spirooxindole fused pyrazolopyridine derivatives (**41**) by the reactions of isatins (**2**), ethyl 2-cyanoacetate (**40**) and 5-amino-3-methylpyrazole (**27**) using a catalytic amount of sodium chloride as catalyst in water under reflux conditions (Scheme 23).

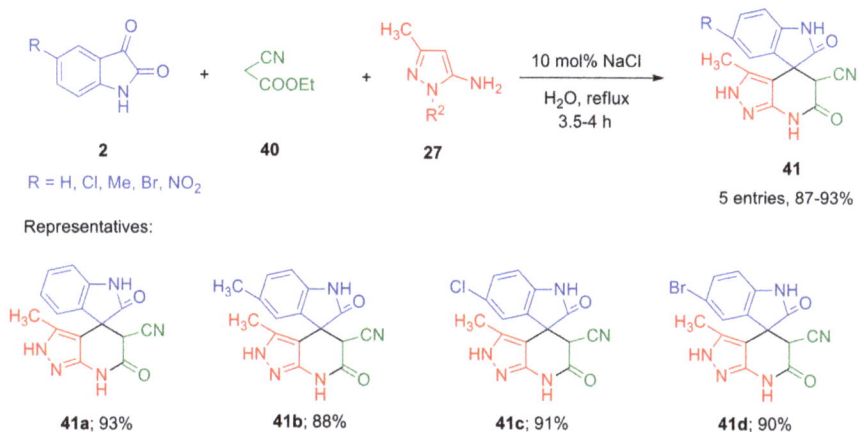

Scheme 23 NaCl-catalyzed aqueous mediated synthesis of pyrazolopyridinyl spirooxindoles

3.   Synthesis of *O*-containing spiroheterocycles

3.1   Synthesis of spirochromenes

A variety of spiroheterocyclic compounds such as spirochromenes (**44**) and spiroacenaphthylene (**45**) were synthesized by the reactions of 4-hydroxycoumarin (**38**), malononitrile (**40a**) and ninhydrin (**42**) acenaphthylene-1,2-dione (**43**) respectively using Amberlite IRA-400 Cl resin as an efficient catalyst in water under reflux conditions (Scheme 24) [60]. The synthesized compounds were found to possess antimicrobial activities.

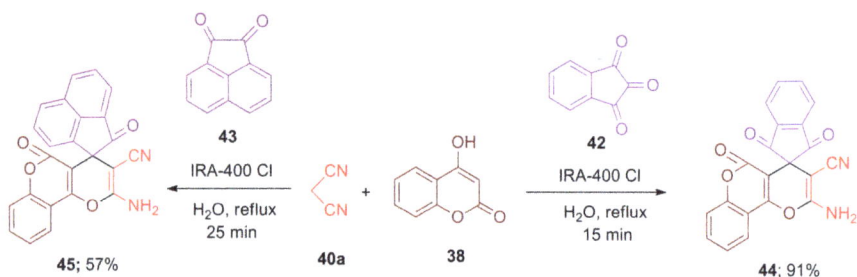

Scheme 24 *Amberlite IRA-400 Cl resin catalyzed synthesis of spirochromenes in an aqueous medium*

4. Synthesis of *N*, *O*-containing spiroheterocycles

4.1 Synthesis of spironaphthopyrano[2,3-*d*]pyrimidine derivatives

Bazgir and his research group [61] developed another *p*-toluene sulfonic acid (*p*-TSA) catalyzed efficient protocol for the synthesis of novel spironaphthopyrano[2,3-*d*]pyrimidine-5,3'-indolines (**47**) *via* one-pot three-component cyclocondensation reaction of isatins (**2**), 2-naphthol (**46**) and barbituric acids (**14,14a,14b**) in aqueous media under reflux conditions (Scheme 25). Scheme 26 shows how two new carbon-carbon and one carbon-oxygen single bonds were formed during this synthesis. When the same reaction was further carried out by Kong et al. [62] employing a catalytic amount of sodium dodecyl sulfate (SDS), a series of spiro[dihydroquinoline-naphthofuranone] derivatives (**48**) was formed instead of spironaphthopyrano[2,3-*d*]pyrimidine-5,3'-indolines (**47**) in water at 80 °C (Scheme 27). After completion of the reaction, the reaction medium containing surfactant was recovered successfully and recycled up to fifth runs. It was proposed that the reaction underwent through the intramolecular ring-opening annulations protocol (Scheme 28)

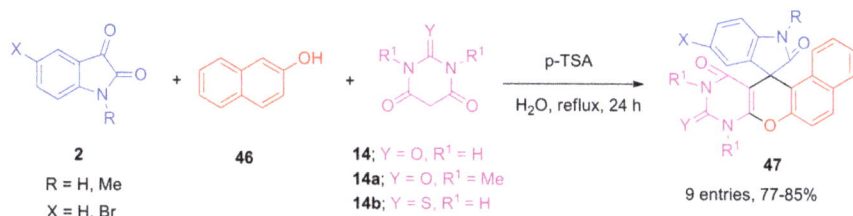

Scheme 25 *p-TSA catalyzed synthesis of spironaphthopyrano[2,3-d]pyrimidine*

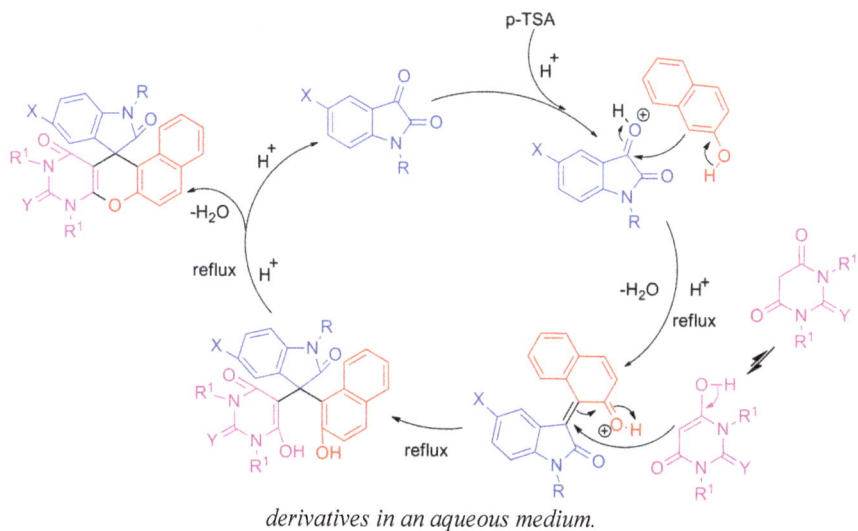

derivatives in an aqueous medium.

Scheme 26 *Plausible mechanism for the synthesis of spironaphthopyrano[2,3-d]pyrimidine-5,3'-indoline derivatives in water.*

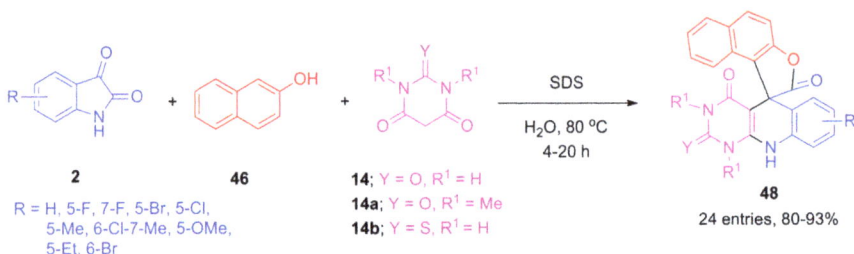

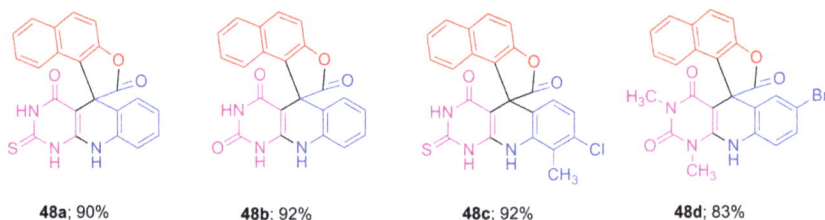

Scheme 27 *SDS catalyzed synthesis of spiro[dihydroquinoline-naphthofuranone] derivatives in an aqueous medium*

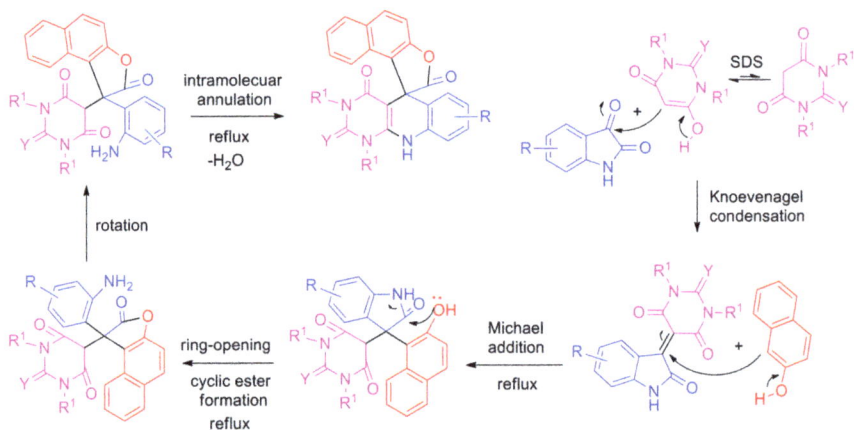

Scheme 28 *Plausible mechanism for the synthesis of spiro[dihydroquinoline-naphthofuranone] derivatives in water.*

4.2 Synthesis of 2-amino-3-cyano-4-indolinon-spiro[pyran or pyran-annulated] heterocycles

In the literature, a number of methods are available for the efficient synthesis of 2-amino-3-cyano-4-indolinon-spiropyran or various pyran-annulated heterocycles *via* one-pot multi-component reactions of isatins, malanonitrile and various C-H activated acid derivatives in aqueous media. Zhao et al. [63] developed a simple, efficient, mild and environment friendly protocol for the three-component synthesis of 2'-amino-3'-cyano-2-oxospiro[indoline-3,4'-pyran] derivatives (**50a,50b**) by the reactions of isatins (**2**), malononitrile (**40a**) and ethyl acetoacetate (**49**) in water under catalyst-free conditions at 60 °C (Scheme 29). Zhu et al. [64] prepared another series of biologically interesting 2-amino-3-cyano-spiro[chromene-4,3'-indoline] derivatives (**51**) starting from isatins (**2**), malononitrile (**40a**) and dimedone (**6a**) in the presence of a catalytic amount of triethylbenzylammonium chloride (TEBA) in aqueous medium under at 60 °C (Scheme 30).

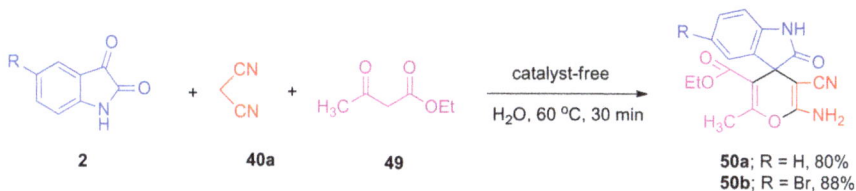

Scheme 29 *Synthesis of 2'-amino-3'-cyano-2-oxospiro[indoline-3,4'-pyran] derivatives in water*

Scheme 30 *TEBA-catalyzed water-mediated synthesis of 2-amino-3-cyano-spiro[chromene-4,3'-indoline] derivatives*

In 2011, Mobinikhaledi et al. [65] reported a simple and efficient water-mediated protocol for the synthesis of 7'-amino-spiro[indoline-3,5'-pyrano[2,3-*d*]pyrimidine]-6'-carbonitrile derivatives (**52**) *via* the condensation reaction of isatins (**2**), malononitrile (**40a**) and barbituric acid (**14**) thiobarbituric acid (**14b**) using tetrabutylammonium bromide (TBAB) as catalyst under reflux conditions (Scheme 31). Later on, in 2016, Singh et al. [66] carried out the same reaction by using thiamine hydrochloride encapsulated silica supported magnetic nanoparticles (Fe$_2$O$_3$@SiO$_2$@vitB1 NPs) in water under ultrasonic-irradiated conditions (Scheme 31). Li et al. [67] demonstrated *L*-proline catalyzed synthesis of 2-amino-3-cyano-spiro[furo[3,4-*b*]pyran-2'-oxo-4,3'-indoline] derivatives (**53**) starting from substituted isatins (**2**), malononitrile (**40a**) and tetronic acid (**34**) in aqueous media at 80 °C (Scheme 32).

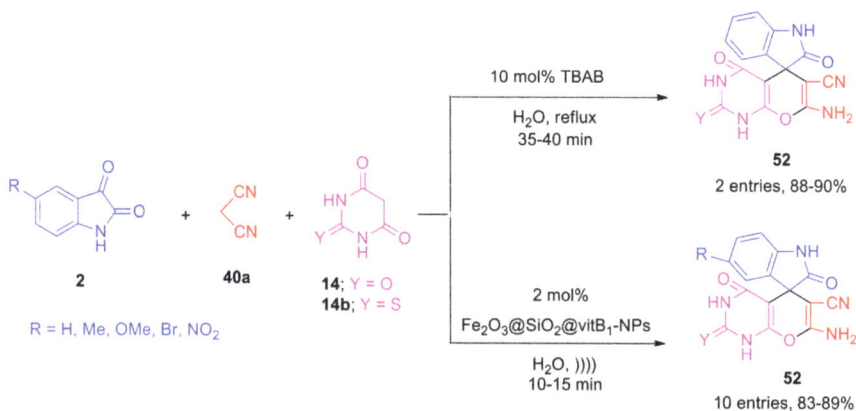

Scheme 31 *Synthesis of 7'-amino-spiro[indoline-3,5'-pyrano[2,3-d]pyrimidine]-6'-carbonitrile derivatives in water*

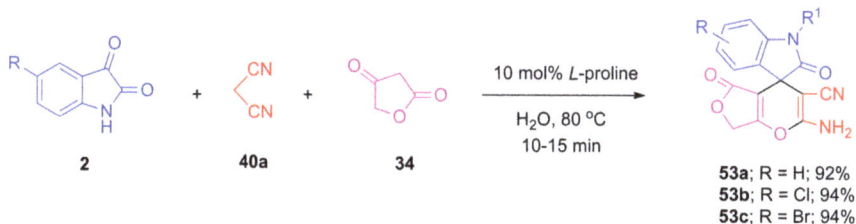

Scheme 32 *L-proline-catalyzed synthesis of 2-amino-3-cyano-spiro[furo[3,4-b]pyran-2'-oxo-4,3'-indoline] derivatives in aqueous medium*

Synthesis of a series of biologically interesting 10'-amino-spiro[indole-3,8'-phenaleno[1,2-*b*]pyran]-9'-carbonitriles (**55**) was achieved by Hari et al. [68] *via* the reactions of isatins (**2**), malononitrile (**40a**) and 3-hydroxy-1*H*-phenalen-1-one (**54**) in the presence of a catalytic amount of ethylenediamine diacetate (EDDA) as catalyst in water at 60 °C (Scheme 33). Ghahremanzadeh et al. [69] synthesized the same series of compounds using *p*-toluene sulfonic acid (*p*-TSA) as a catalyst in aqueous media

(Scheme 33). Karimi et al. [70] have reported the synthesis of 2'-amino-3'-cyano-spiro[indoline-3,4'-pyrano[3,2-*c*]chromene] derivatives (**56**) *via* one-pot three-component reactions of isatins (**2**), malononitrile (**40a**) and 4-hydroxycoumarin (**38**) using alum as a catalyst in water under heating conditions (Scheme 34).

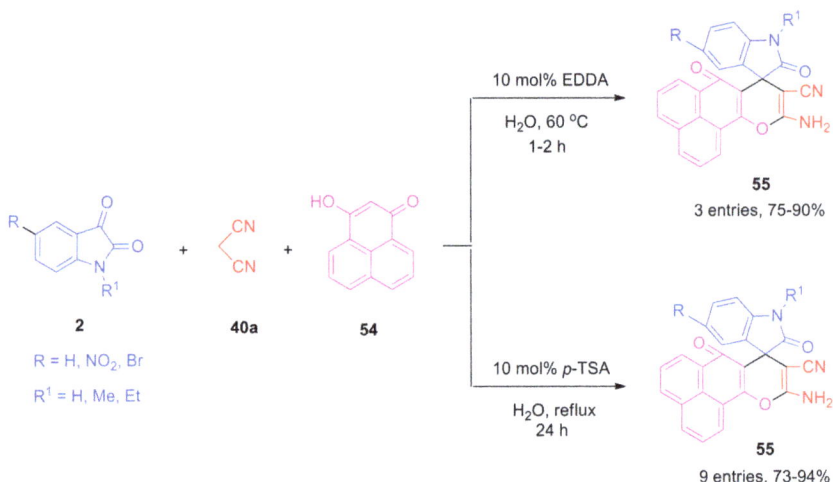

Scheme 33 *Synthesis of 10'-amino-spiro[indole-3,8'-phenaleno[1,2-b]pyran]-9'-carbonitriles in water*

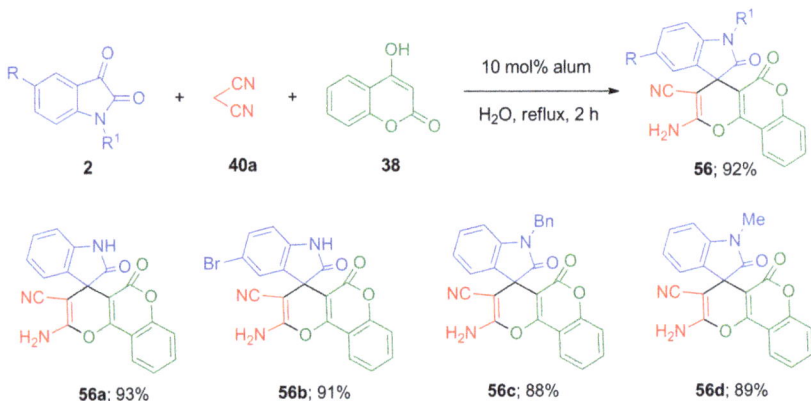

Scheme 34 *Alum-catalyzed synthesis of 2'-amino-3'-cyano-spiro[indoline-3,4'-pyrano[3,2-c]chromenes] in water*

Industrial Applications of Green Solvents I   Materials Research Forum LLC
Materials Research Foundations **50** (2019) 269-319   doi: https://doi.org/10.21741/9781644900239-9

A simple, rapid and ultrasound-assisted protocol was developed for the efficient synthesis of a series of biologically promising spiro[indoline-3,4'-pyrano[3,2-*c*]quinoline derivatives (**57**) starting from isatins (**2**), malononitrile (**40a**) or ethyl cyanoacetate (**40**) and 4-hydroxy-2*H*-quinolin-2-one (**36**) using a 5 mol% piperidine as catalyst in water under ultrasonic irradiation at 50 °C (Scheme 35)[71]. Another ultrasound-assisted, environmentally benign protocol was developed by Dandia et al. [72] for the efficient synthesis of a series of biologically interesting spiro[indoline-3,4'-pyrano[2,3-*c*]pyrazoles (**59**) *via* a one-pot three-component reactions of substituted isatins (**2**), ethyl cyanoacetate (**40**) or malononitrile (**40a**) and 3-methyl-1-phenyl-2-pyrazolin-5-one (**58**) using sodium chloride as catalyst in water at room temperature (Scheme 36). Liu et al. [73] accomplished the same reactions using K_2CO_3 as a catalyst in water at room temperature (Scheme 36).

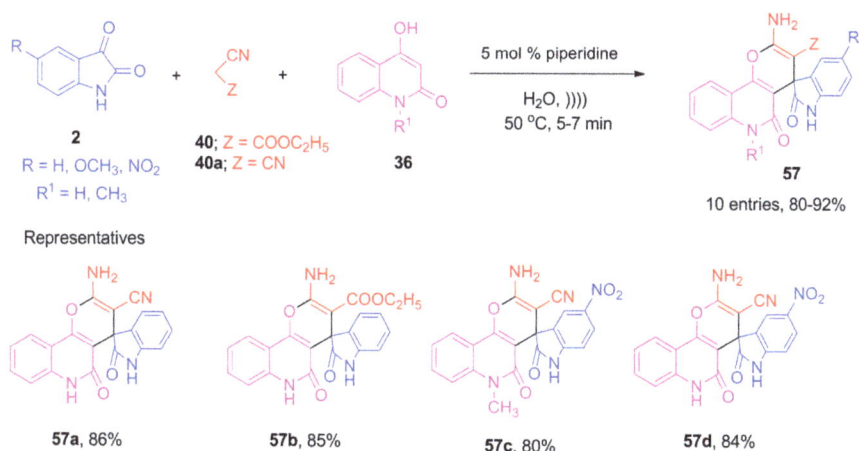

Scheme 35 *Ultrasound-assisted synthesis of spiro[indoline-3,4'-pyrano[3,2-c]quinoline in aqueous media*

Four-component synthesis of spiro[indoline-3,4'-pyrano[2,3-*c*]pyrazoles (**59**) has been achieved by the reactions of β-ketoester (**49**), hydrazine (**30a**), isatins (**2**) and malononitrile (**40a**) using 30 mol% piperidine as catalyst under conventional stirring conditions in water at room temperature (Scheme 37) [74]. Compounds synthesized were found to possess antibacterial activities. Recently, in 2016, catalytic amount of nano-

CeO_2 was employed for the synthesis of another series of spiro[indoline-3,4'-pyrano[2,3-c]pyrazoles (**59**) *via* one-pot four-component reactions of β-ketoester (**49**), aryl hydrazine (**30**), isatins (**2**) and malononitrile (**40a**) in water at 90 °C (Scheme 38) [75]. The compounds synthesized possess antioxidant as well as antibacterial activities.

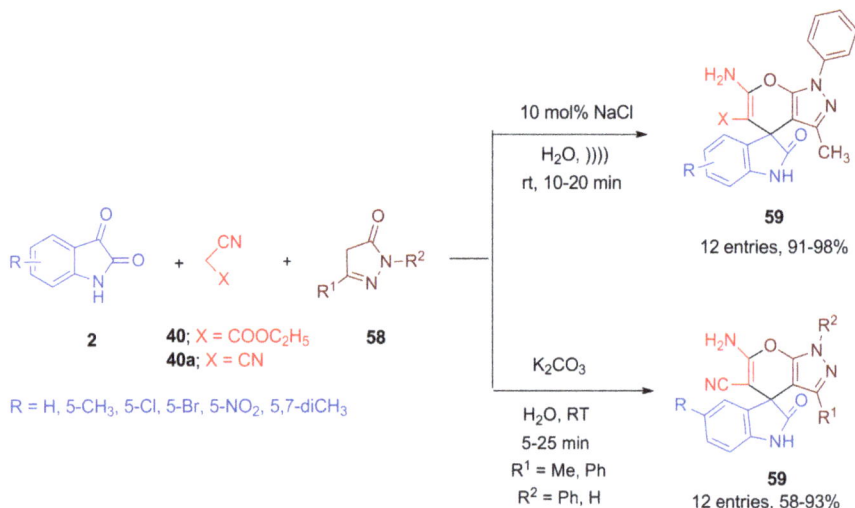

Scheme 36 *Three-component synthesis of spiro[indoline-3,4'-pyrano[2,3-c]pyrazoles in aqueous media*

Scheme 37 *Piperidine-catalyzed four-component synthesis of spiro[indoline-3,4'-pyrano[2,3-c]pyrazoles in aqueous media*

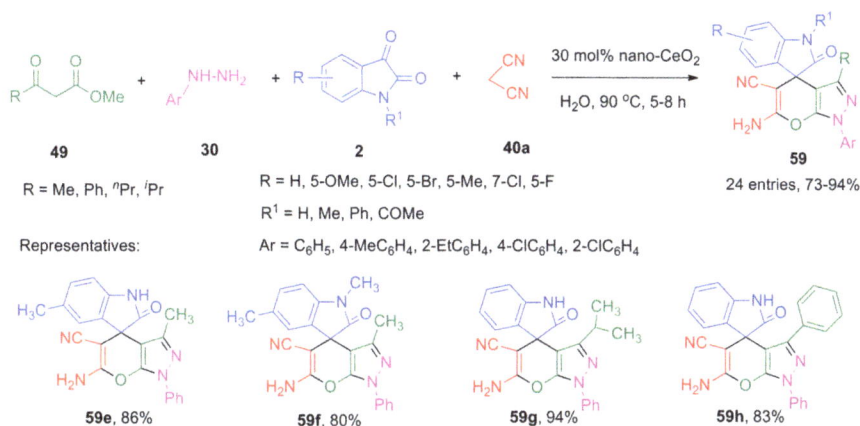

Scheme 38 *Nano-CeO$_2$-catalyzed four-component synthesis of spiro[indoline-3,4-pyrano[2,3-c]pyrazole] derivatives in aqueous media*

4.3 Synthesis of spiro-fused pyrano[2,3-c]pyrazoles

Das and his research group [76] reported the synthesis of highly functionalized diverse tricyclic 4-spiro pyrano[2,3-c]pyrazoles (**62**) through one-pot four-component tandem cyclization reactions among hydrazines (**30**), ethyl acetoacetate (**49**), cyclic ketones (**2,42,60**) and structurally diverse cyclic 1,3-diketones (**6,6a,61**) using catalytic amount of *p*-dodecylbenzenesulphonic acid as an efficient catalyst in water at 90 °C (Scheme 39). *In situ* generated pyrazolone (**58**) was the key intermediate of this reaction. This developed protocol possessed some of the major benefits such as operational simplicity, simple workup procedure, a variety of substrates, excellent yields and use of water as solvent etc.

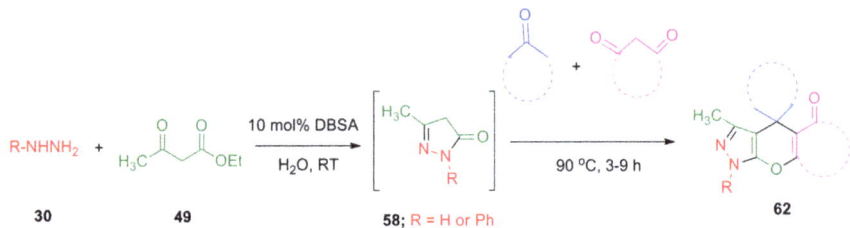

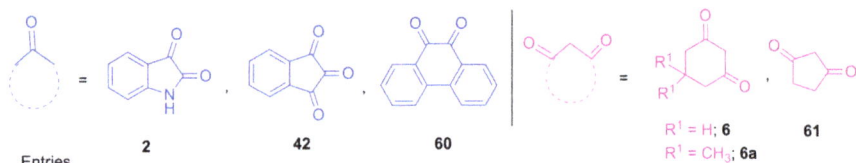

Entries

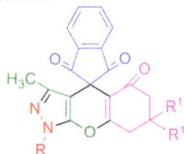

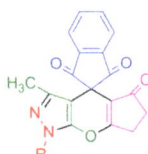

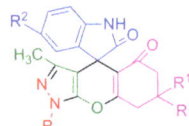

R = H, R¹ = CH₃; **62a**, 96%
R = C₆H₅, R¹ = CH₃; **62b**, 87%
R= 4-NO₂C₆H₅, R¹ = CH₃; **62c**, 82%
R = 4-CNC₆H₄, R¹ = CH₃; **62d**, 86%
R = H, R¹ = H; **62e**, 95%
R = C₆H₅, R¹ = H; **62f**, 88%
R= 4-NO₂C₆H₅, R¹ = H; **62g**, 83%
R = 4-CNC₆H₄, R¹ = H; **62h**, 87%

R = H; **62i**, 94%
R = 4-NO₂C₆H₄; **62j**, 78%

R = H, R¹ = CH₃, R² = Br; **62k**, 93%
R = C₆H₅, R¹ = CH₃, R² = Br; **62l**, 86%
R= 4-NO₂C₆H₅, R¹ = CH₃, R² = Br; **62m**, 80%
R = H, R¹ = H, R² = H; **62n**, 94%
R = C₆H₅, R¹ = H, R² = H; **62o**, 88%
R= 4-NO₂C₆H₅, R¹ = H, R² = H; **62p**, 82%

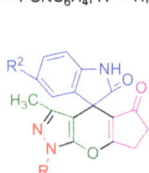

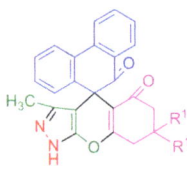

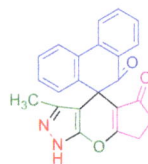

R = H, R² = Br; **62q**, 94%
R = C₆H₅, R² = Br; **62r**, 84%
R = 4-NO₂C₆H₄, R² = Br; **62s**, 78%
R = H, R² = H; **62t**, 92%
R = C₆H₅, R² = H; **62u**, 86%
R = 4-NO₂C₆H₄, R² = H; **62v**, 79%

R¹ = H; **62w**, 78%
R¹ = CH₃; **62x**, 76%

62y, 75%

Scheme 39 *DBSA-catalyzed water mediated synthesis of spiro-fused pyrano[2,3-c]pyrazoles*

Industrial Applications of Green Solvents I   Materials Research Forum LLC
Materials Research Foundations **50** (2019) 269-319   doi: https://doi.org/10.21741/9781644900239-9

4.4   Synthesis of spiro-fused benzo[b]furo[3,4-*e*][1,4]diazepines

Cheng et al. [77] demonstrated rapid, efficient, microwave-assisted domino-cyclization reactions involving one equivalent of ninhydrin (**42**), one equivalent of tetronic acid (**34**) and two equivalents of benzene-1,2-diamines (**63**) to synthesize a series of spiro-substituted benzo[*b*]furo[3,4-*e*][1,4]diazepine derivatives (**64**) in the presence of a catalytic amount of acetic acid in aqueous media at 140 °C (Scheme 40). Scheme 41 shows how five new single bonds were formed in this transformation.

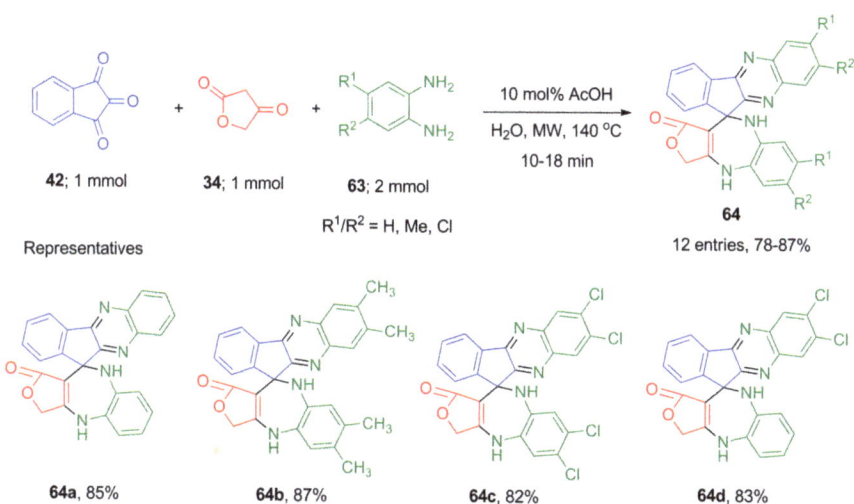

Scheme 40 *Acetic acid-catalyzed synthesis of spiro-fused benzo[b]furo[3,4-e][1,4]diazepines in water.*

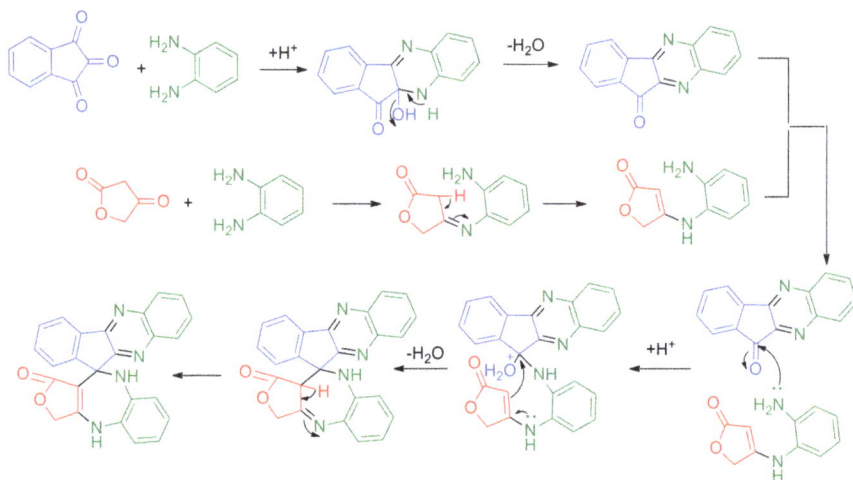

Scheme 41 *Plausible mechanism for the synthesis of spiro-substituted benzo[b]furo[3,4-e][1,4]diazepine derivatives in an aqueous medium*

4.5 Synthesis of spiro{[1,3]dioxanopyridine}-4,6-dione derivatives

A series of biologically interesting spiro[[1,3]dioxane-5,5'-isoxazolo[5,4-*b*]pyridine] derivatives (**67**) was synthesized by Ma et al. [78] *via* the domino-cyclization reactions involving two equivalents of various aldehydes (**23**), one equivalent of Meldrum's acid (**65**) and one equivalent of 3-methylisoxazol-5-amine (**8**) under microwave-assisted catalyst-free conditions at 100 °C (Scheme 42). It was proposed that the transformation followed the hetero Diels-Alder reaction pathway (Scheme 43).

Materials Research Forum LLC
doi: https://doi.org/10.21741/9781644900239-9

Ar = C$_6$H$_5$, 4-FC$_6$H$_4$, 4-ClC$_6$H$_4$, 4-BrC$_6$H$_4$, 4-MeC$_6$H$_4$, 4-NO$_2$C$_6$H$_4$, 2-ClC$_6$H$_4$,
3,4-(OMe)$_2$C$_6$H$_3$, 3,4,5-(OMe)$_3$C$_6$H$_2$, 2-thienyl

10 entries, 77-88%

Representatives

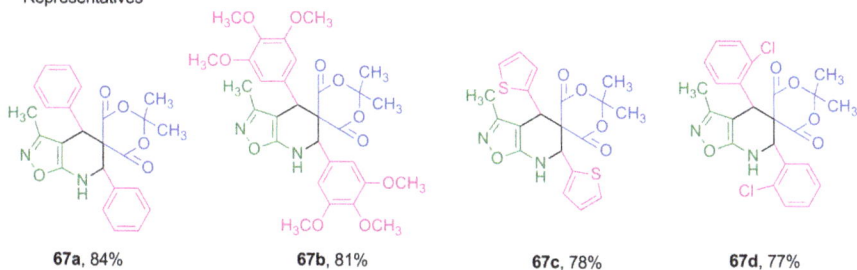

67a, 84%   **67b**, 81%   **67c**, 78%   **67d**, 77%

Scheme 42 *Microwave-assisted synthesis of spiro[[1,3]dioxane-5,5'-isoxazolo[5,4-b]pyridine] derivatives under catalyst-free conditions in water*

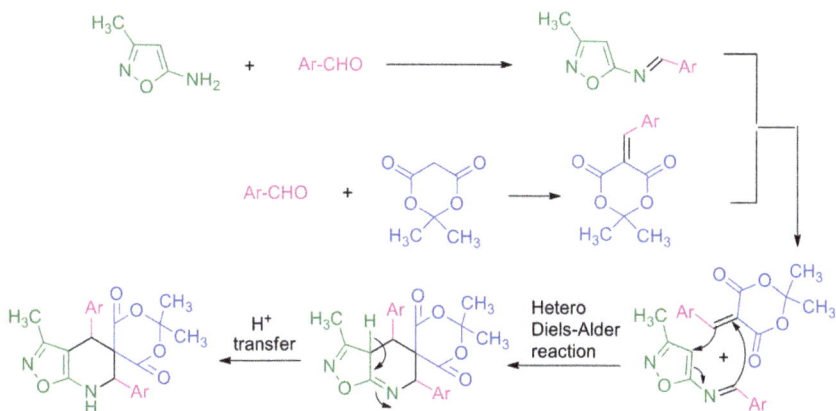

Scheme 43 *Plausible mechanism for the synthesis of spiro[[1,3]dioxane-5,5'-isoxazolo[5,4-b]pyridine] derivatives in an aqueous medium*

5. Synthesis of *N*, *S*-containing spiroheterocycles

5.1 Synthesis of spiro[indole-pyrido[3,2-*e*]thiazine] derivatives

Arya et al. [79] used zeolite supported Brønsted-acid ionic liquid as an efficient catalyst for the synthesis of spiro[indole-pyrido[3,2-*e*]thiazine] derivatives (**69**) *via* three-component reactions of isatins (**2**), substituted anilines (**20**) and 2-mercaptonicotinic acid (**68**) under ultrasonic irradiation in water at 95 °C (Scheme 44). Under the conventional stirring condition, the yields of the desired products were very low. After completion of the reaction, the ionic liquid was recovered quantitatively and reused five times without any loss in its catalytic efficacies.

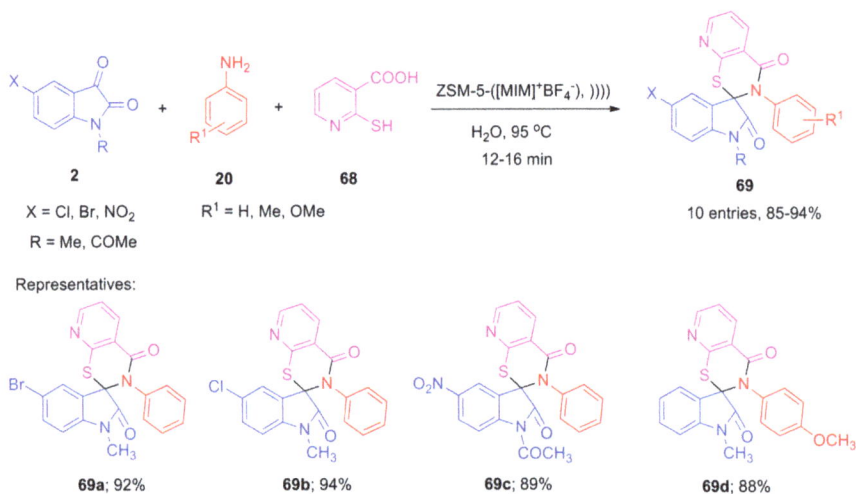

Scheme 44 *ZSM-5 zeolite supported ionic liquid catalyzed synthesis of spiro[indole-pyrido[3,2-e]thiazine] derivatives in an aqueous medium under ultrasonic irradiation*

5.2 Synthesis of spiro[indole-3,4'-pyrazolo[3,4-*e*][1,4]thiazepine] derivatives

Simple, efficient, catalyst-free, ultrasound-assisted three-component condensation reactions of isatins (**2**), 3-amino-5-methylpyrazoles (**27**) and 2-mercaptoacetic acid derivatives (**70**) have been reported by Dandia et al. [80] to afford the corresponding

Industrial Applications of Green Solvents I   Materials Research Forum LLC
Materials Research Foundations **50** (2019) 269-319   doi: https://doi.org/10.21741/9781644900239-9

biologically promising spiro[indole-3,4'-pyrazolo[3,4-*e*][1,4]thiazepine] derivatives (**71**) with good to excellent yields in aqueous medium at ambient temperature (Scheme 45).

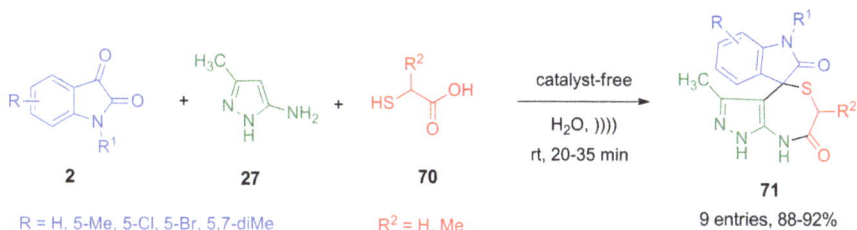

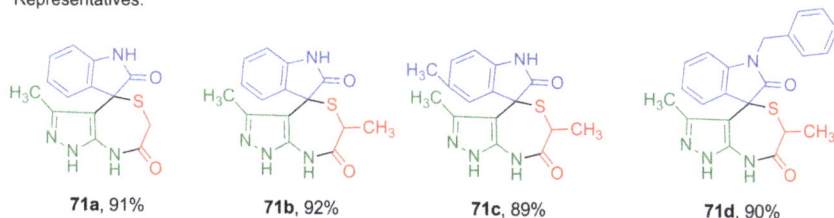

Scheme 45 *Ultrasound-assisted water-mediated synthesis of spiro[indole-3,4'-pyrazolo[3,4-e][1,4]thiazepines]*

5.3   Synthesis of benzothiazole fused spiroheterocyces

Kumar et al. [81] synthesized a series of structurally diverse spiro[pyrimido[2,1-*b*]benzothiazole-3,3'-chromene]-2',4'-diones (**73**), spiro[pyrimido[2,1-*b*]benzothiazole-3,2'-cyclohexane]-1',3'-diones (**74**) and spiro[pyrimido[2,1-*b*]benzothiazole-3,5'-pyrimidine]-2',4',6'-triones (**75**) derivatives *via* pseudo-four component condensation reactions of two equivalents of aldehydes (**23**), one equivalent of 2-aminobenzothiazoles (**72**) and one equivalent of C-H activated acids such as 4-hydroxycoumarin (**38**) or dimedone (**6a**) or 1,3-dimethylbarbituric acid (**14a**) respectively using salfamic acid as a metal-free organocatalyst in water at 80 °C (Scheme 46). The same group also synthesized another series of structurally diverse benzothiazole fused spiroheterocyces

such as spiro[benzothiazolo[2,3-*b*]chromeno[3,4-*e*]pyrimidine-7,3′-indoline]-2′,6-diones
(**77a-c**), spiro[benzothiazolo[2,3-*b*]-pyrimido[5,4-*e*]pyrimidine-5,3′-indoline]-2,2′,4-
triones (**78a-f**), spiro[benzothiazolo[2,3-*b*]-quinazolin-5,3′-indoline]-2′,4-diones (**79a-c**)
and spiro[benzothiazolo[2,3-*b*]pyrano[3,4-*e*]-pyrimidine-5,3′-indoline]-2′,4-diones (**80a-
c**) *via* one-pot three-component reactions of isatins (**2**), 2-aminobenzothiazoles (**72**) and
various C-H activated acids (**38**, **14**, **14a**, **6a**, **76**) using the same sulfamic acid as catalyst
under the same optimized reaction conditions (Scheme 47) [82].

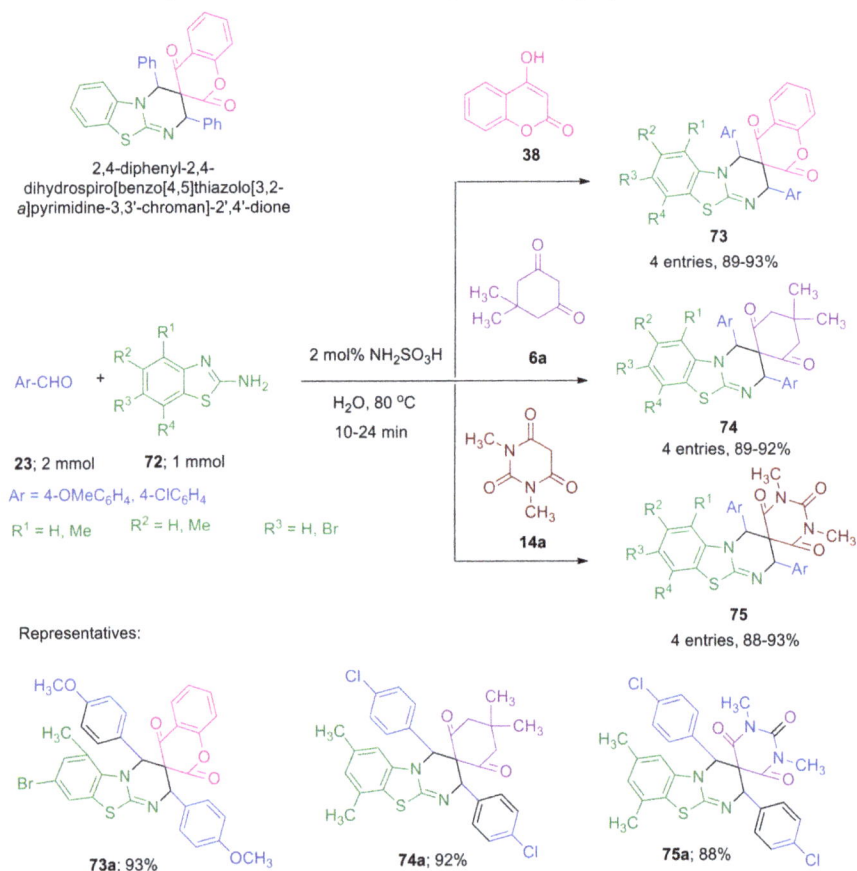

Scheme 46 *NH₂SO₃H-catalyzed synthesis of spirobenzothiazole fused heterocyces in water*

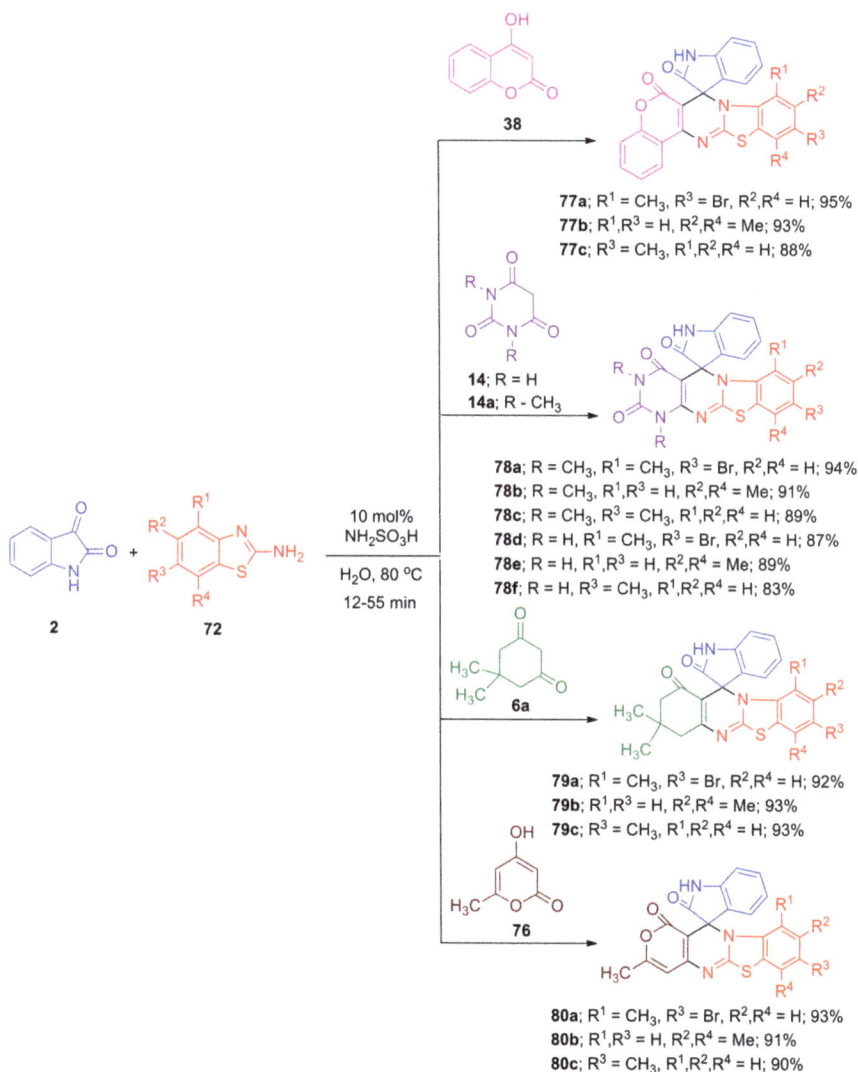

Scheme 47 *NH₂SO₃H-catalyzed synthesis of benzothiazole fused spiroindolinones in water*

5.4   Synthesis of spiro[indoline-3,2′-thiazolidinone] derivatives

Preetam et al. [83] synthesized a number of biologically promising spiro[indoline-3,2′-thiazolidinones] (**81**) through the reactions of amines (**20**), substituted isatins (**2**) and thioglycolic acid (**70**) employing catalytic amount of *p*-dodecylbenzenesulfonic acid (DBSA) as an efficient catalyst in aqueous media at 25 °C (Scheme 48). It was proposed that the reaction underwent *via* the formation of *in situ* generated immine type intermediate **80**. Dandia et al. [84] prepared another series of spiro[indole-3,2′-thiazolidinone]-2,4-dione derivatives (**83**) *via* one-pot three-component tandem reactions of isatins (**2**), 4-amino-1,5-dimethyl-2-phenyl-1*H*-pyrazol-3(2*H*)-one (**82**) and thioglycolic acids (**70**) using a catalytic amount of cetyltrimethylammonium bromide (CTAB) as an efficient phase transfer catalyst in water under ultrasonic irradiation at 45 °C (Scheme 49). Catalyst-free synthesis of a series of novel symmetrical bis-spiro[indoline-3,2′-thiazolidinones] (**84a-f**) was achieved by Panda et al. [85] in water under reflux conditions (Scheme 50).

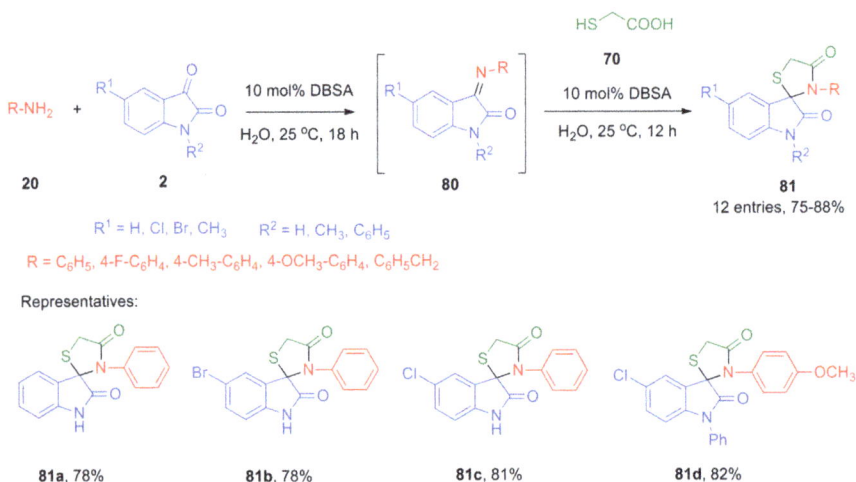

Scheme 48 DBSA-catalyzed synthesis of spiro[indoline-3,2′-thiazolidinone] derivatives in water

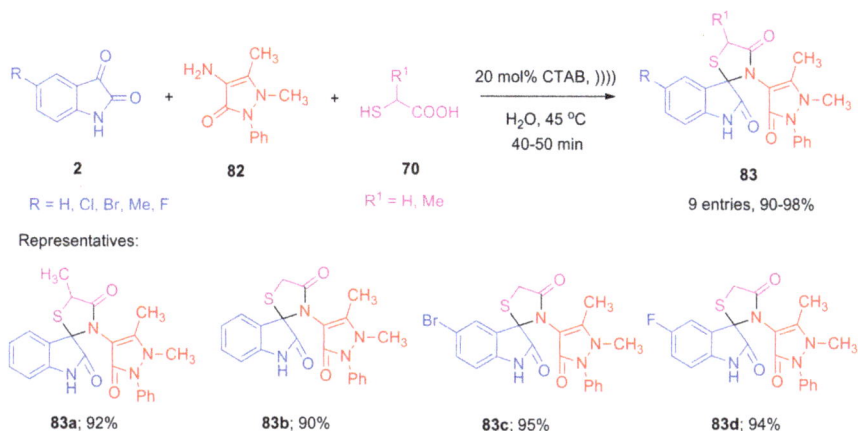

2   **82**   **70**   **83**

R = H, Cl, Br, Me, F   R¹ = H, Me   9 entries, 90-98%

Representatives:

83a; 92%   **83b**; 90%   **83c**; 95%   **83d**; 94%

Scheme 49 *CTAB catalyzed synthesis of spiro[indole-3,2'-thiazolidinone]-2,4-dione derivatives in water*

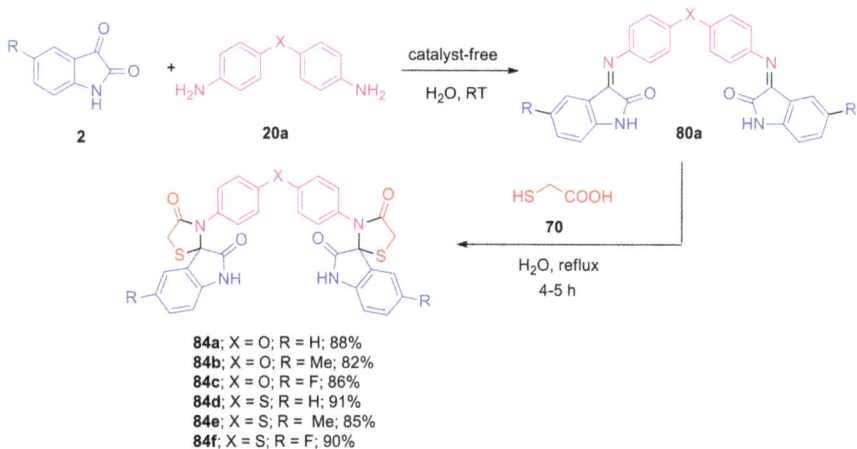

2   **20a**   **80a**

70

84a; X = O; R = H; 88%
84b; X = O; R = Me; 82%
84c; X = O; R = F; 86%
84d; X = S; R = H; 91%
84e; X = S; R = Me; 85%
84f; X = S; R = F; 90%

Scheme 50 *Ultrasound-promoted synthesis of bis-spiro[indoline-3,2'-thiazolidinones] under catalyst-free conditions in an aqueous medium*

5.5 Synthesis of spiro{pyrido[2,1-*b*]benzothiazole-3,3′-indoline derivatives

Hussein et al. [86] reported a simple and efficient protocol for the synthesis of spiro{pyrido[2,1-*b*]benzothiazole-3,3′-indoline derivatives (**86**) *via* one-pot three-component reactions of 2-mercaptoaniline (**84**), malononitrile (**40a**) and various 2-oxoindoline-3-ylidines (**85**) using a catalytic amount of triethylbenzylammonium chloride (TEBACl) in aqueous medium at 80 ºC (Scheme 51).

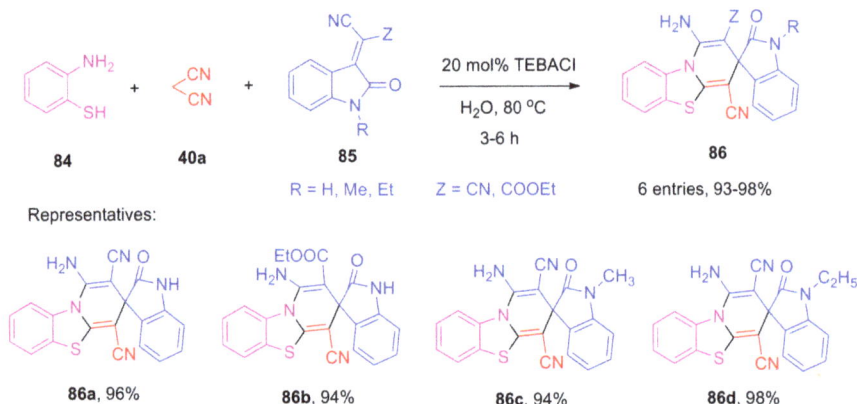

Scheme 51 *TEBACl-catalyzed synthesis of spiro{pyrido[2,1-b]benzothiazole-3,3′-indoline derivatives in water*

Conclusion

Spiroheterocycles with immense biological activities are the backbone of many natural products. Spiroheterocycles are being used as anesthetics, insect antifeedant, antibiotic, antimicrobial, antifungal, cytotoxic, antitumor, inhibitors of microtubule assembly, anti-ulcer, anti-food-deteriorating, CNS stimulant, convallamarogenin, HIV-1 reverse transcriptase inhibitor, cholesterol absorption inhibitor, and antihypertensive agents. Interestingly, to synthesize structurally diverse spiroheterocycles one-pot multi-component reaction strategy has become one of the important tools as it provides a number of advantages such as minimization of the purification processes by reducing the number of work-up steps, operational simplicity, atom economic, energy efficiency, *etc*. On the other hand, water is the safest solvent to carry out organic transformations because of its environmental friendliness as well as it is non-flammable, cheap and abundantly available in nature. Therefore aqueous mediated synthesis of biologically promising spiroheterocycles *via* one-pot multi-component strategy is one of the attractive areas of today's organic chemists. The present chapter covers the "up-to-date" literature related to the aqueous mediated multi-component synthesis of biologically promising spiroheterocycles in a pot.

Acknowledgments

The author is very much grateful to the Indus International University, V.P.O. Bathu, Distt. Una, Himachal Pradesh, India, and the Kartha Education Society, Mumbai, India for the financial help.

References

[1]   G. Brahmachari Handbook of pharmaceutical natural products, 1st Ed. Wiley-VCH, Weinheim (2010).

[2]   G. Brahmachari, Green synthetic approaches for biologically relevant heterocycles, Elsevier, Amsterdam, The Netherland, (2014).

[3]   M. Sannigrahi, Stereocontrolled synthesis of spirocyclics, Tetrahedron 55 (1999) 9007-9071. https://doi.org/10.1016/s0040-4020(99)00482-2

[4]   D.M. James, H.B. Kunze, D.J. Faulkner, Two new brominated tyrosine derivatives from the *Sponge Druinella* (=psammaplysilla) purpurea, J. Nat. Prod. 54 (199) 1137-1140. https://doi.org/10.1021/np50076a040

[5]   J. Kobayashi, M. Tsuda, K. Agemi, H. Shigemiri, M. Ishibashi, T. Sasaki, Y. Mikami, Purealidins B and C, new bromotyrosine alkaloids from the okinawan marine sponge *Psammaplysilla Purea*, Tetrahedron 47 (1991) 6617-6622. https://doi.org/10.1016/s0040-4020(01)82314-0

[6]   N. Deppermann, H. Thomanek, A.H.G.P. Prenzel, W. Maison, Pd-catalyzed assembly of spirooxindole natural products: A short synthesis of horsfiline, J. Org. Chem. 75 (2010) 5994-6000. https://doi.org/10.1021/jo101401z

[7]   E.D. Ananiev, K. Ananieva, G. Abdulova, N. Christova, E. Videnova. Effect of abamectin on protein and RNA synthesis in primary leaves of *Cucurbita pepo L.* (zucchini), Bulg. J. Plant. Physiol. 28 (2002) 85-91.

[8]   M.G. Kulkarni, A.P. Dhondge, S.W. Chavhan, A.S. Borhade, Y.B. Shaikh, D.R. Birhade, M.P. Desai, N.R. Dhatrak, Total synthesis of (±)-coerulescine and (±)-

horsfiline, Beilstein J. Org. Chem. 6 (2010) 876-879.
https://doi.org/10.3762/bjoc.6.103

[9]   D.J. Goldsmith, G. Srouji, C. Kwong. Insect antifeedants. Diels-Alders approach
to the synthesis of ajugarin I, J. Org. Chem. 43 (1978) 3182-3188.
https://doi.org/10.1021/jo00410a019

[10]   G.R. Martinez, P.A. Grieco, E. Williams, K. Kanai, C.V. Srinivasan.
Stereocontrolled total synthesis of antibiotic A-23187 (calcimycin), J. Am. Chem. Soc.
104 (1982) 1436-1438. https://doi.org/10.1021/ja00369a054

[11]   A. Yokosuka, J. Mitsuno, S. Yui, M. Yamazaki, Y. Mimaki. Steroidal glycosides
from *Agave utahensis* and their cytotoxic activity, J. Nat. Prod. 72 (2009) 1399-1404.
https://doi.org/10.1021/np900168d

[12]   J. Yang, X.Z. Wearing, P.W.L. Quesne, J.R. Deschamps, J.M. Cook,
Enantiospecific synthesis of (+)-Alstonisine *via* a stereospecific osmylation process, J.
Nat. Prod. 71 (2008) 1431-1440. https://doi.org/10.1021/np800269k

[13]   M. Miyakoshi, Y. Tamura, H. Masuda, K. Mizutani, O. Tanaka, T. Ikeda, K.
Ohtani, R. Kasai, K. Yamasaki. Steroidal saponins from *Yucca schidigera* (*Mohave
Yucca*), a new antifood-deteriorating agent, J. Nat. Prod. 63 (2000) 332-338.
https://doi.org/10.1021/np9904354

[14]   K. Tanaka, F. Uchiyama, K. Sakamoto. Stereocontrolled total synthesis of (+)-
coriamyrtin, J. Am. Chem. Soc. 104 (1982) 4965-4967.
https://doi.org/10.1021/ja00382a047

[15]   Y. Ma, C. Fan, B. Jia, P. Cheng, J. Liu, Y. Ma, K. Qiao. Total synthesis and
biological evaluation of spirotryprostatin A analogs, Chirality 29 (2017) 737-746.
https://doi.org/10.1002/chir.22746

[16]   D.L. Klass, M. Fiese, L.F. Fieser. Digitogenin, J. Am. Chem. Soc. 77 (1955) 3829-
3833. https://doi.org/10.1021/ja01619a045

[17]   Y. Kita, K. Higuchi, Y. Yoshida, K. Iio, S. Kitagaki, K. Ueda, S. Akai, H. Fujioka.
Enantioselective total synthesis of a potent antitumor antibiotic, fredericamycin A, J.
Am. Chem. Soc. 123 (2001) 3214-3222. https://doi.org/10.1021/ja0035699

[18]   D.F. Taber, T.E. Christos, A.L. Rheingold, I.A. Guzei. Synthesis of (−)-fumagillin,
J. Am. Chem. Soc. 121 (1999) 5589-5590.

[19]   T.-H. Kang, K. Matsumoto, M. Tohda, Y. Murakami, H. Takayama, M. Kitajima,
N. Aimi, H. Watanabe, Pteropodine and isopteropodine positively modulate the

function of rat muscarinic M_1 and 5-HT$_2$ receptors expressed in *Xenopus oocyte*, Eur. J. Pharm. 444 (2002) 39-45. https://doi.org/10.1016/s0014-2999(02)01608-4

[20] M.H. Rønnest, B. Rebacz, L. Markworth, A.H. Terp, T.O. Larsen, A. Krämer, M.H. Clausen. Synthesis and structure-activity relationship of griseofulvin analogues as inhibitors of centrosomal clustering in cancer cells, J. Med. Chem. 52 (2009) 3342-3347. https://doi.org/10.1021/jm801517j

[21] B. Witkop. Gelsemine, J. Am. Chem. Soc. 70 (1948) 1424-1427.

[22] A.A. Mahmoud, A.A. Ahmed, T. Tanaka, M. Iinuma. Diterpenoid acids from grindelia nana, J. Nat. Prod. 63 (2000) 378-380. https://doi.org/10.1021/np9904105

[23] M.P. DeNinno, P.A. McCarthy, K.C. Duplantier, C. Eller, J.B. Etienne, M.P. Zawistoski, F.W. Bangerter, C.E. Chandler, L.A. Morehouse, E.D. Sugarman, R.W. Wilkins, H.A. Woody, L.M. Zaccaro, Steroidal glycoside cholesterol absorption inhibitors, J. Med. Chem. 40 (1997) 2547-2554. https://doi.org/10.1021/jm9702600

[24] J.F.M. Da-Silva, S.J. Garden, A.C. Pinto, The chemistry of isatins: a review from 1975 to 1999, J. Braz. Chem. Soc. 12 (2001) 273-324. https://doi.org/10.1590/s0103-50532001000300002

[25] M. Claire, H. Faraj, G. Grassy, A. Aumelas, A. Rondot, G. Auzou. Synthesis of new 11 beta-substituted spirolactone derivatives. Relationship with affinity for mineralocorticoid and glucocorticoid receptors, J. Med. Chem, 36 (1993) 2404-2407. https://doi.org/10.1021/jm00068a018

[26] M.J. Kornet, A.P. Thio, Oxindole-3-spiropyrrolidines and -piperidines. Synthesis and local anesthetic activity, J. Med. Chem. 19 (1976) 892-898. https://doi.org/10.1021/jm00229a007

[27] C.J. Swain, R. Baker, C. Kneen, R. Herbert, J. Moseley, J. Saunders, E.M. Seward, G.I Stevenson, M. Beer, Novel 5-HT3 antagonists: indol-3-ylspiro(azabicycloalkane-3,5'(4'*H*)-oxazoles), J. Med. Chem. 35 (1992) 1019-1031. https://doi.org/10.1021/jm00084a007

[28] E.A. Laude, D. Bee, O. Crambes, P. Howard, Antitussive and antibronchoconstriction actions of fenspiride in guinea-pigs, Eur. Respir. J. 8 (1995) 1699-1704. https://doi.org/10.1183/09031936.95.08101699

[29] P. Rosenmond, M. Hosseini-Merescht, C. Bub, Ein einfacher Zugang zu den tetracyclischen Vorlaufern der Hetero- und Secoyohimbane, Strychnos-und Oxindolalkaloide, Liebigs Ann. Chem. 2 (1994) 151-154. https://doi.org/10.1002/jlac.199419940208

[30]   M. Carmignani, A.R. Volpe, F.D. Monache, B. Botta, R. Espinal, S.C.D Bonnevaux, C.D. Luca, M. Botta, F. Corelli, A. Tafi, G. Ripanti, G.D. Monache, Novel hypotensive agents from *Verbesina caracasana*. 6. Synthesis and pharmacology of caracasandiamide, J. Med. Chem. 42 (1999) 3116-3125. https://doi.org/10.1021/jm991004l

[31]   J.R. Atack, The benzodiazepine binding site of GABA(A) receptors as a target for the development of novel anxiolytics, Expert Opin. Investig. Drug. 14 (2005) 601-618. https://doi.org/10.1517/13543784.14.5.601

[32]   T. Okita, M. Isobe, Synthesis of the pentacyclic intermediate for dynemicin a and unusual formation of spiro-oxindole ring, Tetrahedron 50 (1994) 11143-11152. https://doi.org/10.1016/s0040-4020(01)89417-5

[33]   R. Sarges, J. Bordner, B.W. Dominy, M.J. Peterson, E.B. Whipple, Synthesis, absolute configuration, and conformation of the aldose reductase inhibitor sorbinil, J. Med. Chem. 28 (1985) 1716-1720. https://doi.org/10.1021/jm00149a030

[34]   I. Inada, H. Satoh, N. Inatomi, H. Nagaya, Y. Maki, Spizofurone, a new anti-ulcer agent, increases alkaline secretion in isolated bullfrog duodenal mucosa, Eur. J. Pharmacol. 124 (1986) 149-155. https://doi.org/10.1016/0014-2999(86)90135-4

[35]   E.M. Smith, G.F. Swiss, B.R. Neustadt, P. McNamara, E.H. Gold, E.J. Sybertz, T. Baum, Angiotensin-converting enzyme inhibitors: spirapril and related compounds, J. Med. Chem. 32 (1989) 1600-1606. https://doi.org/10.1021/jm00127a033

[36]   B. Banerjee, Recent developments on ultrasound-assisted one-pot multicomponent synthesis of biologically relevant heterocycles, Ultrason. Sonochem. 35 (2017) 15-35. https://doi.org/10.1016/j.ultsonch.2016.10.010

[37]   B.H. Rotstein, S. Zaretsky, V. Rai, A.K. Yudin, Small heterocycles in multicomponent reactions, Chem. Rev. 114 (2014) 8323-8359. https://doi.org/10.1021/cr400615v

[38]   B. Banerjee, Recent developments on organo-bicyclo-bases catalyzed multi-component synthesis of biologically relevant heterocycles, Curr. Org. Chem. 22 (2018) 208-233. https://doi.org/10.2174/1385272821666170703123129

[39]   A. Chanda, V.V.Fokin, Organic synthesis "on water", Chem. Rev. 109 (2009) 725.

[40]   R.A. Sheldon, Selective catalytic synthesis of fine chemicals: opportunities and trends, J. Mol. Catal. A Chem. 107 (1996) 75-83.

[41]   R. Ghahremanzadeh, S.C. Azimi, N. Gholami, A. Bazgir, Clean synthesis and antibacterial activities of spiro[pyrimido[4,5-b]-quinoline-5,5'-pyrrolo[2,3-

d]pyrimidine]-pentaones, Chem. Pharm. Bull. 56 (2008) 1617-1620.
https://doi.org/10.1248/cpb.56.1617

[42]   A. Alizadeh, A. Mikaeili, T. Firuzyar, One-pot pseudo five-component synthesis
of spirooxindole derivatives containing fused 1,4-dihydropyridines in water, Synthesis
44 (2012) 1380-1384. https://doi.org/10.1055/s-0031-1290884

[43]   K. Jadidi, R. Ghahremanzadeh, P. Mirzaei, A. Bazgir, Three-component synthesis
of spiro[indoline-3,5'-pyrimido[4,5-b]quinoline]-triones in water, J. Heterocyclic
Chem. 48 (2011) 1014-1018. https://doi.org/10.1002/jhet.655

[44]   Z. Wang, L. Gao, Z. Xu, Z. Ling, Y. Qin, L. Rong, S.-J. Tu, Green synthesis of
novel spiro[indoline-3,4'-pyrazolo[3,4-*b*]pyridine]-2,3'(7'*H*)-dione, spiro[indeno [1,2-
b]pyrazolo[4,3-*e*]pyridine-4,3'-indoline]-2',3-dione and spiro[benzo[*h*] pyrazolo[3,4-
b]quinoline-7,3'-indoline]-2',8(5*H*)-dione derivatives in aqueous medium, Tetrahedron
73 (2017) 385-394. https://doi.org/10.1016/j.tet.2016.12.015

[45]   G.-P. Lu, C. Cai, An efficient, one-pot synthesis of spiro[dihydropyridine-
oxindole]compounds under catalyst-free conditions, J. Chem. Res. 2011 (2011) 547-
551. https://doi.org/10.3184/174751911x13157531279974

[46]   M. Narasimhulu, Y.R. Lee, Ethylenediamine diacetate-catalyzed three-component
reaction for the synthesis of 2,3-dihydroquinazolin-4(1*H*)-ones and their spirooxindole
derivatives, Tetrahedron 67 (2011) 9627-9634.
https://doi.org/10.1016/j.tet.2011.08.018

[47]   A.V. Chate, S.P. Kamdi, A.N. Bhagata, J.N. Sangshetti, C.H. Gill, β-Cyclodextrin
catalyzed one-pot four component auspicious protocol for synthesis of spiro[acridine-
9,3'-indole]-2',4,4'(1'*H*,5'*H*,10*H*)-trione as a potential antimicrobial agent, Synth.
Commun. 48 (2018) 1701-1714. https://doi.org/10.1080/00397911.2017.1421665

[48]   B. Jiang, L.J. Cao, S.J. Tu, W.R. Zheng, H.Z. Yu, Highly diastereoselective
domino synthesis of 6-spirosubstituted pyrido[2,3-*d*]pyrimidine derivatives in water, J.
Comb. Chem. 11 (2009) 612-616. https://doi.org/10.1021/cc900038g

[49]   X.S. Wang, M.M. Zhang, H. Jiang, C.S. Yao, S.J. Tu, Unexpected spiro-
benzoquinolines in the reaction of *N*-(arylidene)naphthalen-2-amine, arylaldehyde, and
1,3-dimethylbarbituric acid in water, Chem. Lett. 36 (2007) 450-451.
https://doi.org/10.1246/cl.2007.450

[50]   S.J. Kalita, B. Das, D.C. Deka, A quick, simple and clean synthesis of
spiro(indoline-3,4'-pyrazolo[4',3':5,6]pyrido[2,3-*d*]pyrimidines) in water through a

novel one-pot multicomponent reaction, Chemistry Select 2 (2017) 5701-5706. https://doi.org/10.1002/slct.201701131

[51]   R. Ghahremanzadeh, M. Sayyafi, S. Ahadi, A. Bazgir, Novel one-pot, three-component synthesis of spiro[indoline-pyrazolo[4′,3′:5,6]pyrido[2,3-*d*]pyrimidine]trione library, J. Comb. Chem. 11 (2009) 393-396. https://doi.org/10.1002/chin.200939166

[52]   R. Ghahremanzadeh, Z. Rashid, A.H. Zarnani, H. Naeimi, Inorganic-organic hybrid silica based tin complex as a novel, highly efficient and recyclable heterogeneous catalyst for the one-pot preparation of spirooxindoles in water, Dalton Trans. 43 (2014) 15791-15797. https://doi.org/10.1039/c4dt02038c

[53]   K. Balamurugan, S. Perumal, J.C. Menéndez, New four-component reactions in water: a convergent approach to the metal-free synthesis of spiro[indoline/acenaphthylene-3,4′-pyrazolo[3,4-*b*]pyridine derivatives, Tetrahedron 67 (2011) 3201-3208. https://doi.org/10.1016/j.tet.2011.03.020

[54]   J. Quiroga, S. Portillo, A. Pérez, J. Gálvez, R. Abonia, B. Insuasty, An efficient synthesis of pyrazolo[3,4-*b*]pyridine-4-spiroindolinones by a three-component reaction of 5-aminopyrazoles, isatin, and cyclic β-diketones, Tetrahedron Lett. 52 (2011) 2664-2666. https://doi.org/10.1016/j.tetlet.2011.03.067

[55]   A. Kamal, K.S. Babu, M.V.P.S. V. Vardhan, S.M.A. Hussaini, R. Mahesh, S.P. Shaika, A. Alarifi, Sulfamic acid promoted one-pot three-component synthesis and cytotoxic evaluation of spirooxindoles, Bioorg. Med. Chem. Lett. 25 (2015) 2199-2202. https://doi.org/10.1016/j.bmcl.2015.03.054

[56]   M. Dabiri, Z.N. Tisseh, M. Nobahar, A. Bazgir, Organic reaction in water: a highly efficient and environmentally friendly synthesis of spiro compounds catalyzed by *L*-proline, Helvetica Chimica Acta 94 (2011) 824-830. https://doi.org/10.1002/hlca.201000307

[57]   S. Ahadi, R. Ghahremanzadeh, P. Mirzaei, A. Bazgir, Synthesis of spiro[benzopyrazolonaphthyridine-indoline]-diones and spiro[chromenopyrazolo-pyridine-indoline]-diones by one-pot, three-component methods in water, Tetrahedron 65 (2009) 9316-9321. https://doi.org/10.1016/j.tet.2009.09.009

[58]   A. Bazgir, S. Ahadi, R. Ghahremanzadeh, H. R. Khavasi, P. Mirzaei, Ultrasound-assisted one-pot, three-component synthesis of spiro[indoline-3,4′-pyrazolo[3,4-*b*]pyridine]-2,6′(1′H)-diones in water, Ultrason. Sonochem. 17 (2010) 447-452. https://doi.org/10.1016/j.ultsonch.2009.09.009

[59]   A. Dandia, A.K. Laxkar, R. Singh, New multicomponent domino reaction on water: highly diastereoselective synthesis of spiro[indoline-3,4'-pyrazolo[3,4-b]pyridines] catalyzed by NaCl, Tetrahedron Lett. 53 (2012) 3012-3017. https://doi.org/10.1016/j.tetlet.2012.03.136

[60]   G. Harichandran, K.S. Devi, P. Shanmugam, M.I. Jesse, K. Kathiravan, Amberlite IRA-400 Cl resin catalyzed multicomponent organic synthesis in water: synthesis, antimicrobial and docking studies of spiroheterocyclic 2-oxindoles and acenaphthoquinone, Current Organocatal. 5 (2018) 13-24. https://doi.org/10.2174/2213337205666180316170023

[61]   R. Ghahremanzadeh, T. Amanpour, M. Sayyafi, A. Bazgir, One-pot, three-component synthesis of spironaphthopyrano[2,3-d]pyrimidine-5,3'-indolines in water, J. Heterocyclic Chem. 47, (2010) 421-424. https://doi.org/10.1002/jhet.331

[62]   D.L. Kong, G.P. Lu, M.S. Wu, Z.F. Shi, Qiang Lin, One-Pot, catalyst-free synthesis of spiro[dihydroquinolinenaphthofuranone] compounds from isatins in water triggered by hydrogen bonding effects, ACS Sustainable Chem. Eng. 5 (2017) 3465-3470. https://doi.org/10.1021/acssuschemeng.7b00145

[63]   L. Zhao, B. Zhou, Y. Li, An efficient one-pot three-component reaction for synthesis of spirooxindole derivatives in water media under catalyst-free condition, Heteroatom Chem. 22 (2011) 673-677. https://doi.org/10.1002/hc.20723

[64]   S.L. Zhu, S.J. Jia, Y. Zhang, A simple and clean procedure for three-component synthesis of spirooxindoles in aqueous medium, Tetrahedron 63 (2007) 9365-9372. https://doi.org/10.1016/j.tet.2007.06.113

[65]   A. Mobinikhaledi, N. Foroughifar, M.A.B. Fard, Simple and efficient method for three-component synthesis of spirooxindoles in aqueous and solvent-free media, Synth. Commun. 41 (2011) 441-450. https://doi.org/10.1080/00397911003587507

[66]   N.G. Singh, M. Lily, S.P. Devi, N. Rahman, A. Ahmed, A.K. Chandra, R. Nongkhlaw, Synthetic, mechanistic and kinetic studies on the organo-nano catalyzed synthesis of oxygen and nitrogen containing Spiro compounds under ultrasonic condition, Green Chem. 18 (2016) 4216-4227. https://doi.org/10.1039/c6gc00724d

[67]   Y. Li, H. Chen, C. Shi, D. Shi, S. Ji, Efficient one-pot synthesis of spirooxindole derivatives catalyzed by L-proline in aqueous medium, J. Comb. Chem. 12 (2010) 231-237. https://doi.org/10.1021/cc9001185

[68] G.S. Hari, Y.R. Lee, Efficient one-pot synthesis of spirooxindole derivatives by ethylenediamine diacetate catalyzed reactions in water, Synthesis 2010 (2010) 0453-0464. https://doi.org/10.1055/s-0029-1217116

[69] R. Ghahremanzadeh, T. Amanpour, A. Bazgir, Clean synthesis of spiro[indole-3,8'-phenaleno[1,2-*b*]pyran]-9'-carbonitriles and spiro[indole-3,4'-pyrano[4,3-b]pyran]-3'-carbonitriles by one-pot, three-component reactions, J. Heterocyclic Chem. 47 (2010) 46-49. https://doi.org/10.1002/jhet.247

[70] A.R. Karimi, F. Sedaghatpour, Novel mono- and bis(spiro-2-amino-4*H*-pyrans): alum-catalyzed reaction of 4-hydroxycoumarin and malononitrile with isatins, quinones, or ninhydrin, Synthesis 2010 (2010) 1731-1735. https://doi.org/10.1055/s-0029-1219748

[71] S. Gholizadeh, K. Radmoghadam, Ultrasound-assisted three-component synthesis of spiro[4*H*-pyrano[3,2-*c*]quinolin-4,3'-indoline]-2',5(6*H*)-diones in water, Orient. J. Chem. 29 (2013) 1637-1641. https://doi.org/10.13005/ojc/290450

[72] A. Dandia, A.K. Jain, D.S. Bhati, NaCl as a novel and green catalyst for the synthesis of biodynamic spiro heterocycles in water under sonication, Synth. Commun. 41 (2011) 2905-2919. https://doi.org/10.1080/00397911.2010.515365

[73] Y. Liu, D Zhou, Z. Ren, W. Cao, J. Chen, H. Deng, Q. Gu, A green efficient synthesis of spiro[indoline-3,4'(1*H*)-pyrano[2,3-*c*]pyrazol]-2-one derivatives, J. Chem. Soc. 2009 (2009) 154-156. https://doi.org/10.3184/030823409x416875

[74] S. Ahadi, Z. Yasaei, A. Bazgir, A clean and one-pot synthesis of spiroindoline-pyranopyrazoles, J. Heterocyclic Chem. 47 (2010) 1090-1094. https://doi.org/10.1002/jhet.437

[75] R. Shrestha, K. Sharma, Y.R. Lee, Y.-J. Wee, Cerium oxide-catalyzed multicomponent condensation approach to spirooxindoles in water, Mol. Divers 20 (2016) 847-858. https://doi.org/10.1007/s11030-016-9670-2

[76] P. Mukherjee, S. Paul, A.R. Das, Expeditious synthesis of functionalized tricyclic 4-spiro pyrano[2,3-*c*]pyrazoles in aqueous medium using dodecylbenzenesulphonic acid as a Brønsted acid-surfactantcombined catalyst, New J. Chem. 39 (2015) 9480-9486. https://doi.org/10.1039/c5nj01728a

[77] C. Cheng, B. Jiang, S.J. Tu, G. Li, [4+2+1] Domino cyclization in water for chemo- and regioselective synthesis of spiro-substituted benzo[*b*]furo[3,4-*e*][1,4]diazepine derivatives, Green Chem. 13 (2011) 2107-2115. https://doi.org/10.1039/c1gc15183e

[78]   N. Ma, B. Jiang, G. Zhang, S.-J. Tu, W. Wever, G. Li, New multicomponent
domino reactions (MDRs) in water: highly chemo-, regio- and stereoselective
synthesis of spiro{[1,3]dioxanopyridine}-4,6-diones and pyrazolo[3,4-*b*]pyridines,
Green Chem. 12 (2010) 1357-1361. https://doi.org/10.1039/c0gc00073f

[79]   K. Arya, D.S. Rawat, H. Sasai, Zeolite supported Brønsted-acid ionic liquids: An
eco approach for synthesis of spiro[indole-pyrido[3,2-*e*]thiazine] in water under
ultrasonication, Green Chem. 14 (2012) 1956-1963.
https://doi.org/10.1039/c2gc35168d

[80]   A. Dandia, R. Singh, J. Joshi, S. Maheshwari, P. Soni, Ultrasound promoted
catalyst-free and selective synthesis of spiro[indole-3,49-pyrazolo[3,4-
e][1,4]thiazepines] in aqueous media and evaluation of their anti-hyperglycemic
activity, RSC Adv. 3 (2013) 18992-19001. https://doi.org/10.1039/c3ra43745k

[81]   M. Kumar, A.K. Arya, J, George, K. Arya, R.T. Pardasani, DFT studied hetero-
diels–alder cycloaddition for the domino synthesis of spiroheterocycles fused to
benzothiazole and chromene/pyrimidine rings in aqueous media, J. Heterocyclic
Chem. 54 (2017) 3418-3426. https://doi.org/10.1002/jhet.2964

[82]   A.K. Arya, M. Kumar, An efficient green chemical approach for the synthesis of
structurally diverse spiroheterocycles with fused heterosystems, Green Chem. 13
(2011) 1332-1338. https://doi.org/10.1039/c1gc00008j

[83]   A. Preetam, M. Nath, Ambient temperature synthesis of spiro[indoline-3,2'
thiazolidinones] by a DBSA-catalyzed sequential reaction in water, Tetrahedron Lett.
57 (2016) 1502-1506. https://doi.org/10.1016/j.tetlet.2016.02.079

[84]   A. Dandia, R. Singh, S. Bhaskaran, S.D. Samant, Versatile three component
procedure for combinatorial synthesis of biologically relevant scaffold spiro[indole-
thiazolidinones] under aqueous conditions, Green Chem. 13 (2011) 1852-1859.
https://doi.org/10.1039/c0gc00863j

[85]   S.S. Panda, N. Jain, N. Jehan, S. Bhagat, S.C. Jain, An eco-friendly synthesis of
some novel symmetrical bis spiro-indoles, Phosphorus, Sulfur, and Silicon, 187 (2012)
101-111. https://doi.org/10.1080/10426507.2011.582057

[86]   E.M. Hussein, A.M. El-Khawaga, Simple and clean procedure for three-
component syntheses of spiro{pyrido[2,1-*b*]benzothiazole-3,3'-indolines} and
spiro{thiazolo[3,2-*a*]pyridine-7,3'-indolines} in aqueous medium, J. Heterocyclic
Chem. 49 (2012) 1296-1301. https://doi.org/10.1002/jhet.908

Industrial Applications of Green Solvents I
Materials Research Foundations 50 (2019) 320-344

Materials Research Forum LLC
doi: https://doi.org/10.21741/9781644900239-10

Chapter 10

Application of Ionic Liquids in Gas Separation Membranes

M. Zia-ul-Mustafa[1], Hafiz Abdul Mannan[1], Hilmi Mukhtar[1,*], Rizwan Nasir[2], Dzeti Farhah Mohshim[3], NAHM Nordin[1]

[1]Chemical Engineering Department, Universiti Teknologi PETRONAS, Bandar Seri Iskandar, 32610, Perak, Malaysia

[2]Department of Chemical Engineering, University of Jeddah, Jeddah, Kingdom of Saudi Arabia

[3]Petroleum Engineering Department, Universiti Teknologi PETRONAS, Bandar Seri Iskandar, 32610, Perak, Malaysia

hilmi_mukhtar@utp.edu.my*

Abstract

Gas emission is a direct result of huge industrial progresses since the last century. To overcome the hazardous effects of these acid gases, it is crucial to separate and capture these unwanted gases. Ionic liquids owing to negligible vapor pressure, high thermal resistance and widespread electrochemical stability have found their application in gas separation membranes. In this chapter, a comprehensive summary of the applications of ionic liquids in gas separation membranes is described. The main classifications of ionic liquid membranes (ILMs) such as supported ionic liquid membranes (SILMs), ionic liquid polymeric membranes (ILPMs) and ionic liquid mixed matrix membranes (ILMMMs) and their applications for the separation of various mixed gases systems have been discussed in detail.

Keywords

Ionic Liquids, Acid Gas Separation Membranes, Membrane Types, Rubbery ILPMs, ILMs

List of Abbreviations

1-ethyl-3-methylimidazolium	[emim]
Bis(trifluoromethanesulfonyl)imide	Tf$_2$N
Bistriflimide	TFSI
Dichloromethane	DCM
Ionic liquid membranes	ILMs

Industrial Applications of Green Solvents I   Materials Research Forum LLC
Materials Research Foundations **50** (2019) 320-344   doi: https://doi.org/10.21741/9781644900239-10

Ionic liquid mixed matrix membranes	ILMMs
Ionic liquid polymeric membranes	ILPMs
Poly(vinylidene fluoride)-(hexafluoropropyl)	PVDF-HFP
Polyacrylonitrile	PAN
polyamide	PA
Polyethersulfone	PES
Polyionic liquids	PILs
Polysulfone	PSF
Pressure swing absorption	PSA
Room-temperature ionic liquids	RTILs
Supported ionic liquid membranes	SILMs
Task-specific ionic liquids	TSILs
Tetrafluoroborate	BF_4

Contents

1.   Introduction

CO_2 emission, acid rain [1], and fog and haze [2] are causing global warming issues. International energy agency has reported that CO_2 emission is increasing by 6% annually due to the utilization of fossil fuels as an energy source [3] and these fossil fuels will remain a major source of electric power generation. Rapid growing human civilization demands on controlling these emissions to facilitate a clean-living environment for humans. Therefore, it is a basic requirement to capture these harmful gas emissions with appropriate methods.

Physical and chemical solvents scrubbing, pressure swing absorption (PSA), cryogenic distillation, amine absorption, and membrane separation are mainly techniques in use, to treat industrial trail gases or process gases.

However, membrane technology in gas separation application is getting importance due to its advantages over other conventional processes, such as ease of operation, low energy requirements, compact size, low operating and capital cost, and environmental friendly separation. Polymeric materials are fascinating and getting importance due to the brilliant film forming nature and mechanical stability and lower price in gas separation applications. Many polymers have been reported in membrane fabrication such as polysulfone (PSF), polyethersulfone (PES), polyacrylonitrile (PAN), and polyamide etc.

What are ionic liquids (ILs)?

ILs are organic salts that exist in liquid form at room temperature. ILs remain liquid at low temperature due to the irregularity of the cation, attached with resonance-stabilized anions. Ionic liquids due to their negligible vapor pressure and higher thermal stability are feasible in gas separation processes.

Classification of ionic liquids is reported as room-temperature ILs (RTILs) [4], task-specific ILs (TSILs) [5] and polyionic liquids (PILs) [6]. All types of ionic liquids are made up of a combination of cations and anions. These cations and anions with proper functionalization can alter the properties of the material in which they are incorporated such as hydrophobicity/hydrophilicity and specific chemical interactions. In polymers, the ionic liquids act as wetting agents and produce flexibility in the polymer chains, this characteristic of the ionic liquids opens a way to use them in polymer gas separation membranes.

Pual W. et al. [7] in 1914 reported the use of ionic liquid and nowadays use of ILs has become important in many applications, as well as in gas separation membranes through dispensing in the polymer materials. Particularly, ILs were recommended as CO_2 separation medium by Blanchard et al. [8] in 2001. At present, in the last two decades, usage of ionic liquids has been expanded tremendously, in many multidisciplinary fields such as chemistry, material science, chemical engineering and environmental science [9]. Many studies have been reported in gas separation membranes as well [10-12]. However, the usage of ILs has also limitations because of its high cost, uncertain toxicities, and environmental effects.

ILs have great potential to substitute conventional organic solvents because of their exceptional characteristics such as negligible vapor pressure, non-flammability and higher ionic conductivity [13]. Firstly, ionic liquids were incorporated in supported ionic liquid membranes (SILMs) to substitute the conventional solvents which have the limitation of solvent loss because of volatilization, even though conventional supported liquid membranes have high permeability values because of higher liquid diffusivities [14].

Types of ionic liquids

Mainly three types of ionic liquids are described in the literature, task-specific ILs (TSILs) [5] room-temperature ILs (RTILs) [4] and polyionic liquids (PILs) [6]. In tasks specific ionic liquids, ILs act as a complexing agent, basically IL is added with maximum loading in the base material. In this manner, such ILs absorb higher quantities of the target gas. Such as, Davis et al. [15] reported the development of IL-based amine functionality, in which IL efficiently absorbed 0.5 mol of CO_2 per mol of IL. More, complexing agents are also incorporated in ILs to increase solubility uptake of many target gases [16, 17]. Relatively, RTILs due to modular nature are polymerized. Applications of RTILs are green solvents for reactions, in membranes separations and electrochemical systems [18]. Imidazolium and pyridinium based ionic liquids have been reported as reasonable solvents for CO_2 separation due to tunable cation or anion properties to achieve better gas separation performance [19]. Furthermore, polymeric ionic liquids, because of higher intake capacity of CO_2 and good mechanical strength [20] values in comparison to straight RTIL [21], have also been incorporated in gas separation membranes. Polymerization of RTIL monomer by altering the n-alkyl length has been helpful to enhance the permeability of target gases i.e. CO_2, N_2, and methane (CH_4) [22].

2.   Gas separation through ionic liquid polymer membranes

Separation of gas mixtures has been achieved using porous and non-porous membranes. The driving force for separation is the difference in partial pressure of gas between two edges of the membrane. Permeation is associated with rate control and separation is measured by the selectivity of the membrane at a specific pressure, temperature, flow rate and membrane area [23]. Gas separation occurs only if a specific constituent of gas

Industrial Applications of Green Solvents I   Materials Research Forum LLC
Materials Research Foundations **50** (2019) 320-344   doi: https://doi.org/10.21741/9781644900239-10

passes through the membrane more quickly compared to others. Based on pore sizes of the membrane matrix, few important transport mechanisms in gas separation membranes have been stated in literature such as Knudsen diffusion, solution-diffusion, molecular sieving, and Poiseuille flow. Porous membrane includes Poiseuille flow, Knudsen diffusion, and molecular sieving, in which the separation process is based on molecular size through the small pores in the membrane matrix whereas the common commercial applications are based on nonporous via solution-diffusion mechanism [24].

Global warming is the key issue for CO_2 capture from industrial gas mixtures. RTILs provides a highly adjustable and tunable medium for the progress of new processes and materials designed to the removal of CO_2 from power plant emitted gases as well as in CH_4 sweetening processes [25]. Membrane technology is an attractive method to CO_2 separation from N_2 and CH_4 in industrial processes. Higher selective polymer membranes have been proposed as a long-term substitute for other separation processes for certain CO_2 separation processes [26]. Ionic liquid membranes are considered highly useful for selective separation of CO_2 from gaseous mixtures.

Initially, ionic liquids were incorporated into porous polymer supports where ionic liquids were injected into the pores. The pores were wetted with ILs with the help of capillary forces. These supported membranes provided higher permeability values of the target gas. However, loss of ionic liquids at higher pressure was the main issue in the expansion of these membranes for long term usage. As a replacement of these membranes, physical blending was considered appropriate to resolve the solvent loss issue. In this method, the reinforcement of ionic liquid into polymeric membranes is done [27]. Recently, a lot of work has been reported by researchers on the blending of IL in the polymer matrix. From the literature survey, it has been observed that CO_2 has more solubility in imidazolium-based RTILs. Therefore, 1-R-3-methyl imidazolium-based RTILs have gained special attention in CO_2 separations. Their low viscosity as compared to other RTILs is also another factor for their selection [28]. Both glassy and rubbery polymers have been used as a host polymer for the development of ILPMs as discussed in the following sections. In addition, block copolymer-based polymers have also been used in the development of ILPMs.

Rubbery ionic liquid polymeric membranes

Initially, the proof-of-concept was demonstrated by Bara et al. [29] for poly(RTIL)/RTIL blend membranes. They prepared ILPMs by blending poly(RTIL)/([emim][Tf₂N]) due to the fact that poly(RTIL) demonstrated typically low diffusivity than pure SILMs. The objective was to provide enough free IL to facilitate the transport of target gas. CO_2 permeability of composite ILPM was 44 barrer, while the permeability of poly(RTIL) membrane was 9.2 barrer only. However, a reduction in CO_2/CH_4 selectivity from 39 to 27 was observed due to increased diffusion of both gases.

Later, the same group studied the effects of variation in anions and cations in RTILs on the transport properties of composite poly(RTIL)/RTIL ILPMs [30, 31]. It was observed that [Tf_2N] anion has the highest CO_2 permeability due to higher molar volume compared to other anions. Similarly, imidazolium-based cations having different functionalities were tested and it was found that imidazolium cation with fluoroalkyl group was more selective towards CO_2/CH_4 separation due to the better combination of CO_2 diffusivity and solubility.

Zarca et al. [32] reported the gas transport properties of poly(RTIL)/([C_4mim][Cl]) in the presence of copper salt. The presence of copper salt reduced CO_2 permeability from 226.7 barrer for poly(RTIL)/([C_4mim][Cl]) to 8.4 barrer for poly(RTIL)/([C_4mim][Cl]) containing 8wt. % copper salt due to reduction in gas solubility. However, CO_2/N_2 selectivity increased from 29 for poly(RTIL)/([C_4mim][Cl]) to 34 for poly(RTIL)/([C_4mim][Cl]) containing 8wt. % copper salt due to increase in CO_2 gas diffusivity.

Recently, Tome et al. [33] explored the film forming abilities of poly(RTIL)/([C_2mim][C(CN)$_3$]) ILPMs with various molecular weights of poly(RTILs). Stable membranes were not obtained for lower molecular weights of poly(RTILs). Medium and higher molecular weight poly(RTILs) containing 60 % of free ([C_2mim][C(CN)$_3$]) IL displayed outstanding performance (CO_2 permeability = 542 and 439 barrer respectively, CO_2/N_2 selectivity = 54 and 64.4 respectively) and surpassed the Robeson upper bound limit. However, the synthesized membranes had gone through a severe loss in mechanical strength with the addition of IL in the membrane matrix.

Following the approach of using poly(RTIL) polymers, Hong et al. [34] reported the synthesis of poly(vinylidene fluoride)-(hexafluoropropyl) (PVDF-HFP) copolymer/1-ethyl-3-methylimidazolium tetrafluoroborate ([emim][BF_4]) membrane. The copolymer was selected due to better mechanical strength and miscibility with the IL. There was no particular interaction among the polymer and the IL and the components were mixed by physical mixing only. The CO_2 permeability of ILPM reached 400 barrer at 2 atm pressure coupled with CO_2/N_2 selectivity of 60 under mixed gas feed conditions (50/50 %). This performance was beyond the Robeson upper bound line for CO_2/N_2 separation.

Following the success of this strategy, PVDF-HFP copolymer was later tested with various combinations of imidazolium-based ILs having different anions and cations. For example, Uchytil et al. [35] synthesized ILPMs with PVDF-HFP copolymer along with two different ILs i.e. ([emim][Tf$_2$N]) and ([hmim][Tf$_2$N]). Better performances in terms of CO$_2$ permeability and CO$_2$/CH$_4$ selectivity were obtained with ([emim][Tf$_2$N]) based ILPMs than ([hmim][Tf$_2$N]) based ILPMs. The authors concluded that this difference could be due to the difference in the chain length of imidazolium-based cations. The CO$_2$ permeability of PVDF-HFP/([emim][Tf$_2$N]) and PVDF-HFP/([hmim][Tf$_2$N]) ILPMs were 1620 and 877 barrer respectively, compared to 1.3 barrer for pure PVDF-HFP polymer membrane. Similarly, CO$_2$/CH$_4$ selectivities of PVDF-HFP/([emim][Tf$_2$N]) and PVDF-HFP/([hmim][Tf$_2$N]) ILPMs were 15.3 and 11.3 respectively, compared to 7.7 for pure PVDF-HFP polymer membrane.

Friess et al. [36] also explored the opportunity of PVDF-HFP/([emim][Tf$_2$N]) ILPM for gas and vapor separation and found that CO$_2$ permeability increased significantly in the presence of ([emim][Tf$_2$N]) IL. However, the decreasing trend in CO$_2$/CH$_4$ selectivity was observed. Due to a decrease in mechanical strength of ILPMs, the authors suggested the change in transport mechanism from diffusion-controlled transport to the solubility-controlled mechanism. The authors identified a severe loss in mechanical properties especially Young's modulus of ILPM at higher IL loading (3MPa) compared to pure polymer membrane (1500MPa).

Jansen et al. [37] compared the effects of ([emim][Tf$_2$N]) and ([Hdmim][Tf$_2$N]) ILs on PVDF-HFP based ILPMs at varying ionic liquid concentrations. The blends of ILs ([emim][Tf$_2$N]) and ([Hdmim][Tf$_2$N]) in various ratios were found to provide better performance due to complementary properties of both ILs. Fluoropolymer and fluorinated anion i.e. bis(trifluoromethylsulfonyl)imide demonstrated improved compatibilities in the membrane matrix. The elastic modulus and break strength of ILPMs decreased vividly as the IL loading was increased. ILPM with 40 wt. % loading of ([emim][Tf$_2$N]) and ([Hdmim][Tf$_2$N]) ILs in 50/50 blend ratio demonstrated a stable performance at 60°C. CO$_2$ permeability was found to be 212 barrer and the corresponding CO$_2$/CH$_4$ selectivity was 8.48.

Table 1 [29, 32-35, 37-39] presents the literature summary of rubbery ILPMs developed in recent years. The [emim][Tf$_2$N] IL has been extensively used as free IL due to its higher CO$_2$ permeability and suitable CO$_2$/CH$_4$ selectivity. A comparison of various SILMs shows that [emim][Tf$_2$N] has the highest reported CO$_2$ permeability with a moderate CO$_2$/CH$_4$ selectivity and low viscosity as shown in Table 2 [40].

Industrial Applications of Green Solvents I   Materials Research Forum LLC
Materials Research Foundations **50** (2019) 320-344   doi: https://doi.org/10.21741/9781644900239-10

Table 1 Literature summary of rubbery ILPMs

Polymer	Ionic Liquid	PCO$_2$ (barrer)	CO$_2$/CH$_4$ Selectivity	Reference
Poly(RTIL)	[emim][Tf$_2$N]	44	27	[29]
P[vbim][NTf$_2$]	[emim][NTf$_2$]	300.1	14.8	[38]
	[emim][B(CN)$_4$]	365.4	15.8	
	[emim][BF$_4$]	233.2	17.1	
poly(RTIL)	[C$_4$mim][Cl] with CuCl salt	8.4	34*	[32]
poly(RTIL)	[C$_2$mim][C(CN)$_3$]	439	64.4*	[33]
p(VDF-HFP)	[emim][BF$_4$]	400	60*	[34]
p(VDF-HFP)	[emim][Tf$_2$N]	1620	15.3	[35]
	[hmim][Tf$_2$N]	877	11.3	
PVDF	[emim][B(CN)$_4$]	1778	41.4*	[39]
PVDF-HFP	Mixture of [emim][Tf$_2$N] and [Hdmim][Tf$_2$N]	212	8.48	[37]

* CO$_2$/N$_2$ selectivity

Table 1 Permeability and selectivity of imidazolium-based RTIL [40]

Ionic Liquid	CO$_2$ Single Gas Permeability (barrer)	CO$_2$/CH$_4$ Mixed Gas Selectivity	Viscosity ($\times$10-3 Pa.s) (cP))
[emim][TF$_2$N]	1702 $\pm$ 13	17 $\pm$ 0.9	26
[emim][DCA]	1237 $\pm$ 30	24 $\pm$ 1.4	21
[emim][CF$_3$SO$_3$]	1171$\pm$ 16	22 $\pm$ 1.2	45
[C$_6$mim][TF$_2$N]	1136 $\pm$ 20	9.9 $\pm$ 0.9	55
[emim][BF$_4$]	968 $\pm$ 10	27 $\pm$ 0.8	34
[emim][BF$_4$]	939 $\pm$ 10	12.9 $\pm$ 1.1	25

Glassy ionic liquid polymeric membranes

Poly(RTIL), PVDF and PVDF-HFP polymers are rubbery in nature that could be a reason for lower selectivity of ILPMs. Therefore, glassy polyimide (PI) polymer was also evaluated as a host polymer for ILPMs. Kanehashi et al. [41] studied the performance of PI/[bmim][Tf$_2$N] ILPMs under a wide range of IL loading (0-81 %). At lower IL loading, reduction in CO$_2$ permeability was observed due to the blocking-effect of IL. The presence of IL blocked the diffusion path of CO$_2$ gas by the plasticizing effect. However, at higher IL loading (51-81 %), a linear increase in the CO_2 permeability of ILPMs was observed due to the formation of separate IL domain in the membrane matrix that provided excessive freedom for gas diffusion. However, the permeability of ILPM at 81 % IL loading (501 barrer) was lower than base PI polymer (1156 barrer) due to a decrease in gas diffusivity and solubility in ILPMs. At the same time, the CO_2/CH_4 selectivity of PI/[bmim][Tf$_2$N] ILPM was reported as 14.5 which was also lower than base PI showing a selectivity of 18.6.

Liang et al. [42] also explored the opportunity of glassy polyamide (PI) and polybenzimidazole (PBI) polymers with [bmim][Tf$_2$N] IL in a composite arrangement by synthesizing composite ionic liquid and polymer membrane (CILPM) for gas separation. The membranes were tested at the elevated condition of temperature (35-200°C) and lower pressure range (2-6 bar). PBI based CILPMs were found to be more suitable for H_2 separation while PI-based CILPMs were found to be suitable for CO_2/CH_4 separation. At higher temperatures, the membrane exhibited higher permeability at the expense of selectivity. The best performance (CO_2 permeability = 7.9 barrer, CO_2/CH_4 selectivity = 54.9) was achieved with PI/[bmim][Tf$_2$N] at 30 % IL loading at 35°C which showed the potential of glassy polymers as a membrane matrix for ILPMs.

Mohshim et al. [43] compared the effect of anions on the performance of polyethersulfone PES based ILPMs. ILPMs were synthesized at lower loadings of IL (10-20 %) in PES/[emim][Tf$_2$N] and PES/[emim][CF$_3$SO$_3$] ILPMs. The ILPMs based on [emim][Tf$_2$N] IL showed higher CO_2 permeability due to lower viscosity and higher affinity towards CO_2 gas. However, due to lower permeability and blocking the effect of [emim][CF$_3$SO$_3$] as a plasticizer, the CO_2/CH_4 selectivity of PES/[emim][CF$_3$SO$_3$] ILPM was found to be higher than PES/[emim][Tf$_2$N] ILPM. The synthesized ILPMs displayed stable performance even at a higher feed pressure of 30 bar indicating their suitability as a potential material for CO_2 separation at elevated pressures. However, lower IL concentration was used by physical blending of ionic liquid in solvent first and the polymer was added after that in the solution. A very high viscous solution was developed by this method resulting in difficulty in solubility of the polymer in the solvent and ionic liquid mixture. Consequently, the drying process was lowered. This resulted in irregularity in the membrane structure. Moreover, the membranes were not fully characterized in terms of IL distribution, mechanical stability, the effect of IL on the morphology, interaction between polymer and IL. Additionally, the synthesized ILPMs were tested under single gas conditions and no information was reported on the long-term stability of the membranes.

Polysulfone (PSF) was also used as a base polymer for ILPM synthesis due to its thermal and mechanical stability. Lu et al. [44] incorporated various loadings of a few ILs in the

PSF matrix and evaluated the membranes for CO_2/N_2 separation. Instead of a dense membrane structure, the synthesized ILPMs demonstrated porous structures which might be due to fast evaporation of dichloromethane (DCM) solvent leaving behind pores in membrane structure. However, interestingly, the synthesized ILPMs maintained their selectivity with higher CO_2 permeability. Table 3 presents the literature summary of glassy ILPMs [41-45].

Table 3 Literature summary of glassy ILPMs [41-45]

Polymer	Ionic Liquid	PCO_2 (barrer)	CO_2/CH_4 Selectivity	Reference
PI	[bmim][Tf$_2$N]	501	14.5	[41]
PI	[bmim][NTf$_2$]	7.9	54.9	[42]
PES	[emim][Tf$_2$N] [emim][CF$_3$SO$_3$]	45 25	11 24	[43]
PSF	[bmim][TFSI] [bdim][TFSI] [dcim][TFSI] [dems][TFSI] [Hdph][TFSI] [Hdph][DC]	3000* 3600* 3400* 5800* 5200* 4900*	28** 27.7** 24.6** 30** 25** 25**	[44]
PSF	TEAF	69.39***	53	[45]

* × 10^5 GPU   ** CO_2/N_2 selectivity   *** GPU

Block copolymer based ILPMs

In recent years, elastomeric multi-block copolymer poly(ether-block-amide) has been extensively studied as a host polymer for ILPMs. It has an unstructured rubbery segment, polyether (PE), and a hard-glassy semi-crystalline segment, polyamide (PA). It is commercially available with the trademark of Pebax in different grades. Pebax has a phase separated microstructure where hard PA segments provide mechanical stability and soft PE segments facilitate high permeability due to high chain mobility. In addition, PE segments demonstrate a higher attraction to the quadrupolar carbon dioxide, and thus, provide higher CO_2 selectivity over nonpolar gases (CH_4, N_2, and H_2).

Bernardo et al. [46] first explored the opportunity of using Pebax as a base polymer with [bmim][CF$_3$SO$_3$] IL at higher IL loadings (80 wt. %). They used two grades of Pebax namely, Pebax 2533 and Pebax 1657 for membrane synthesis. Reduction in mechanical strength of ILPMs was observed as IL loadings increased due to the suspension of IL in the polyether phase. Pebax 1657 was found to be stiffer, less permeable, and higher selective than Pebax 2533. Pebax 2533 didn't show any significant change in gas permeability, diffusivity, or solubility after the incorporation of [bmim][CF_3SO_3] IL. However, Pebax 1657 showed a substantial increment in CO_2 permeability and reduction in selectivity after the incorporation of [bmim][CF_3SO_3] IL.

Later, Qiu et al. [47] synthesized Pebax 1657/[Bmim][Tf_2N] ILPMs at various loadings of [Bmim][Tf_2N] IL. Ionic liquid reduced the glass transition of the composite membranes since ionic liquid acted as a plasticizer in the matrix. Initially, at lower IL loading, reduction in gas permeability was observed due to the blocking effect of IL. However, as the concentration of IL increased, higher gas permeability was observed. At 40 wt. % IL loading, CO_2 permeability was 286 barrer, three times higher than neat Pebax membrane. However, a slight reduction in CO_2/CH_4 selectivity was observed due to softening of the matrix and the passing of larger gas species such as CH_4 and N_2. This observation is commonly reported for ILPMs. Moreover, Pebax 1657/[Bmim][Tf_2N] ILPMs demonstrated CO_2-induced plasticization phenomena even at lower pressures (3 bar).

Estahbanati et al. [48] investigated the effect of [bmim][BF_4] IL on Pebax 1657 copolymer. Membrane structure became more irregular and complex on blending IL in Pebax matrix due to the plasticization effect of IL on Pebax polymer. This effect was more severe at higher IL loadings because IL reduced intermolecular hydrogen bonding between copolymer segments. The irregular membrane structure caused higher gas permeability. However, CO_2 permeability was enhanced due to the high affinity of CO_2 in both polymer and IL. Thus, CO_2 permeability increased from 110 barrer for neat Pebax to 190 barrer for Pebax 1657/[bmim][BF_4] ILPM at 50 % loading. The corresponding CO_2/CH_4 selectivity improved from 20.8 to 24.4.

Pebax 1074/[bmim][PF_6] ILPM system was studied by Mahdavi et al. [49] at various loadings of IL (0-80 wt.%). ILPMs exhibited complex morphology due to reduced inter-chain bonding caused by the occurrence of ionic liquid in the membrane matrix. The incorporation of IL in Pebax matrix increased the gas permeability and decreases the selectivity. The authors concluded that IL had enhanced the chain mobility of the blend membrane, thereby, increased the permeability of smaller and larger gases which caused a reduction in the selectivity.

Pebax 1657/[emim][BF_4] polymer-IL pair was again explored by Fam et al. [50] in dense as well as multi-layer composite hollow fiber membrane configuration. In both cases, improved permeability was obtained with IL embedded membranes. However, CO_2/CH_4 selectivity was better in dense configuration than hollow fiber configuration due to the possibility of defects in hollow fiber membranes. The membranes maintained the rubbery characteristics in the presence of IL which was characterized by an increase in CO_2 permeability with respect to operating pressure.

Table 4 [46-54] presents the description of the recently reported studies on block copolymer and ionic liquid pairs for the development of ILPMs. Despite the high performance of ILPMs, a decrease in mechanical properties of ILPMs was observed with increases in IL weight %. Plasticization happens at higher ionic liquid loading, resulting in loss in mechanical properties. Due to this reason, ILPMs studies were mostly reported at lower operating pressure.

Table 4 Selected studies on block copolymer based ILPMs

Polymer	Ionic Liquid	PCO$_2$ (barrer)	CO$_2$/CH$_4$ Selectivity	Reference
Pebax 2533 Pebax 1657	[bmim][CF$_3$SO$_3$]	260 300	10 15	[46]
Pebax1657	[emim][BF$_4$]	550	18	[51]
Pebax1657	[bmim][Tf$_2$N]	286	12.5	[47]
Pebax1657	[bmim][BF$_4$]	190	24.4	[48]
Pebax 1074	[bmim][PF$_6$]	104.26	18.51	[49]
Pebax1657	[emim][BF$_4$]	270.1	27.3	[50]
Pebax1657	[bmim][Tf$_2$N] [bmim][DCA] [bmim][BF$_4$]	157.3 126.4 90.8	12.5 13.8 17.6	[52]
Pebax1657	[DnBM][Cl]	135	11	[53]
PS-PEO-PS PS-PMMA-PS PS-PIL-PS	[emim][TFSA]	710 840 980	22 13 19	[54]

3. Gas separation in supported ionic liquid membranes (SILMs)

Ionic liquids can be used in gas separation membrane in various ways. A common way to incorporate ionic liquids in membranes is using them as separation agents in SILMs [18]. In SILMs ionic liquid is injected in polymer support, the pores are made wetted with ILs [55] and ILs remain stuck within polymer due to capillary forces. SILMs provide higher permeability values of the target gas, in gas separation performance the rate of diffusion, dissolution, and desorption of target gas play an important role [56].

Barghi et al. [57] studied the effect of [bmim][PF$_6$] ionic liquid on inorganic membrane support illustrating almost CO_2 diffusivity 30 times higher than CH_4. Santos et al. [58] have reported an enhanced CO_2 permeability in [emim][TFSI]/polyvinylidene fluoride(PVDF) supported ionic liquid membranes from 2.7 to 325 barrer. Due to negligible vapor pressure, ionic liquids are not lost by evaporation and hence [emim][Ac] has been used in SILMs for CO_2/N_2[58, 59]. Albo et al. [59] reported different ways to insert [emim][Ac] ionic liquid in porous Al_2O_3/TiO_2 tubes and confirmed long-lasting stability. TiO_2 coated with [emim][Ac] ionic liquid showed P_{CO_2} = 2.78 ± 0.11 × 10^{-8} $mol/m^2s.Pa$ giving selectivity value of 30.72 for CO_2 and N_2 at 4 bar pressure. Santos et al. [58] reported that SILMs containing [emim][Ac], [bmim][Ac] and [vbtma][Ac] immobilized in poly(vinylidene fluoride) resulted in improvement of selectivity with rise in operating temperature because of high activation energy of N_2. However, SLMs and SILMs are limited to perform at lower pressure. At higher pressure, ionic liquid gets pushed away from the porous supports resulting in a decrease in membrane performance [42, 60].

The pressure difference across SILMs rarely exceeds 2 bar which shows the limitation of SLMs and SILMs [32]. Despite the fact that SILMs, having a good impact on permeability enhancement of target gas, these membranes have shown the main drawback in losing ILs from polymeric supports at higher pressures. Much work has been reported on SILMs [27, 61,62] but a big issue in these membranes has been the displacement of ionic liquid upon applying high pressure. In order to overcome this issue and keeping used of ionic liquids in polymers for gas separation, the trend has changed to use IL-based mixed matrix membranes. Recently, Rizwan et al. [12] reported IL-based mixed matrix membranes using EDA modified SAPO-34 into the PES matrix along with the addition of [emim][Tf_2N] and observed the improvement in CO_2/CH_4 selectivities by~37 folds as compared to pure PES membrane. Table 5 presents a summary of permeability and selectivity data for gas separation in supported ionic liquid membranes.

4.   Gas separation in ionic liquid mixed matrix membranes (ILMMMs)

MMMs are heterogeneous membranes having a combination of inorganic materials reinforced in polymeric materials. The MMMs have been reported by Kulprathipanja et al. [68] as a better option for gas separation as compared to polymeric membranes. MMMs are considered a revolutionary method to enhance permeability and selectivity. In these membranes, gas separation performance is contributed by inorganic filler having molecular sieving ability. These fillers may be porous or nonporous such as carbon molecular sieves, zeolites are porous and silica is non-porous.

Industrial Applications of Green Solvents I   Materials Research Forum LLC
Materials Research Foundations **50** (2019) 320-344   doi: https://doi.org/10.21741/9781644900239-10

Table 5: Literature summary of supported ionic liquid membranes (SILMs)

Support	Ionic Liquid	PCO2 Permeability (barrer)	CO2/CH4 Selectivity	References
PES with 80% porosity	[emim][DCA]	1237	24	[63]
PVDF	[emim][CF3SO3]	1171	17	[63]
PES	[emim][BF4]	968	27	[63]
PES	[emim][Tf2N]	1702	17	[63]
PES	[C6mim][Tf2N]	1136	9.9	[63]
PES	[emim][CF3SO3]	1771	22	[63]
PES	[EtMepy][(PFBu)SO3]	897	6.6	[64]
PTFE	[emim][C(CN)3]	667	19.4	[65]
PAN	[emim][DCA]	41.5	58*	[66]
PS-b-P4VP	[emim][DCA]	600	65*	[67]
PTFE	[emim][C(CN)3]	667	57*	[65]
PVDF	[emim][AC]	878.8	33.7*	[58]
PES	[EtMepy][(PFBu)SO3]	897	12.3*	[64]
poly(tetrafluoroethylene) PTFE	[emim][B(CN)4]	742	49*	[65]
PTFE	[emim][C(CN)3	667	57*	[65]

* CO_2/N_2 selectivity

In ionic liquid mixed matrix membranes, usually, three components are incorporated together to develop a membrane i.e. polymer, inorganic filler, and an ionic liquid. The addition of ionic liquid gives more advantages in the separation process, making ionic liquid mixed matrix membranes. ILMMMs have been studied by many researchers [62, 69, 70]. Oral et al. [71] studied the incorporation of ionic liquid emim[Tf2N] and emim[CF3SO3] into polyimide-zeolite (SAPO-34) and observed that addition of these ionic liquids into MMM has improved the filler-polymer interface as well as improved the permeability and selectivity. The use of ionic liquid in mixed matrix membranes also increases polymer and inorganic filler compatibility. For example, Zhou et al. [72] observed that gas separation performance of novel ionic cross-linked polyether membranes based on poly(ionic liquid)s and poly(trimethylene ether)glycol units. This membrane showed CO_2 permeability 86–113 barrer and CO_2/N_2 selectivity 41–19 depending upon the density of the cross-linking used. It was observed that permeability for CO_2 was improved along with CO_2/N_2 enhancement. Table 6 [67, 70, 73] provides a summary of permeability and selectivity data for gas separation in supported ionic liquid membranes

Table 6 Selected studies on ionic liquid mixed matrix membranes (ILMMMs)

Ionic liquid	Polymer	CO_2 Single Gas Permeability (barrer)	CO_2/CH_4 Selectivity	Reference
[emim][Tf$_2$N]	PIL-based	44	27	[67]
[emim][Tf$_2$N]	Styrene	72.1	32.2	[70]
[emim][Tf$_2$N]	PIL-based	44	39*	[67]
[emim][Tf$_2$N]	Styrene	72.1	42.4*	[70]
[hmim][Tf$_2$N]	PTFE	600	18*	[73]

* CO_2/N_2 selectivity

5.   Future recommendations in research and development

At present, membranes for gas separations are lacking in durability and performance. The plasticization has been the major issue concerning with polymeric membranes. Also, the poor interactions between polymers and fillers have resulted in non-ideal filler-polymer interfacial morphology, e.g. interface voids, a rigid polymer layer around the inorganic fillers and particle pore blockage (in case of porous fillers) [74]. When glassy polymers are used as base material, poor polymer-filler interaction arises, that gives rise to the development of interfacial voids and hence poor selectivity. One more, big issue during membranes synthesis has been filler agglomeration, as it produces phase separation between filler and polymer, leading to the weakening of the composite and forming non-selective defects [75]. Consequently, these issues are causing challenges for improved gas separation performances and there is a need for exploring efficient technology to address these issues.

The immobilization of ionic liquids in the polymer by different methods produces numerous types of ionic liquid membranes. ILs tunability and the features of the polymer suggest new opportunities for efficient gas separations. The synergy of ILs and polymers helps to achieve the required properties to gain the requirements of the desired applications. In this chapter, the literature review of IL-based membranes used for gas separation applications has been reviewed focusing on PILMs, SILMs, and ILMMMs.

Although ILs based membranes for gas separation provide many advantages, however, there are still challenges present for improving gas separation membranes. Mainly the gas separation performance is restricted by permeability/selectivity trade-off; therefore, synthesizing new ionic liquid-based membranes with enhanced permeability and selectivity is important. One effective way is to blend ionic liquid with a polymer. It has been reported as the presence of ionic liquids in polymers increases the plasticization pressure of membrane [11]. In the case of CO_2/CH_4 separation, IL helps CO_2 absorption in the polymer, increasing CO_2 permeance and selectivity of CO_2/CH_4 gas [4]. A tertiary complex system including polymer, inorganic filler, and the ionic liquid is significant and can be helpful to get desired gas separation application, however, a lot of expertise and efforts are required to deal with such complex system.

Additionally, the blending of ILs in the tertiary component system is helpful to form defect-free interfacial morphology. It functionalizes the fillers present, minimizing the interfacial morphology issue. Besides, it improves polymer-filler adhesion and helps to form defect-free interfacial morphology. A comparison of the performance of SILMs, PILMs, and ILMMMs has shown that trend is shifting toward ILMMMs due to the less ionic loss in ILMMMs as shown in Fig 1. Additionally, it has been suggested that ionic liquids, new solvents, new fillers as well as surface-modified fillers should [76] be incorporated in the polymer to achieve better gas separation performance.

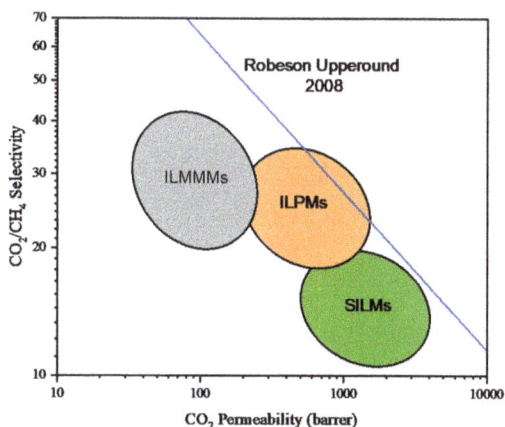

Fig. 1 Robeson's upper bound curves showing the general trade-off between permeability and selectivity for CO_2/CH_4 separation from this literature review data

Furthermore, it is necessary to perform systematic and broad investigations on the structural properties of IL, polymer and inorganic fillers; so that membranes for desired gas separation applications can be developed.

Conclusion

To enhance gas separation performance, the addition of various additives into the existing polymer membranes have been found to be a very beneficial approach. Few advantages and disadvantages of rubbery ionic liquid polymeric membranes (ILPMs), glassy ionic liquid polymeric membranes (ILPMs), supported ionic liquid membranes (SILMs), and ionic liquid mixed matrix membranes (ILMMMs) have been reported in this chapter with specific incorporation of ionic liquids into gas separation membranes to enhance the gas separation. Novel combinations of polymers and ionic liquids has been recommended to attain high selectivity and high permeabilities of target gases. Ionic liquids with the combination of many anions and cations along with suitable polymers and solvents are attracting the researchers to meet the desired gas separation applications. In addition, the combined effect of ionic liquids with inorganic fillers in polymeric materials has been found to improve gas separation performance without compromising the mechanical strength of the membranes. In short, ionic liquids are expected to attain more attention in gas separation membranes in the future.

Acknowledgment

The authors acknowledge the support given by the Universiti Teknologi PETRONAS.

References

[1]   J.L. Anderson, J.K. Dixon, E.J. Maginn, J.F. Brennecke, Measurement of SO_2 solubility in ionic liquids, The J. Phys. Chem. B 110 (2006) 15059-15062. https://doi.org/10.1021/jp063547u

[2]   J. Erisman, A. Bleeker, J. Galloway, M. Sutton, Reduced nitrogen in ecology and the environment, Environ. Pollut. 150 (2007) 140-149. https://doi.org/10.1016/j.envpol.2007.06.033

[3]   I. Statistics, CO_2 emissions from fuel combustion-highlights, IEA, Paris http://www. iea. org/CO_2 highlights/CO_2 highlights. pdf. Cited July (2011). https://doi.org/10.1787/data-00433-en

[4]   J.E. Bara, T.K. Carlisle, C.J. Gabriel, D. Camper, A. Finotello, D.L. Gin, R.D. Noble, Guide to CO_2 separations in imidazolium-based room-temperature ionic liquids, Ind. Eng. Chem. Res. 48 (2009) 2739-2751. https://doi.org/10.1021/ie8016237

[5]   K.N. Ruckart, R.A. O'Brien, S.M. Woodard, K.N. West, T.G. Glover, Porous solids impregnated with task-specific ionic liquids as composite sorbents, The J. Phys. Chem. C 119 (2015) 20681-20697. https://doi.org/10.1021/acs.jpcc.5b04646

[6]   W. Qian, J. Texter, F. Yan, Frontiers in poly (ionic liquid) s: syntheses and applications, Chem. Soc. Rev. 46 (2017) 1124-1159. https://doi.org/10.1039/c6cs00620e

[7]   P. Walden, "On the Molecular Size and Electrical Conductivity of Some Molten Salts," News of the Imperial Academy of Sciences. VI serial, 8: 6 (1914), 405-422.

[8]   L.A. Blanchard, Z. Gu, J.F. Brennecke, High-pressure phase behavior of ionic liquid/CO_2 systems, The J. Phys. Chem. B 105 (2001) 2437-2444. https://doi.org/10.1021/jp003309d

[9]   Z. Lei, B. Chen, Y.M. Koo, D.R. MacFarlane, Introduction: Ionic liquids, Chem. Rev. 117 (2017) 6633-6635. https://doi.org/10.1021/acs.chemrev.7b00246

[10]   H.A. Mannan, H. Mukhtar, M.S. Shahrun, M.A. Bustam, Z. Man, M.Z.A. Bakar, Effect of [EMIM][Tf$_2$N] ionic liquid on ionic liquid-polymeric membrane (ILPM) for CO_2/CH_4 separation, Procedia Eng. 148 (2016) 25-29. https://doi.org/10.1016/j.proeng.2016.06.477

[11]   D.F. Mohshim, H. Mukhtar, Z. Man, Composite blending of ionic liquid–poly (ether sulfone) polymeric membranes: Green materials with potential for carbon dioxide/methane separation, J. Appl. Polym. Sci. 133 (2016). https://doi.org/10.1002/app.43999

[12]   R. Nasir, N.N.R. Ahmad, H. Mukhtar, D.F. Mohshim, Effect of ionic liquid inclusion and amino–functionalized SAPO-34 on the performance of mixed matrix membranes for CO_2/CH_4 separation, J. Environ. Chem. Eng. 6 (2018) 2363-2368. https://doi.org/10.1016/j.jece.2018.03.032

[13]   L.C. Tome, I.M. Marrucho, Ionic liquid-based materials: a platform to design engineered CO_2 separation membranes, Chem. Soc. Rev. 45 (2016) 2785-2824. https://doi.org/10.1039/c5cs00510h

[14]   K. Simons, K. Nijmeijer, J.E. Bara, R.D. Noble, M. Wessling, How do polymerized room-temperature ionic liquid membranes plasticize during high pressure

CO_2 permeation?, J. Membr. Sci. 360 (2010) 202-209.
https://doi.org/10.1016/j.memsci.2010.05.018

[15]   E.D. Bates, R.D. Mayton, I. Ntai, J.H. Davis, CO_2 capture by a task-specific ionic liquid, J. Am. Chem. Soc. 124 (2002) 926-927. https://doi.org/10.1021/ja017593d

[16]   D. Camper, J.E. Bara, D.L. Gin, R.D. Noble, Room-temperature ionic liquid-amine solutions: Tunable solvents for efficient and reversible capture of CO_2, Ind. Eng. Chem. Res. 47 (2008) 8496-8498. https://doi.org/10.1021/ie801002m

[17]   J.E. Bara, C.J. Gabriel, S. Lessmann, T.K. Carlisle, A. Finotello, D.L. Gin, R.D. Noble, Enhanced CO_2 separation selectivity in oligo (ethylene glycol) functionalized room-temperature ionic liquids, Ind. Eng. Chem. Res. 46 (2007) 5380-5386. https://doi.org/10.1021/ie070437g

[18]   R.D. Noble, D.L. Gin, Perspective on ionic liquids and ionic liquid membranes, J. Membr. Sci. 369 (2011) 1-4.

[19]   D. Camper, J. Bara, C. Koval, R. Noble, Bulk-fluid solubility and membrane feasibility of Rmim-based room-temperature ionic liquids, Ind. Eng. Chem. Res. 45 (2006) 6279-6283. https://doi.org/10.1021/ie060177n

[20]   H. Ohno, M. Yoshizawa, W. Ogihara, Development of new class of ion conductive polymers based on ionic liquids, Electrochim. Acta, 50 (2004) 255-261. https://doi.org/10.1016/j.electacta.2004.01.091

[21]   J. Tang, W. Sun, H. Tang, M. Radosz, Y. Shen, Enhanced CO_2 absorption of poly (ionic liquid)s, Macromolecules, 38 (2005) 2037-2039. https://doi.org/10.1021/ma047574z

[22]   J.E. Bara, S. Lessmann, C.J. Gabriel, E.S. Hatakeyama, R.D. Noble, D.L. Gin, Synthesis and performance of polymerizable room-temperature ionic liquids as gas separation membranes, Ind. Eng. Chem. Res. 46 (2007) 5397-5404. https://doi.org/10.1021/ie0704492

[23]   R. Spillman, M. Sherwin, Gas separation membranes: the first decade, Chem. Tech. 20 (1990) 378-384.

[24]   G. Lu, J.D. Da Costa, M. Duke, S. Giessler, R. Socolow, R. Williams, T. Kreutz, Inorganic membranes for hydrogen production and purification: a critical review and perspective, Journal of Colloid and Interface Science 314 (2007) 589-603. https://doi.org/10.1016/j.jcis.2007.05.067

[25]   J.E. Bara, D.E. Camper, D.L. Gin, R.D. Noble, Room-temperature ionic liquids and composite materials: platform technologies for CO_2 capture, Acc. Chem. Res. 43 (2009) 152-159. https://doi.org/10.1021/ar9001747

[26]   R.W. Baker, K. Lokhandwala, Natural gas processing with membranes: an overview, Ind. Eng. Chem. Res. 47 (2008) 2109-2121. https://doi.org/10.1021/ie071083w

[27]   Z. Dai, R.D. Noble, D.L. Gin, X. Zhang, L. Deng, Combination of ionic liquids with membrane technology: A new approach for CO_2 separation, J. Membr. Sci. 497 (2016) 1-20.

[28]   C. Cadena, J.L. Anthony, J.K. Shah, T.I. Morrow, J.F. Brennecke, E.J. Maginn, Why is CO_2 so soluble in imidazolium-based ionic liquids?, J. Am. Chem. Soc. 126 (2004) 5300-5308. https://doi.org/10.1021/ja039615x

[29]   J.E. Bara, E.S. Hatakeyama, D.L. Gin, R.D. Noble, Improving CO_2 permeability in polymerized room-temperature ionic liquid gas separation membranes through the formation of a solid composite with a room-temperature ionic liquid, Polym. Adv. Technol. 19 (2008) 1415-1420. https://doi.org/10.1002/pat.1209

[30]   J.E. Bara, D.L. Gin, R.D. Noble, Effect of anion on gas separation performance of polymer−room-temperature ionic liquid composite membranes, Ind. Eng. Chem. Res. 47 (2008) 9919-9924. https://doi.org/10.1021/ie801019x

[31]   J.E. Bara, R.D. Noble, D.L. Gin, Effect of "Free" cation substituent on gas separation performance of polymer−room-temperature ionic liquid composite membranes, Ind. Eng. Chem. Res. 48 (2009) 4607-4610. https://doi.org/10.1021/ie801897r

[32]   G. Zarca, W.J. Horne, I. Ortiz, A. Urtiaga, J.E. Bara, Synthesis and gas separation properties of poly(ionic liquid)-ionic liquid composite membranes containing a copper salt, J. Membr. Sci. 515 (2016) 109-114. https://doi.org/10.1016/j.memsci.2016.05.045

[33]   L.C. Tomé, D.C. Guerreiro, R.M. Teodoro, V.D. Alves, I.M. Marrucho, Effect of polymer molecular weight on the physical properties and CO_2/N_2 separation of pyrrolidinium-based poly(ionic liquid) membranes, J. Membr. Sci. 549 (2018) 267-274. https://doi.org/10.1016/j.memsci.2017.12.019

[34]   S.U. Hong, D. Park, Y. Ko, I. Baek, Polymer-ionic liquid gels for enhanced gas transport, Chem. Comm. 0 (2009) 7227-7229. https://doi.org/10.1039/b913746g

[35]   P. Uchytil, J. Schauer, R. Petrychkovych, K. Setnickova, S.Y. Suen, Ionic liquid membranes for carbon dioxide–methane separation, J. Membr. Sci. 383 (2011) 262-271. https://doi.org/10.1016/j.memsci.2011.08.061

[36]   K. Friess, J.C. Jansen, F. Bazzarelli, P. Izák, V. Jarmarová, M. Kačírková, J. Schauer, G. Clarizia, P. Bernardo, High ionic liquid content polymeric gel membranes: correlation of membrane structure with gas and vapour transport properties, J. Membr. Sci. 415 (2012) 801-809. https://doi.org/10.1016/j.memsci.2012.05.072

[37]   J.C. Jansen, G. Clarizia, P. Bernardo, F. Bazzarelli, K. Friess, A. Randová, J. Schauer, D. Kubicka, M. Kacirková, P. Izak, Gas transport properties and pervaporation performance of fluoropolymer gel membranes based on pure and mixed ionic liquids, Sep. Purif. Tech. 109 (2013) 87-97. https://doi.org/10.1016/j.seppur.2013.02.034

[38]   L. Hao, P. Li, T. Yang, T.S. Chung, Room temperature ionic liquid/ZIF-8 mixed-matrix membranes for natural gas sweetening and post-combustion CO_2 capture, J. Membr. Sci. 436 (2013) 221-231. https://doi.org/10.1016/j.memsci.2013.02.034

[39]   H.Z. Chen, P. Li, T.-S. Chung, PVDF/ionic liquid polymer blends with superior separation performance for removing CO_2 from hydrogen and flue gas, Int. J. Hydrogen Energy 37 (2012) 11796-11804. https://doi.org/10.1016/j.ijhydene.2012.05.111

[40]   P. Scovazzo, Determination of the upper limits, benchmarks, and critical properties for gas separations using stabilized room temperature ionic liquid membranes (SILMs) for the purpose of guiding future research, J. Membr. Sci. 343 (2009) 199-211. https://doi.org/10.1016/j.memsci.2009.07.028

[41]   S. Kanehashi, M. Kishida, T. Kidesaki, R. Shindo, S. Sato, T. Miyakoshi, K. Nagai, CO_2 separation properties of a glassy aromatic polyimide composite membranes containing high-content 1-butyl-3-methylimidazolium bis(trifluoromethylsulfonyl)imide ionic liquid, J. Membr. Sci. 430 (2013) 211-222. https://doi.org/10.1016/j.memsci.2012.12.003

[42]   L. Liang, Q. Gan, P. Nancarrow, Composite ionic liquid and polymer membranes for gas separation at elevated temperatures, J. Membr. Sci. 450 (2014) 407-417. https://doi.org/10.1016/j.memsci.2013.09.033

[43]   D.F. Mohshim, H. Mukhtar, Z. Man, Composite blending of ionic liquid–poly(ether sulfone) polymeric membranes: Green materials with potential for carbon dioxide/methane separation, J. Appl. Polym. Sci. 133 (2016) 43999. https://doi.org/10.1002/app.43999

[44]   S.C. Lu, A.L. Khan, I.F.J. Vankelecom, Polysulfone-ionic liquid based membranes for CO_2/N_2 separation with tunable porous surface features, J. Membr. Sci. 518 (2016) 10-20. https://doi.org/10.1016/j.memsci.2016.06.031

[45]   R. Ur Rehman, S. Rafiq, N. Muhammad, A.L. Khan, A. Ur Rehman, L. Ting Ting, M. Saeed, F. Jamil, M. Ghauri, X. Gu, Development of ethanolamine-based ionic liquid membranes for efficient CO_2/CH_4 separation, J. Appl. Polymer. Sci. 134 (2017) 45395. https://doi.org/10.1002/app.45395

[46]   P. Bernardo, J.C. Jansen, F. Bazzarelli, F. Tasselli, A. Fuoco, K. Friess, P. Izák, V. Jarmarová, M. Kačírková, G. Clarizia, Gas transport properties of Pebax®/room temperature ionic liquid gel membranes, Sep. Purif. Technol. 97 (2012) 73-82. https://doi.org/10.1016/j.seppur.2012.02.041

[47]   Y. Qiu, J. Ren, D. Zhao, H. Li, M. Deng, Poly(amide-6-b-ethylene oxide)/[Bmim][Tf$_2$N] blend membranes for carbon dioxide separation, Journal of Energy Chemistry 25 (2016) 122-130. https://doi.org/10.1016/j.jechem.2015.10.009

[48]   E. Ghasemi Estahbanati, M. Omidkhah, A. Ebadi Amooghin, Preparation and characterization of novel Ionic liquid/Pebax membranes for efficient CO_2/light gases separation, Journal of Industrial and Engineering Chemistry 51 (2017) 77-89. https://doi.org/10.1016/j.jiec.2017.02.017

[49]   H.R. Mahdavi, N. Azizi, M. Arzani, T. Mohammadi, Improved CO_2/CH_4 separation using a nanocomposite ionic liquid gel membrane, Journal of Natural Gas Science and Engineering 46 (2017) 275-288. https://doi.org/10.1016/j.jngse.2017.07.024

[50]   W. Fam, J. Mansouri, H. Li, V. Chen, Improving CO_2 separation performance of thin film composite hollow fiber with Pebax®1657/ionic liquid gel membranes, J. Membr. Sci. 537 (2017) 54-68. https://doi.org/10.1016/j.memsci.2017.05.011

[51]   H. Rabiee, A. Ghadimi, T. Mohammadi, Gas transport properties of reverse-selective poly(ether-b-amide6)/[Emim][BF4] gel membranes for CO_2/light gases separation, J. Membr. Sci. 476 (2015) 286-302. https://doi.org/10.1016/j.memsci.2014.11.037

[52]   M. Li, X. Zhang, S. Zeng, L. bai, H. Gao, J. Deng, Q. Yang, S. Zhang, Pebax-based composite membranes with high gas transport properties enhanced by ionic liquids for CO_2 separation, RSC Adv. 7 (2017) 6422-6431. https://doi.org/10.1039/c6ra27221e

[53] A. Jomekian, B. Bazooyar, R.M. Behbahani, T. Mohammadi, A. Kargari, Ionic liquid-modified Pebax® 1657 membrane filled by ZIF-8 particles for separation of CO_2 from CH_4, N_2 and H_2, J. Membr. Sci. 524 (2017) 652-662. https://doi.org/10.1016/j.memsci.2016.11.065

[54] Y. Gu, E.L. Cussler, T.P. Lodge, ABA-triblock copolymer ion gels for CO_2 separation applications, J. Membr. Sci. 423-424 (2012) 20-26. https://doi.org/10.1016/j.memsci.2012.07.011

[55] L.C. Branco, J.G. Crespo, C.A. Afonso, Studies on the selective transport of organic compounds by using ionic liquids as novel supported liquid membranes, Chem. Eur. J. 8 (2002) 3865-3871. https://doi.org/10.1002/1521-3765(20020902)8:17<3865::aid-chem3865>3.0.co;2-l

[56] R. Kreiter, J.P. Overbeek, L.A. Correia, J.F. Vente, Pressure resistance of thin ionic liquid membranes using tailored ceramic supports, J. Membr. Sci. 370 (2011) 175-178. https://doi.org/10.1016/j.memsci.2010.12.024

[57] S. Barghi, M. Adibi, D. Rashtchian, An experimental study on permeability, diffusivity, and selectivity of CO_2 and CH_4 through [bmim][PF_6] ionic liquid supported on an alumina membrane: Investigation of temperature fluctuations effects, J. Membr. Sci. 362 (2010) 346-352. https://doi.org/10.1016/j.memsci.2010.06.047

[58] E. Santos, J. Albo, A. Irabien, Acetate based supported ionic liquid membranes (SILMs) for CO_2 separation: influence of the temperature, J. Membr. Sci. 452 (2014) 277-283. https://doi.org/10.1016/j.memsci.2013.10.024

[59] J. Albo, T. Yoshioka, T. Tsuru, Porous Al_2O_3/TiO_2 tubes in combination with 1-ethyl-3-methylimidazolium acetate ionic liquid for CO_2/N_2 separation, Sep. Purif. Tech. 122 (2014) 440-448. https://doi.org/10.1016/j.seppur.2013.11.024

[60] L.Z. Liang, Q. Gan, P. Nancarrow, A study on permeabilities and selectivities of small-molecule gases for composite ionic liquid and polymer membranes, App. Mech. Mater. 448-453 (2013) 765-770. https://doi.org/10.4028/www.scientific.net/amm.448-453.765

[61] L.C. Tomé, D.J. Patinha, C.S. Freire, L.P.N. Rebelo, I.M. Marrucho, CO_2 separation applying ionic liquid mixtures: the effect of mixing different anions on gas permeation through supported ionic liquid membranes, RSC Adv. 3 (2013) 12220-12229. https://doi.org/10.1039/c3ra41269e

[62] Y.C. Hudiono, T.K. Carlisle, J.E. Bara, Y. Zhang, D.L. Gin, R.D. Noble, A three-component mixed-matrix membrane with enhanced CO_2 separation properties based

Industrial Applications of Green Solvents I	Materials Research Forum LLC
Materials Research Foundations **50** (2019) 320-344	doi: https://doi.org/10.21741/9781644900239-10

on zeolites and ionic liquid materials, J. Membr. Sci. 350 (2010) 117-123. https://doi.org/10.1016/j.memsci.2009.12.018

[63]   P. Scovazzo, D. Havard, M. McShea, S. Mixon, D. Morgan, Long-term, continuous mixed-gas dry fed CO_2/CH_4 and CO_2/N_2 separation performance and selectivities for room temperature ionic liquid membranes, J. Membr. Sci. 327 (2009) 41-48. https://doi.org/10.1016/j.memsci.2008.10.056

[64]   A.B. Pereiro, L.C. Tomé, S. Martinho, L.P.N. Rebelo, I.M. Marrucho, Gas permeation properties of fluorinated ionic liquids, Ind. Eng. Chem. Res. 52 (2013) 4994-5001. https://doi.org/10.1021/ie4002469

[65]   L.C. Tomé, C. Florindo, C.S. Freire, L.P.N. Rebelo, I.M. Marrucho, Playing with ionic liquid mixtures to design engineered CO_2 separation membranes, Phys. Chem. Chem. Phys. 16 (2014) 17172-17182. https://doi.org/10.1039/c4cp01434k

[66]   J. Grünauer, V. Filiz, S. Shishatskiy, C. Abetz, V. Abetz, Scalable application of thin film coating techniques for supported liquid membranes for gas separation made from ionic liquids, J. Membr. Sci. 518 (2016) 178-191. https://doi.org/10.1016/j.memsci.2016.07.005

[67]   J. Grünauer, S. Shishatskiy, C. Abetz, V. Abetz, V. Filiz, Ionic liquids supported by isoporous membranes for CO_2/N_2 gas separation applications, J. Membr. Sci. 494 (2015) 224-233. https://doi.org/10.1016/j.memsci.2015.07.054

[68]   S. Kulprathipanja, R.W. Neuzil, N.N. Li, Separation of fluids by means of mixed matrix membranes, Google Patents, 1988.

[69]   J.E. Bara, D.L. Gin, R.D. Noble, Effect of anion on gas separation performance of polymer– room-temperature ionic liquid composite membranes, Ind. Eng. Chem. Res. 47 (2008) 9919-9924. https://doi.org/10.1021/ie801019x

[70]   Y.C. Hudiono, T.K. Carlisle, A.L. LaFrate, D.L. Gin, R.D. Noble, Novel mixed matrix membranes based on polymerizable room-temperature ionic liquids and SAPO-34 particles to improve CO_2 separation, J. Membr. Sci. 370 (2011) 141-148. https://doi.org/10.1016/j.memsci.2011.01.012

[71]   R.D.N. C. A. Oral, S. B. Tantekin-Ersolmaz,, Ternarymixed-matrix membranes containing room temperature ionicliquids, Proceedings of the North Am.Membr.Soc.Conference (NAMS '11), (2011).

[72]   X. Zhou, M.M. Obadia, S.R. Venna, E.A. Roth, A. Serghei, D.R. Luebke, C. Myers, Z. Chang, R. Enick, E. Drockenmuller, Highly cross-linked polyether-based

Industrial Applications of Green Solvents I   Materials Research Forum LLC
Materials Research Foundations **50** (2019) 320-344   doi: https://doi.org/10.21741/9781644900239-10

1,2,3-triazolium ion conducting membranes with enhanced gas separation properties, Eur. Polym. J. 84 (2016) 65-76. https://doi.org/10.1016/j.eurpolymj.2016.09.001

[73]   P.T. Nguyen, B.A. Voss, E.F. Wiesenauer, D.L. Gin, R.D. Noble, Physically gelled room-temperature ionic liquid-based composite membranes for CO_2/N_2 separation: effect of composition and thickness on membrane properties and performance, Ind. Eng. Chem. Res. 52 (2012) 8812-8821. https://doi.org/10.1021/ie302352r

[74]   H. Vinh-Thang, S. Kaliaguine, Predictive models for mixed-matrix membrane performance: a review, Chem. Rev. 113 (2013) 4980-5028. https://doi.org/10.1021/cr3003888

[75]   A.C. Balazs, T. Emrick, T.P. Russell, Nanoparticle polymer composites: where two small worlds meet, Science 314 (2006) 1107-1110. https://doi.org/10.1126/science.1130557

[76]   T.H. Bae, J.S. Lee, W. Qiu, W.J. Koros, C.W. Jones, S. Nair, A high-performance gas-separation membrane containing submicrometer-sized metal–organic framework crystals, Angew. Chem. Int. Ed. 49 (2010) 9863-9866. https://doi.org/10.1002/anie.201006141

Keyword Index

About the Editors

Dr. Inamuddin is currently working as Assistant Professor in the Chemistry Department, Faculty of Science, King Abdulaziz University, Jeddah, Saudi Arabia. He is a permanent faculty member (Assistant Professor) at the Department of Applied Chemistry, Aligarh Muslim University, Aligarh, India. He obtained Master of Science degree in Organic Chemistry from Chaudhary Charan Singh (CCS) University, Meerut, India, in 2002. He received his Master of Philosophy and Doctor of Philosophy degrees in Applied Chemistry from Aligarh Muslim University (AMU), India, in 2004 and 2007, respectively. He has extensive research experience in multidisciplinary fields of Analytical Chemistry, Materials Chemistry, and Electrochemistry and, more specifically, Renewable Energy and Environment. He has worked on different research projects as project fellow and senior research fellow funded by University Grants Commission (UGC), Government of India, and Council of Scientific and Industrial Research (CSIR), Government of India. He has received Fast Track Young Scientist Award from the Department of Science and Technology, India, to work in the area of bending actuators and artificial muscles. He has completed four major research projects sanctioned by University Grant Commission, Department of Science and Technology, Council of Scientific and Industrial Research, and Council of Science and Technology, India. He has published 138 research articles in international journals of repute and eighteen book chapters in knowledge-based book editions published by renowned international publishers. He has published forty-two edited books with Springer, United Kingdom, Elsevier, Nova Science Publishers, Inc. U.S.A., CRC Press Taylor & Francis Asia Pacific, Trans Tech Publications Ltd., Switzerland and Materials Research Forum LLC, U.S.A. He is the member of various editorial boards of the journals and serving as associate editor for journals such as Environmental Chemistry Letter, Applied Water Science, Euro-Mediterranean Journal for Environmental Integration, Springer-Nature, Frontiers Section Editor of Current Analytical Chemistry, published by Bentham Science Publishers, editorial board member for Scientific Reports-Nature and editor for Eurasian Journal of Analytical Chemistry. He has attended as well as chaired sessions in various international and national conferences. He has worked as a Postdoctoral Fellow, leading a research team at the Creative Research Initiative Center for Bio-Artificial Muscle, Hanyang University, South Korea, in the field of renewable energy, especially biofuel cells. He has also worked as a Postdoctoral Fellow at the Center of Research Excellence in Renewable Energy, King Fahd University of Petroleum and Minerals, Saudi Arabia, in the field of polymer electrolyte membrane fuel cells and computational fluid dynamics of polymer electrolyte membrane fuel cells. He is a life member of the Journal of the Indian

Chemical Society. His research interest includes ion exchange materials, a sensor for heavy metal ions, biofuel cells, supercapacitors and bending actuators.

Mohd Imran Ahamed is a Research Scholar at Department of Chemistry, Aligarh Muslim University, Aligarh, India. He is working towards his Ph.D. thesis entitled Synthesis and characterization of inorganic-organic composite heavy metals selective cation-exchangers and their analytical applications. He has published several research and review articles in the journals of international recognition. He has also edited various books published by Springer and Materials Research Forum LLC, U.S.A. He has completed his B.Sc. (Hons) Chemistry from Aligarh Muslim University, Aligarh, India, and M.Sc. (Organic Chemistry) from Dr. Bhimrao Ambedkar University, Agra, India. His research work includes ion exchange chromatography, wastewater treatment, and analysis, bending actuator and electrospinning.

Prof. Abdullah M. Asiri is the Head of the Chemistry Department at King Abdulaziz University since October 2009 and he is the founder and the Director of the Center of Excellence for Advanced Materials Research (CEAMR) since 2010 till date. He is the Professor of Organic Photochemistry. He graduated from King Abdulaziz University (KAU) with B.Sc. in Chemistry in 1990 and a Ph.D. from University of Wales, College of Cardiff, U.K. in 1995. His research interest covers color chemistry, synthesis of novel photochromic and thermochromic systems, synthesis of novel coloring matters and dyeing of textiles, materials chemistry, nanochemistry and nanotechnology, polymers and plastics. Prof. Asiri is the principal supervisors of more than 20 M.Sc. and six Ph.D. theses. He is the main author of ten books of different chemistry disciplines. Prof. Asiri is the Editor-in-Chief of King Abdulaziz University Journal of Science. A major achievement of Prof. Asiri is the discovery of tribochromic compounds, a new class of compounds which change from slightly or colorless to deep colored when subjected to small pressure or when grind. This discovery was introduced to the scientific community as a new terminology published by IUPAC in 2000. This discovery was awarded a patent from European Patent office and from UK patent. Prof. Asiri involved in many committees at the KAU level and on the national level. He took a major role in the advanced materials committee working for KACST to identify the national plan for science and technology in 2007. Prof. Asiri played a major role in advancing the chemistry education and research in KAU. He has been awarded the best researchers from KAU for the past five years. He also awarded the Young Scientist Award from the Saudi Chemical Society in 2009 and also the first prize for the distinction in science from the Saudi Chemical Society in 2012. He also received a recognition certificate from the American Chemical Society (Gulf region Chapter) for the advancement of chemical science in the Kingdome. He received a Scopus certificate for the most publishing scientist in Saudi Arabia in chemistry in 2008. He is also a member of the editorial board of various journals of international repute. He is the Vice- President of Saudi Chemical Society (Western Province Branch). He holds four USA patents, more than one thousand publications in international journals, several book chapters and edited books.

www.ingramcontent.com/pod-product-compliance
Lightning Source LLC
Chambersburg PA
CBHW071321210326
41597CB00015B/1300